I0056259

Catalogue of
Intermediate Hosts of
Animal Parasites in India

THE AUTHORS

Dr. K. Muraleedharan specialized in Veterinary Parasitology, served as teacher and research scientist of University of Agricultural Sciences, Bangalore in the cadre of Associate Professor and Professor. Associated with the functioning of PL-480 Research Project, two ICAR Research Schemes, UAS-KDDC Diagnostic Laboratories, establishment of Parasitology Department at Bidar Veterinary College and at Konehalli Research Station. Published more than 100 research papers and received many gold medals and awards for technical capabilities and outstanding contributions. A persistent and prolific writer interested in publishing and meticulous research observations.

Dr.M.S.Jagannath is a distinguished Veterinary Parasitologist, teacher, researcher and an able administrator. He had advanced training from London School of Hygiene and University of Edinburgh. UK. His career spanning over 41 years held various positions including as Founder Director of Centre of Advanced studies at the Veterinary College, Bangalore, India and Prof of Medical Parasitology, University of Addis Ababa, Ethiopia. He has published 250 research papers and findings are cited in 40 text books. Research included first description of parasites, new host records, new taxonomy, developing battery of modern diagnostics, studies on molecular biology and immunology of parasites. He has guided several Master and Doctoral students. Recipient of 3 Fellowships and 5 gold medals for academic excellence and research. Completed 3 national, one state and 30 other research projects.

Dr. Placid Eugene D'Souza has put in 33 years of service as a teacher and researcher in Veterinary Parasitology. He has guided 19 M.V.Sc and 5 Ph.D students, involved in conducting 30 training programmes for teachers/scientists/veterinary officers since 1996. Course Director for 11, twenty one days ICAR National training programmes for faculty. He is the recipient of two fellowships, two gold medals and seven National and state awards.

He has been involved in Editorial activities of popular monthly journals such as Poultry Adviser, Livestock Adviser, Veterinarian and a referee for large number of National and International journals besides Compendia of Trainings and conferences.

He has significantly contributed to the study of helminth, protozoan and arthropod parasites as evidenced by 39 research projects, 142 research papers and 96 presentations to his credit. His expertise has been availed by many National, State level bodies and Universities all over India.

Catalogue of Intermediate Hosts of Animal Parasites in India

K. MURALEEDHARAN
Former Professor and University Head of Veterinary Parasitology,
University of Agricultural Sciences, Bangalore

M.S. JAGANNATH
Former Professor and Founder Director, Centre of Advanced Faculty Training,
Department of Veterinary Parasitology,
University of Agricultural Sciences, Bangalore.
Former Professsor of Medical Parasitology, Faculty of Medicine,
Addis Ababa University, Addis Ababa, Ethiopia

PLACID E. D'SOUZA
Professor and Head cum Director, Centre of Advanced Faculty Training,
Department of Veterinary Parasitology,
Karnataka Veterinary, Animal and Fisheries Sciences University Regional Campus,
Hebbal, Bangalore

2014
Daya Publishing House®
A Division of
Astral International Pvt. Ltd.
New Delhi – 110 002

© 2014 AUTHORS
ISBN 9789351242963

Publisher's note:
Every possible effort has been made to ensure that the information contained in this book is accurate at the time of going to press, and the publisher and author cannot accept responsibility for any errors or omissions, however caused. No responsibility for loss or damage occasioned to any person acting, or refraining from action, as a result of the material in this publication can be accepted by the editor, the publisher or the author. The Publisher is not associated with any product or vendor mentioned in the book. The contents of this work are intended to further general scientific research, understanding and discussion only. Readers should consult with a specialist where appropriate.
Every effort has been made to trace the owners of copyright material used in this book, if any. The author and the publisher will be grateful for any omission brought to their notice for acknowledgement in the future editions of the book.
All Rights reserved under International Copyright Conventions. No part of this publication may be reproduced, stored in a retrieval system, or transmitted in any form or by any means, electronic, mechanical, photocopying, recording or otherwise without the prior written consent of the publisher and the copyright owner.

Published by   : **Daya Publishing House®**
A Division of
Astral International Pvt. Ltd.
– ISO 9001:2008 Certified Company –
4760-61/23, Ansari Road, Darya Ganj
New Delhi-110 002
Ph. 011-43549197, 23278134
E-mail: info@astralint.com
Website: www.astralint.com

Laser Typesetting   : **Classic Computer Services**, Delhi - 110 035

Printed at   : **Replika Press Pvt. Ltd.**

PRINTED IN INDIA

Karnataka Veterinary, Animal & Fisheries Sciences University

P.B. No. 6, Nandinagar, Bidar – 585 401, Karnataka

Dr. C. RENUKAPRASAD
M.V.Sc. Ph.D.
Vice Chancellor

Phone: 08482 245264
Fax: 08482 245107
E-mail: vckvafsu@yahoo.co.in
c.renukaprasad@gmail.com
www.kvafsu.edu.in

Foreword

There was a need felt for a comprehensive document on intermediate hosts of animal parasites in India. I am extremely happy to place on record that two retired professors and the current professor of the Department of Veterinary Parasitology of our University have made a maiden attempt to prepare a comprehensive treatise on intermediate hosts of animal Parasites in India in a catalogue from. In this book the up to date information on intermediate and susceptible experimental hosts arranged chronologically along with the related references would significantly lessen the laborious and time consuming task of accessing relevant literature. I am sure that this catalogue will enable the aspiring parasitologists to readily acquaint themselves with the quantum of work done in the past till date in identifying the intermediate hosts and their role towards controlling parasites. This single window portal will help the researchers to take up advanced studies within this area. In addition this book will be beneficial to teachers, biologists and reviewers in the subject. I feel that everyone who has interest in different aspects of Parasitology like biology, therapeutics, biochemistry, immunology and molecular biology should possess a personal copy for ready reference from the Indian perspective.

Dr. C. Renukaprasad
Vice Chancellor

Preface

Domestic animals harbour a wide variety of internal parasites which cause great economic loss to the exchequer. Many of these parasites are transmissible from animals to animals through various intermediate hosts. Large quantum of work involving intermediate hosts of animal parasites has been undertaken in India for long. Such information is well documented in the annals of various research publications as well as in unpublished forms in souvenirs, reports and theses. Immediate access or sifting specific information from them is a little difficult. A compilation of up to date information gathered on intermediate hosts of animal parasites in India to fill these lacunae has been felt for quite some time. Therefore an attempt has been made to collect such findings to prepare a catalogue on the intermediate hosts of parasites arranged chronologically along with authority concerned with year.

The catalogue will enable to locate the required details from the institutional libraries for further in-depth or detailed study as needed. Thus time consuming and laborious task of reference collections has been minimized so that researchers could find more time to focus on the selected topic for research.

Undoubtedly, this compact book serves as a ready reckoner to teachers, text-book writers of parasitology and scientists of veterinary and medical profession, practicing veterinarians, meat inspectors, physicians, surgeons and public health authorities, epidemiologists, zoologists, students and those engaged in controlling parasitic problems.

For the contents of the Catalogue, we are indebted to many authors who have provided reprints of their research findings. We acknowledge the facilities given for r◄►ing the journals in various institutional libraries and the internet. Finally, we ⊆✕ss our gratitude to M/s Astral International Private Limited, New Delhi -110 ◗◖ndia) for publishing this catalogue.

K. Muraleedharan

M.S. Jagannath

Placid E. D'Souza

Contents

Abbreviations

Abstr.: Abstract

Acad.: Academy

Adv.: Advances

Agri.: Agriculture

Ann.: Annual

Biol.: Biology

Coll.: College

Conf.: Conference

Centr.: Center

Cong.: Congress

Dis.: Disease/Diseases

E: Experimental

Ed./Eds.: Editor/Editors

FH: Final host

Hlth.: Health

ICAR: Indian Council of Agricultural Research

ICMR: Indian Council of Medical Research

IH: Intermediate host

Inaug.: Inaguration

Inst.: Institution

Internatl.: International

Med.: Medical/Medicine

N: Natural

Natl: National

NDDB: National Dairy Devevlopment Board

Nuclr.: Nuclear

p./pp.: Page/pages

PH: Paratenic host

Parasitol.: Parasitology

Proc.: Proceedings

Pt.: Part

Pub.: Public

Reg.: Regional

Rep.: Report

Res.: Research

Sect.: Section

Sci.: Science

RH: Reservoir host

Semr.: Seminar

Stud.: Studies

Symp.: Symposium

Syn.: Synonym

TH: Transport host

Vet.: Veterinary

Zool.: Zoology

Zoon.: Zoonosis

Introduction

The parasites of domestic animals are of great concern as they affect their health resulting in huge financial loss mostly due to reduction in milk, meat, eggs, wool and other by-products as well as working ability of draught animals in addition to higher mortality rate. Some of these parasites are transmissible to man or vice versa producing mild to serious health hazards. To avoid such maladies of high magnitude, necessary treatment has to be extended to the affected animals after timely detection of the presence of the parasites to eliminate them from the body. Implementation of appropriate measures for control of these parasites will reduce their occurrence.

Sound knowledge of the biology and life cycle of these parasites provides many clues to understand certain vulnerable points where we can check their growth and propagation which are very essential for controlling and eradicating parasitism. Many of the helminth and protozoan parasites are having either direct or indirect life cycles. The parasites with indirect life cycle require the presence of external agencies called intermediate hosts or vectors for their larval stages to develop and attain the infective stage which is capable of entering the definitive or final hosts. The transmission of infective larvae to final hosts occurs by various means such as penetration, ingestion and through their bites. From the portal of entry, further development takes place to juvenile stages which reach their site of predilection and attain sexual maturity.

Usually the intermediary agencies are derived from phylum Annelida, Arthropoda, Mollusca, Pisces, Reptilia and Mammalia which are involved in transmitting parasites of domestic animals of India. Some of the mammals act as reservoir or transport hosts in which no development of parasites takes place.

Trematodes require one or two intermediate hosts to complete their life-cycle. The fresh-water snails or rarely land snails act as first intermediate hosts. After completion of development of different larval forms in snails, the motile cercariae emerged from the infected snail, encyst either on vegetations or enter inside a second intermediate host and encyst to form metacercariae which are infective to final hosts. The second intermediate hosts may be snails, crabs, dragonfly naiads or fishes as in different species of trematodes. But in blood flukes no cercarial encystment takes place. Instead, cercariae released from the snails actively swim in water and they ultimately enter the final hosts.

With respect to cestodes and nematodes, the infection to final hosts takes place by ingestion of variety of intermediate hosts such as oribatid mites, grass-hoppers, beetles, ants, house-flies, fleas, earthworms, copepods, frogs, rats, lizards, squirrels, fishes, snakes, birds, mosquitoes and ticks. Various food animals act as intermediate hosts of taeniid tapeworms and from the consumption of their infected meat containing encysted larvae, man and animals like cats and dogs get this infection. Even certain rodents are known to act as intermediate hosts of certain tapeworms. The diphyllobothrid tapeworm requires two intermediate hosts as in trematodes and cyclops as well as fishes act as the first and second intermediate host respectively. The nematode, *Trichinella spiralis* is unique as the first stage larvae laid by adult females in final host without any free-living existence, encyst in the muscles of the same host. The encapsulated larvae later become infective to another host on ingestion the affected meat. Thereby the same infected animal serve as final host at one stage and intermediate host at a later stage.

Biting arthropods like tabanids, mosquitoes or ticks transmit the infective stage of blood protozoans which include *Trypanosoma, Plasmodia, Babesia* and *Theileria* through their bites. Various food animals like cattle, sheep and goats act as intermediate hosts whose flesh is eaten by carnivorous animals which serve as final hosts in *Sarcocystis*. While in the case of ingestion of oocysts shed in excreta by definitive or reservoir hosts, cats both domestic and wild, on sporulation or congenital means transmit infection to man and food animals.

Various herbivores act as intermediate hosts for the tongue worm, an arthropod parasite of domestic and wild carnivores, and rarely man.

Information on these intermediate hosts, mode of transmission of parasites to the final hosts, on different larval stages and on transmission experiments employing laboratory animals and vulnerable young host animals or development of immunological means will be available from this purview of this Catalogue arranged in nine parts.

Part I- Tables provide information by class or family-wise on intermediate hosts of parasites. Part II – Bibliography. Section III includes consolidated information on parasitic zoonosis involving intermediate hosts and Section IV deals with books providing information on intermediate hosts published by Indian authors. Part III provides page-wise index of parasites and Part IV provides page-wise index of research scientists.

In short, the Catalogue is a check list of nearly 450 intermediate hosts found in India related to 180 animal parasites which belong to 50 families. The works of about 1700 scientists, published in more than 200 journals, both in India and abroad or released research findings from the proceedings of conferences, symposia and reports research projects for the past one century were consulted for compiling this Catalogue.

PART I

INTERMEDIATE HOSTS OF PARASITES

Section I
HELMINTH PARASITES

I.A PHYLUM: PLATYHELMINTHES
1. CLASS: TREMATODA

Intermediate Host	Larva	Final Host	Reference
Final hosts include both primary and secondary hosts; E- Experimental infection; N-Natural infection			
1.1 FAMILY: DICROCOELIIDAE			
Dicrocoelium dendriticum (Rudolphi, 1819) Loss, 1899			
1st IH: Land snail (N)	Cercariae	—	Srivastava, 1944
1st IH: *Macrochlamys cassida Euaustenia monticola*	Cercariae	—	Bhalerao, 1946
1st IH: *Macrochlamys indica Zootecus insularis*	Cercariae	—	Anon (1944, 1946) (vide Chowdhury, 2001)
Eurytrema pancreaticum (Janson, 1889)			
Bradybaenia similaris	—	—	Gomathinayagam *et al.*, 2004 (cited)

1.2 FAMILY: OPISTHORCHIIDAE

Opisthorchis sp.

Intermediate Host	Larva	Final Host	Reference
1st IH F. W. snails: *Digoniostoma pulchella* (N) *Melania tuberculata* (N)	Cercariae	—	Mohandas, 1974
Melania tuberculata (N) *Melania tuberculata* (N) *Melania scabra* (N) *Digoniostoma pulchella* (N)	*Cercariae indicae* VIII Sewell, 1922 *Cercaria* sp. III Kerala Mohandas, 1976 *Cercaria* sp. IV Kerala	—	Mohandas, 1976
2nd IH Fish: *Puntius sophore* (N)	Metacercariae		Rekha Devi *et al.*, 2005

Opisthorchis caninus (Lewis and Cunningham, 1872) Baker, 1911

Intermediate Host	Larva	Final Host	Reference
1st IH: *Melanoides tuberculata* (N)	Cercariae		Anon, 1963 (vide Chowdhury, 2001)
Fish 2nd IH: Murrel *Ophiocephalus punctatus* (N) Carp *Cirrhina mrigala* (N) *Labeo rohita* (N)	Metacercariae	Guinea pig (E)	Rai and Pande, 1965

Intermediate Host	Larva	Final Host	Reference
2nd IH: *Ophiocephalus punctatus* (N) *Cirrhina mrigala* (N) *Labeo rohita* (N)	Metacercariae	Pups (E)	Rai, 1966
2nd IH: *Channa* (*Ophiocephalus*) *punctatus* (N)	Metacercariae	Albino mice (E) Rat (E)	Pande and Premvati, 1974
2nd IH: *Channa* (*Ophiocephalus*) *punctatus* (N)	Metacercariae	Hamster (E) Rhesus monkey (E)	Pande and Shukla, 1974
Ophiocephalus punctatus (N)	Metacercariae+	Pups (E)	Pramod Kumar *et al.*, 1983
Ophiocephalus punctatus (N) *Ophiocephalus striatus* (N) *Ophiocephalus marulius* (N)	Metacercariae	Rabbit (E) Guinea pig (E) Albino mice (E) Pups (E)	Vijay Sachadeva, 1994
Opisthorchis noverca Braun, 1902			
Melanoides (*Tarebia*) *lineatus* (N)	Cercaria, probably, pleurolophocercous	—	Sahai, 1969-70
Melanoides (*Tarebia*) *lineatus* (N) 1st IH	Cercariae	—	Sahai and Srivastava, 1972
Snail: 1st IH *Melanoides* (*Tarebia*) *lineatus* (N) Fish: 2nd IH *Cirrhinus reba* (N) (E) *Labeo bata* (N) (E)	Sporocyst Redia Cercaria, pleurolophocercous Metacercariae	Pups (E) Dogs (N) Pigs (N)	Sahai and Srivastava, 1978

1.3 FAMILY: FASCIOLIDAE

Fasciola sp.

Intermediate Host	Larva	Final Host	Reference
Lymnaea auricularia (N)	Cercariae	—	Sanai, 1969-70
Indoplanorbis exustus (N) *Lymnaea acuminata* (N)	*Cercariae*	—	Sandey, 1982
Lymnaea (Pseudosuccinea) acuminata	Cercariae	—	Pokhriyal et al., 1996
Thiara (M.) tuberculata (N)	Cercariae	—	Pokhriyal et al., 1998
Thiara (Melanoides) tuberculata (N) *Lymnaea luteola* (N) *Bellamya bengalensis* (N)	Cercariae	—	Tamloorkar et al., 2001
Liver fluke and amphistome (combined information): *Lymnaea acuminata* (N) *Lymnaea luteola* (N) *Thiara (Melanoides) tuberculata* (N) *Indoplanorbis exustus* (N) *Paludomus obesus* (N) *Ferrissia tenuis* (N) *Ballamya bengalensis* (N) *Bellamya dissimilis* (N)	Cercariae	—	Pemola Devi and Jauhari, 2008

Fasciola gigantica Cobbold, 1885

Intermediate Host	Larva	Final Host	Reference
Lymnaea acuminata (N)	Cercariae Metacercariae	Bull (E) Goats (E)	Bhalerao, 1933

Intermediate Host	Larva	Final Host	Reference
Lymnaea acuminata (N)	Cercariae		Bhalerao, 1943
Lymnaea acuminata (N) Lymnaea luteola (N)	Cercariae		Srivastava, 1944a
Lymnaea acuminata (N) Lymnaea (psuedosuccinea) acuminata (N)	Cercariae Metacercariae	Guinea pig (E)	Thapar and Tandon, 1952
Lymnaea (psuedosuccinea) acuminata (N) Lymnaea F. patula (N)	Cercariae	—	Singh and Malaki, 1963
Lymnaea auricularia var rufescens (N)	Cercariae Metacercariae	Lambs (E)	Deo et al., 1967
Lymnaea auricularia var rufescens (N)	Cercariae	—	Patnaik, 1968
Lymnaea auricularia var rufescens (N)	Cercariae Metacercariae	Guinea pig (E) Rabbits (E)	Patnaik and Ray, 1968
Lymnaea auricularia var rufescens (N)	Cercariae	—	Patnaik and Acharya, 1971
	Cercariae Metacercariae	Rabbits (E)	Singh, 1971a
	Cercariae Metacercariae	Guinea pig (E)	Singh, 1971b
Snails (N)	Cercariae Metacercariae	—	Chandra, 1972
Lymnaea acuminata (N)	Cercariae Metacercariae	Guinea pig (E)	Srivastava and Singh, 1972a

Intermediate Host	Larva	Final Host	Reference
Lymnaea acuminata (N)	Cercariae Metacercariae	Guinea pig (E)	Srivastava and Singh, 1972 b
	Metacercariae	Rhesus monkey (E)	Anon, 1971/1972 (vide Chowdhury, 2001)
	Cercariae Metacercariae	Laboratory animals (E)	Srivastava and Singh, 1974
Lymnaea auricularia var rufescens (N)	Cercariae	—	Das, 1976
Lymnaea acuminata (N)	Larval stages		Yadav and Karyakarte, 1979
Lymnaea auricularia var rufescens (N)	Cercariae	—	Dutt and Bali, 1980
Lymnaea auricularia var rufescens (N)	Cercariae	—	Das and Subramanian, 1981
Lymnaea acuminata (N)	Cercariae	—	Sardey, 1982
	Cercariae Metacercariae	Cow calves (E)	Bhatia and Roy Choudhury, 1985
Lymnaea acuminata (N)			Rao et al., 1985
Lymnaea auricularia (N)	Cercariae Metacercariae	—	Gupta et al., 1986
Lymnaea auricularia (E)	Cercariae Metacercariae	Guinea pig (E) Rabbits (E)	Gupta and Chandra, 1987
Lymnaea acuminata (N)	Cercariae	Farm animals (N)	Gupta and Paul, 1987
Lymnaea auricularia rufescens (N)	Cercariae Metacercariae	Sheep (E)	Khajuria and Bali, 1987a

Intermediate Host	Larva	Final Host	Reference
Lymnaea auricularia rufescens (N)	Metacercariae		Khajuria and Bali, 1987b
	Metacercariae	Rabbits (E)	Yadav and Gupta, 1988
Lymnaea auricularia (E)	Cercariae	Sheep (E)	Gupta *et al.*, 1989
	Metacercariae	Goat (E)	
		Buffalo calves (E)	
		Rabbits (E)	
Lymnaea auricularia race *rufescens* (N)	Cercariae (N)	—	Prasad, 1989
	Metacercariae (N)		
Lymnaea auricularia sensustricto (N)	Metacercariae (N)	Guinea pig (E)	Sharma *et al.*, 1989
	Metacercariae	Lambs (N)	Banerjee and Singh, 1991
Lymnaea auricularia (E)	Cercariae	Rabbits (E)	Gupta and Yadav, 1992
	Metacercariae	Goats (E)	
		Buffaloes (E)	
	Cercariae	Kids (E)	Yadav and Gupta, 1992
	Metacercariae		
Lymnaea auricularia race *rufescens* (N)	Cercariae		Chaudhari *et al.*, 1993
Lymnaea auricularia (E)	Metacercariae	Rabbits (E)	Gupta and Yadav, 1993a
Lymnaea auricularia (N)	Cercariae	—	Gupta and Yadav, 1993b
	Cercariae	Rabbits (E)	Yadav and Gupta, 1993
	Metacercariae		
Lymnaea auricularia (N)	Cercariae	—	Gupta and Yadav, 1994a

Intermediate Host	Larva	Final Host	Reference
	Metacercariae	Rabbits (E)	Gupta and Yadav, 1994b
	Metacercariae	Rabbits (E)	Gupta and Yadav, 1994c
		Goats (E)	Yadav and Gupta, 1994
Lymnaea auricularia (N)	Cercariae	—	Bhatnagar and Gupta, 1995
Lymnaea auricularia (E)	Cercariae Metacercariae	Rabbits (RB1 to Rb4)	Gupta and Yadav, 1995a
Lymnaea auricularia (E)	Cercariae Metacercariae	Rabbits (RB1 to Rb4)	Gupta and Yadav, 1995b
Lymnaea auricularia (E)	Cercariae Metacercariae	Goats (E)	Yadav and Gupta, 1995a
Lymnaea auricularia (E)	Cercariae Metacercariae	Goats (E)	Yadav and Gupta, 1995b
Lymnaea auricularia (N)	Cercariae	—	Rai et al., 1996
	Metacercariae	Calves, male (B. taurus x B. indicus) (E) Buffalo calves (E)	Sanyal, 1996
Lymnaea auricularia (E)	Metacercariae	Buffalo	Sanyal and Gupta, 1996a
Lymnaea auricularia (E)	Metacercariae	Buffalo	Sanyal and Gupta, 1996b
Lymnaea auricularia (E)	Cercariae Metacercariae	Kids (E)	Yadav and Gupta, 1996a
	Cercariae Metacercariae	Rabbits (E) Goats (E)	Yadav and Gupta, 1996b

Intermediate Host	Larva	Final Host	Reference
Lymnaea auricularia (N)	Cercariae	—	Yadav *et al.*, 1997
Lymnaea auricularia (E)	Metacercariae	Sheep (N)	Mandal *et al.*, 1998
Lymnaea auricularia (E)	Metacercariae	Calves (cattle, buffaloes) (E)	Sanyal, 1998
Lymnaea auricularia (E)	Metacercariae	Calves (cattle, buffaloes) (E)	Sanyal and Gupta, 1998
	Metacercariae	—	Yadav and Gupta, 1998a
	Metacercariae	Rabbits (E)	Yadav and Gupta, 1998b
	Metacercariae	Buffalo calves (E)	Mehra *et al.*, 1999
	Metacercariae (viability)	—	Prasad *et al.*, 1999
Lymnaea auricularia rufescens (N)	Metacercariae	Lambs (E)	Santra *et al.*, 1999a
Lymnaea auricularia rufescens (N)	Metacercariae	Lambs (E)	Santra *et al.*, 1999b
Lymnaea auricularia (N)	Cercariae Metacercariae	Rabbits (E)	Shikari *et al.*, 1999
Lymnaea auricularia (E)	Cercariae Metacercariae	Calves, buffalo, male (E)	Yadav *et al.*, 1999
Lymnaea auricularia (N)	Cercariae	—	Bhatnagar and Prasad, 2001
Lymnaea auricularia (N)	Cercariae	—	Yadav *et al.*, 2001
Lymnaea auricularia (E)	Metacercariae	Sheep (E)	Dixit *et al.*, 2002

Intermediate Host	Larva	Final Host	Reference
Lymnaea auricularia (N)	Cercariae	Calves, CB (E) Metacercariae	Velusamy, 2002
Lymnaea auricularia (E)	Cercariae Metacercariae	Calves, buffalo (E) Sheep (E)	Dixit et al., 2003
	Metacercariae	Buffaloes (N)	Ganga et al., 2003
Lymnaea auricularia (N)	Cercariae Metacercariae	Cattle (N)	Gupta et al., 2003
	Metacercariae	Buffaloes (N)	Dixit et al., 2004
	Metacercariae	Buffaloes (N)	Ganga et al., 2004
	Cercariae Metacercariae	Sheep (E) Cattle (E) Buffaloes (E)	Raina, et al., 2004
	Metacercariae		Upadhyay, and Kumar, 2004
Lymnaea auricularia (N)	Cercariae Metacercariae	—	Velusamy et al., 2004a
Lymnaea auricularia (N)	Cercariae Metacercariae	Calves, H.F (E)	Velusamy et al., 2004b
Lymnaea auricularia (N)	Cercariae Metacercariae	—	Velusamy et al., 2004c
Lymnaea auricularia (N)	Cercariae Metacercariae	Calves, CB (E)	Velusamy et al., 2004d
	Metacercariae	Cattle (E)	Ghosh et al., 2005a

Intermediate Host	Larva	Final Host	Reference
	Metacercariae	Cattle, CB (E) Rabbits (E)	Ghosh et al., 2005b
	Metacercariae	Calves, H.F., male (E)	Jayraw et al., 2005a
	Metacercariae	Calves, H.F., male (E)	Jayraw et al., 2005b
	Metacercariae	Calves, H. F., male (E)	Jayraw et al., 2005c
	Metacercariae	Calves, buffalo (E)	Nambi Azhahia et al., 2005
	Metacercariae	Calves, buffalo (E)	Pal et al., 2005
		Cattle (E) Buffalo (E) Sheep (E)	Raina et al., 2005
Lymnaea auricularia (N)	Cercariae Metacercariae		Singh et al., 2005
Lymnaea auricularia (N)	Cercariae Metacercariae	Sheep (E) Buffalo calves (E)	Yadav et al., 2005
	Metacercariae	Cattle (E) Buffalo (E) Sheep (E) Rabbit (E)	Yokananth et al., 2005
Lymnaea auricularia (N)	Cercariae Metacercariae	New Zealand white rabbits (E) Buffalo calves (E)	Ghosh et al., 2006
	Metacercariae	Buffalo calves (E)	Gupta et al., 2006

Intermediate Host	Larva	Final Host	Reference
	Metacercariae	Buffalo calves (E)	Raina et al., 2006
	Metacercariae	Cattle (E)	Sriveny, et al., 2006
Lymnaea auricularia (N)	Cercariae Metacercariae	Calves, cross bred, male (E)	Velusamy et al., 2006
Lymnaea auricularia (N)	Cercariae Metacercariae		Yadav et al., 2006a
Lymnaea auricularia (N)	Cercariae Metacercariae	—	Yadav et al., 2006b
	Metacercariae	Buffalo yearlings, male (E)	Ganga et al., 2007
Lymnaea auricularia (N)	Metacercariae	Buffalo (E)	Yadav et al., 2007
	Metacercariae	Buffalo calves (E)	Pal and Das Gupta, 2007
	Metacercariae	Calves, CB, Bos taurus x Bos indicus (E)	Ingale et al., 2008
	Metacercariae (Lab. maintained)	Buffalo calves (E) New Zealand white rabbits (E)	Niranjan Kumar et al., 2008a
	Metacercariae (Lab. maintained)	Buffalo calves (E) New Zealand white rabbits (E)	Niranjan Kumar et al., 2008b
	Metacercariae (Lab. maintained)	Buffaloes (N) (E)	Ghosh et al., 2009

Intermediate Host	Larva	Final Host	Reference
	Metacercariae	Calves, male, H. F cross	Jayaraw *et al.*, 2009
	Metacercariae	Buffaloes (N) (E)	Jedappa *et al.*, 2009
	Metacercariae	Mice (E) Rabbit (E)	Raina *et al.*, 2009
Lymnaea auricularia (N)	Cercariae	Cattle (N) Buffaloes (N) Sheep (N) Goats (N)	Rajat Garg *et al.*, 2009
Lymnaea auricularia (N)	Cercariae	—	Singh *et al.*, 2009
		Mice (E)	Smitha *et al.*, 2009
	Metacercariae	Mice (E)	Tripathi *et al.*, 2009
Lymnaea auricularia (N)	Cercariae Metacercariae	Calves, CB, male (E)	Velusamy *et al.*, 2009
Lymnaea auricularia (N)	Cercariae	—	Vohra and Agrawal, 2009
Lymnaea auricularia (N)	Cercariae	—	Vohra *et al.*, 2009
Lymnaea auricularia (N)	Cercariae	—	Yadav, *et al.*, 2009
Lymnaea auricularia (E)	Metacercariae (Bubaline origin)	Buffaloes, Murrah yearlings, riverine (E)	Edith *et al.*, 2010a
Lymnaea auricularia (E)	Metacercariae (Bubaline origin)	Buffaloes, Murrah yearlings, riverine (E)	Edith *et al.*, 2010b
Lymnaea auricularia (E)	Metacercariae (Bubaline origin)	Buffaloes, Murrah yearlings, riverine (E)	Edith *et al.*, 2010c

Intermediate Host	Larva	Final Host	Reference
	Metacercariae	Buffaloes, Murrah yearlings, riverine (E)	Ganga et al., 2010
	Metacercariae	Buffaloes (N) (E) Cattle (N) (E)	Jedappa et al., 2010
Lymnaea auricularia (N)	Metacercariae		Nagar Gaurav et al., 2010
		Mice (E)	Smitha et al., 2010
Lymnaea auricularia (E)	Metacercariae (Bubaline origin)	Buffalo yearlings, male (N) (E)	Edith et al., 2011a
Lymnaea auricularia (E)	Metacercariae (Bubaline origin)	Buffaloes, Murrah yearlings, riverine (E)	Edith et al., 2011b
Lymnaea auricularia (N)	Metacercariae (200)	Cross bred calf (E)	Ani Bency Jacob et al., 2011
Lymnaea auricularia (E)	Metacercariae	Buffalo calves (N) (E)	Niranjan Kumar et al., 2011
	Metacercariae (bubaline origin)	Buffaloes (N) (E)	Raina et al., 2011
	Metacercariae	Buffalo, male Murrah breed	Singh et al., 2011
Lymnaea acuminata (N)	Sporocysts Rediae Cercariae	—	Sunitha et al., 2011
Lymnaea auricularia (N)	Metacercariae		Ani Bency Jacob et al., 2012
Lymnaea auricularia (E)	Metacercariae (Bubaline origin)	Buffaloes, Murrah yearlings, riverine (E)	Edith et al., 2012

Intermediate Host	Larva	Final Host	Reference
Lymnaea auricularia (N)			Gupta *et al.*, 2012
Indoplanorbis exustus (N)	Larval stages	—	Neeha Singh *et al.*, 2012
Lymnaea acuminata (N)	Sporocysts Rediae Cercariae	—	Sunita *et al.*, 2012
	Metacercariae (fresh)	Buffalo, Murrah male	Singh *et al.*, 2012
Lymnaea acuminata (N)	Sporocysts Rediae Cercariae	—	Sunita *et al.*, 2013a
Lymnaea acuminata (N)	Sporocysts Rediae Cercariae	—	Sunita *et al.*, 2013b
Lymnaea acuminata (N)	Sporocysts Rediae Cercariae	—	Sunita *et al.*, 2013c
Lymnaea acuminata (N)	Rediae Cercariae	—	Sunita *et al.*, 2013d
	Fasciola hepatica Linnaeus, 1758		
Lymnaea acuminata (E) *Lymnaea luteola* (E)	Cercariae	—	Srivastava, 1944b
		Mice (E)	Anon, 1987 (vide Chowdhury, 2001)

Intermediate Host	Larva	Final Host	Reference
Fasciola indica Varma, 1953			
Lymnaea auricularia race rufescens (N), probable	Cercariae	—	Varma, 1954
Lymnaea acuminata (E)	Cercariae Metacercariae	Guinea pigs (E)	Tandon, 1970
Fasciolopsis buski (Lankester, 1857) Buckley, 1939			
Segmentina trochoideus (N)	Cercariae	—	Buckley, 1939
Helicorbis coenosus (Syn: Segmentina coenosus) (E)	Cercariae Metacercariae	Pig (E)	Tripathi et al., 1973
Helicorbis coenosus (E)	Cercariae Metacercariae	Guinea pigs (E) Rabbits (E) Pig (E)	Malaviya, 1985
Lymnaea auricularia sensu stricto (N)		—	Anon, 1985, 1988 (vide Chowdhury, 2001)
Hippeutis coenosus (N)		—	Malaviya and Varma, 1989
Indoplanorbis exustus (N)	Cercariae	—	Preet and Prakash, 2001a
Indoplanorbis exustus (N)	Cercariae	—	Preet and Prakash, 2001b
Gyraulus convexiusculus (N)		—	Hafeez, Md., 2003 (citation)
Segmentina hemispharerula (N) Hippeutis umbilicalis (N)		—	Vena Tandon et al., 2012 (citation)

1.4 FAMILY: ECHINOSTOMATIDAE

Intermediate Host	Larva	Final Host	Reference
Artyfechinostomum malayanum (Leiper, 1911) Mendheim, 1943 (Syn: Echinostoma malayanum Leiper, 1911, see below)			
Frog: *Rana cyanophlyctis* 2nd IH (N)	Metacercariae	Pigs (N) Rats, Albino (E)	Premvati and Pande, 1974
Artyfechinostomum oraoni Bandyopadhyay, Manna and Nandy, 1989			
F. W. Snail: *Lymnaea* spp. (N) 1st IH	Rediae developed from miracidia		Maji et al., 1995
		Pigs (N), RH	Bandyopadhyay et al., 1995
Artyfechinostomum sufrartyfex Lane, 1915 (See also Echinostoma malayanum Leiper, 1911)			
Indoplanorbis exustus (N)	Cercariae		Peter, 1955
F. W. Snails: 1st IH *Indoplanorbis exustus* *Lymnaea luteola f. australis* *Digoniostoma pulchella*	Metacercariae	White rats (E) Piglets (E)	Matta and Pande, 1966
Rana cyanophlyctis 2nd IH (N)	Metacercariae	Pup (E) Rabbits (E)	Dharmendra Nath, 1969
Indoplanorbis exustus 2nd IH (N) *Rana cyanophlyctis* 2nd IH (N)	Metacercariae	Piglets (E)	Agrawal, 1971

Intermediate Host	Larva	Final Host	Reference
Indoplanorbis exustus 2nd IH (N); *Rana cyanophlyctis* (adults and tadpoles) 2nd IH (N)	Metacercariae	Piglets (E)	Agrawal and Pande, 1972
Pond frog: *Rana cyanophlyctis* 2nd IH (N)	43- collar spined metacercariae	Probably, Pigeon (E); Chick (E); Duck (E); Albino rat (E)	Dharmendra Nath, 1972a
Pond frog: *Rana cyanophlyctis* 2nd IH (N)	43- collar spined metacercariae	Spiny- tailed lizard: *Uromastix hardwickii* (E)	Dharmendra Nath, 1972b
Lymnaea luteola (N); *Lymnaea acuminata* (N); *Indoplanorbis exustus* (N); *Digoniostoma pulchella* (N); *Vivipara bengalensis* (N)	Metacercarial cyst, echinochasmid 43- spined	Albino rat (E); Albino mice (E); Rabbit (E); Pup (E); Spiny tailed lizard (E)	Dharmendra Nath, 1973a

Echinochasmus sp.

Intermediate Host	Larva	Final Host	Reference
Fish: *Aplocheilus panchax* (N); *Aplocheilus megastigma* (N)	Metacercariae		Madhavi, 1980

Echinochasmus baugulai, Verma, 1935

Intermediate Host	Larva	Final Host	Reference
Snails: Marine *Natica marochiensis* (N); Bivalve	Cercariae; Metacercariae	Birds (N); Gull-billed tern (*Gelochelidon nilotica*) (N)	Ramalingam, 1960

Intermediate Host	Larva	Final Host	Reference
Katelysia opima (N)		Black headed gull (Larus ridibundus) (N) Curlew (Numenius arquata) (N) Common or fan tailed snipe (Capella gallinago) (N) Little egret (Egreta garzetta) (N) Pond heron (Ardeola grayii) (N) Domestic duck (Anas domestica) (E)	
Alocinma travancorica (N)	Aegilis group of Gymnocephalus cercariae; identical to Cercariae indicae XLI	Ardeola grayii (N) Chicks, White Leghorn (E)	Madhavi et al., 1989
Fishes: Gambusia affinis (E)	Metacercariae		
Snail: Thiara tuberculata (N) Fishes: Aplocheilus panchax (N) (E) Oryzias melastigma (N) Gambusia affinis (N) Channa punctata (N)	Cercariae Gymnocephalus type Metacercariae	Ardeola grayi (N) White Leghorn Chicks (E)	Dhanumkumari et al., 1991

Intermediate Host	Larva	Final Host	Reference
	Echinochasmus corvus Bhalerao, 1926		
Pond Fish: *Ophiocephalus punctatus* (N)	Metacercariae	Chicks , White Leghorn (E) Rhode Island Red (E) Pigeons (E)	Dr-armendra Nath and Pande, 1970
Ophiocephalus punctatus (N)	Metacercariae, 24 spined	Domestic poultry (E)	Dr-armendra Nath, 1973b
Ophiocephalus punctatus (N)	Metacercariae, 24 spined	Domestic poultry (E)	Dr-armendra Nath, 1974a
Ophiocephalus punctatus (N)	Metacercariae, echino-chasmid, 24 spined	Chicks, WLH (E) Pigeons (E)	Dr-armendra Nath, 1974b
	Echinoparyphium sp.		
Lymnaea luteola f. australis (N) *Inoplanorbis exustus* (N) -1st IH (N) *Lymnaea luteola* *Indoplanorbis exustus* Tadpoles-2nd IH (N)	Cercariae Metacercariae	Duckings (E) Chicken (E) Pigeon (E)	Singh, 1978
	Echinoparyphium dunni Lei and Umathevy, 1965		
Lymnaea luteola 1st and 2nd IH (E) *Indoplanorbis exustus* 2nd IH (N)	Cercariae and metacercariae	Duck (E)	Sreekumar, 1966
Lymnaea luteola f. impura 2nd IH (N) *Indoplanorbis exustus* 2nd IH (N)	Metacercariae	Duck (E)	Sreekumar and Peter, 1973

Intermediate Host	Larva	Final Host	Reference
Echinoparyphium flexum (Linton, 1892) Dietz, 1960			
Vivipara bengalensis 2nd IH (N)	Metacercariae	Chicks (E) Pigeons (E)	Matta and Pande, 1967
Vivipara bengalensis 2nd IH (N)	Metacercariae	—	Rai and Panda, 1967
Vivipara bengalensis race mandiensis 2nd IH (N)	Metacercariae	Chicks (E)	Muraleedharan and Pande, 1967
2nd IH *Vivipara bengalensis* (N) metacercarial cyst	Echinostomatid, 45-spined	Pigeon (E) Chick (E) Domestic duck (E) Albino rat (E)	Dharmendra Nath, 1973a
2nd IH *Vivipara bengalensis* (N)	Echinostomatid, 45-spined metacercarial cyst	Pigeon (E) Chick (E) Domestic duck (E) Albino rat (E)	Dharmendra Nath, 1973b
2nd IH *Vivipara bengalensis* (N)	Echinostomatid, 45-spined metacercaria	Chicks (E)	Dharmendra Nath, 1975
Echinostomes / Echinostomatoid			
Lymnaea luteola (N) *Gyraulus convexiusculus* (N)	Cercariae	—	Mukherjee, 1966
Lymnaea auricularia (N) *Lymnaea luteola* (N) *Indoplanorbis exustus* (N)	Cercariae	—	Sahai, 1969-70

Intermediate Host	Larva	Final Host	Reference
2nd IH *Rana cyanophlyctis* (N)	Metacercariae, echinostomatid, 27- spined echinostomatid, 51- spined echinostomatid, 55- spined		Dharmendra Nath, 1972a
2nd IH *Vivipara bengalensis* (N) *Rana cyanophlyctis* (N) *Ophiocephalus punctatus* (N).	Metacercariae, echinostomatid, 55 spined		Dharmendra Nath, 1973c
2nd IH *Ophiocephalus punctatus* (N)	1.Metacercariae, echinostomatid 2. Metacercariae, echinostomatid, 55 spined		Dharmendra Nath, 1974b
Digoniostoma pulchella (N) *Idiopoma dissimilis* (N) *Lymnaea luteola* (N) *Indoplanorbis exustus* (N) *Melania tuberculata* (N)	Cercariae		Mohandas, 1974a
Indoplanorbis exustus (N) *Lymnaea* sp. (*Lymnaea acuminata* and *Lymnaea luteola*) (N)	Cercariae	—	Muraleedharan et al., 1973
Vivipara bengalensis (N)	Larval stages		Borkatoky and Das, 1980
Lymnaea luteola (N) *Lymnaea auricularia rufescens* (N)	Cercariae	—	Dutt and Bali, 1980

Intermediate Host	Larva	Final Host	Reference
Idiopom 76a dissimilis (N), 1st &2nd IH	Cercariae Metacercariae		Mohandas, 1981
L. luteola f. typica (N) *Indoplanorbis exustus* (N) *Gyraulus convexiusculus* (N)	Cercariae	—	Chaudhri *et al.*, 1982
Lymnaea acuminata (N) *Lymnaea auricularia* (N) *Lymnaea luteola* (N) *Indoplanorbis exustus* (N) *Gyraulus convexiusculus* (N)	Cercariae	—	Choubisa, 2002
Lymnaea acuminata f. patula (N) *Lymnaea acuminata f. chlamys* (N) *Lymnaea acuminata f. typical* (N) *Lymnaea luteola f. rufescens* (N) *Indoplanorbis exustus* (N) *Gyraulus convexiusculus* (N) *Vivipara bengalensis race gigantea* (N) *Vivipara bengalensis race mandiensis* (N)	Cercariae	—	Choubisa, 2008
Freshwater molluscs (N) (*Indoplanorbis exustus* and *Bellamyia bengalensis*)	Cercariae	—	Johnpaul *et al.*, 2010

Intermediate Host	Larva	Final Host	Reference
Echinostoma ivaniosi Mohandas, 1973			
Lymnaea luteola f. typica (N), E) Lymnaea acuminata f. gracilor (N, E) Lymnaea luteola (E), both 1st and 2nd IH Lymnaea acuminata (E) Indoplanorbis exustus (E) Idiopoma dissimilis (E)	Echinostomatid 37 spined cercaria and metacercaiae	Fowl (*Gallus gallus domesticus*) (E)	Mohandas, 1973
Idiopoma dissimilis (N)	Cercariae		Mchandas, 1981
Echinostoma luteola Singh, 1977			
Lymnaea luteola (N)	Cercariae		Singh, 1977
Echinostoma malayanum Leiper, 1911 **(Syn: Artyfechinostomum sufratyfex, Lane, 1915;** Artyfechinostomum malayanum (Leiper, 1911) Mendheim, 1943)			
Indoplanorbis exustus (N), 1st IH	Cercaria with 43-45 collar spines	—	Mohandas, 1971
Indoplanorbis exustus (N) (E), 2nd IH Lymnaea luteola f. typica 2nd IH (N) (E)	Metacercariae Metacercariae	White rats (E) White mouse (E) White rats (E) White mouse (E)	
Indoplanorbis exustus (N), 1st IH	Sporocysts Rediae Cercariae		Mchandas, 1974a

Intermediate Host	Larva	Final Host	Reference
Snail (N)	Cercaria	—	Mohandas, 1974b
Indoplanorbis exustus (N) (E) Lymnaea luteola f. typica 2nd IH (E)	Cercaria Metacercaria	White rats (E)	Mohandas, 1974c
Indoplanorbis exustus 2nd IH (N)	Metacercaria	Albino rats (Rattus norvegicus) (E)	Mohandas and Nadakal, 1978
Echinostoma revolutum (Frolich, 1802) Looss, 1899			
Lymnaea 1st IH (N)	Cercariae Echinostoma revoluti, Beaver	—	Peter, 1954
Lymnaea luteola f. succinea 1st IH (N)	Cercariae Echinostoma revoluti, Beaver, 1937	—	Peter, 1955
Lymnaea luteola f. succinea 1st IH (N)	Cercariae Echinostoma revoluti, Beaver, 1937	—	Peter, 1957
Lymnaea auricularia var. rufescens (N)			Patnaik, 1968
Lymnaea auricularia var. rufescens 1st & 2nd IH (E) Lymnaea luteola 2nd IH (N)	Cercariae Metacercriae	Fowl (E) Pigeons (E)	Patnaik and Ray, 1966
Lymnaea auricularia var. rufescens 1st & 2nd IH (E)	Cercariae Metacercriae	—	Patnaik and Ray, 1967
Lymnaea auricularia var. rufescens (N) 1st and 2nd IH	Cercariae Metacercriae	—	Patnaik and Ray, 1968
Vivipara bengalensis (N)	Echinostmatid, 37- spined metacercarial cyst	Pigeon (E) Chick (E) Domestic duck (E)	Dharmendra Nath, 1973a

Intermediate Host	Larva	Final Host	Reference
Lymnaea luteola f. impura, 1st IH (E) Indoplanorbis exustus, 2nd IH Lymnaea luteola f. impura, 2nd IH Tadpoles, 2nd IH	Cercariae Metacercariae	Duck (E)	Sreekumar, 1966
Lymnaea luteola –1st IH (E) Indoplanorbis exustus Lymnaea luteola var. impura Tadpoles, 2nd IH (E)	Cercariae Metacercariae	Duck (E)	Sreekumar and Peter, 1973
Idiopoma dissimilis (N)	Cercariae	—	Mohandas, 1971
Snail (N)	Cercariae	—	Mohandas, 1974b
Idiopoma dissimilis (N)	Cercariae	—	Mohandas, 1981
Euparyphium malayanum Leiper, 1911			
Lymnaea acuminata 1st IH (N) Lymnaea ovata 1st IH (N)	Cercariae indicae XXIII	—	Sewell, 1922
Lymnaea luteola 1st IH (E)	Cercariae indicae XXIII, Sewell, 1922	Dog (E) Cat (E)	Rao, 1933
Lymnaea luteola (N)	Cercariae indicae XXIII, Sewell, 1922	—	Mudaliyar and Alwar, 1947
Lymnaea luteola f. typica (N)	Cercariae indicae XXIII, Sewell, 1922	—	Peter, 1955
Lymnaea luteola f. succinea (N)	Cercariae indicae XXIII, Sewell, 1922	—	Peter, 1955
Idiopoma dissimilis (N)	Cercariae		Mohandas, 1981

Intermediate Host	Larva	Final Host	Reference
Hypoderaeum conoideum (Bloch, 1782)			
Lymnaea luteola f. impura 1st IH (E) *Indoplanorbis exustus* (E) 2nd IH	Cercaria Metacercaria	Duckling (E)	Sreekumar, 1966
Lymnaea luteola f. impura 1st IH (E) *Indoplanorbis exustus* (E) 2nd IH *Lymnaea luteola,* 2nd IH	Cercaria Metacercaria	Ducking (E)	Sreekumar and Peter, 1973
Stephanoprora pandei Nath, 1973			
Freshwater fish: *Puntius sophore* (N)	Metacercaria, 22 spined	White leghorn Chicks (E)	Dharmendra Nath, 1971
Puntius sophore (N)	Metacercaria, echinochasmid, 22 spined	Chicks (E)	Dharmendra Nath, 1974
1.5 FAMILY: HETEROPHYIDAE			
Centrocestus formosanus (Nishigori, 1924)			
Cirrhina reba (N) *Amblypharyagodon mola* (N) *Labeo bata* (N) *Puntius* sp. (N)	*Centrocestus* type metacercaria	Albino rats (E) Guinea pig (E) Pups (E) Cockerels, WLH (E) Pigeon (E)	Dharmendra Nath and Pande, 1970
Cirrhina reba (N) *Amblypharyagodon mola* (N) *Labeo bata* (N) *Puntius sarana* (N)	*Centrocestus* type metacercaria	Pigeons (E) Chick (E) Domestic ducks (E)	Dharmendra Nath, 1972a

Intermediate Host	Larva	Final Host	Reference
Cirrhina reba (N) *Amblypharyagodon mola* (N) *Labeo bata* (N) *Puntius sarana* (N)	*Centrocestus* type metacercaria	Pigeons (E)	Dharmendra Nath, 1972b
Fishes (13 species): *Esomus danricus* (N) *Puntius sophore* (N) *Puntius chola* (N) *Puntius tincto* (N) *Nandus nandus* (N) *Osteobrame cotio* (N) *Xenentodon cancila* (N) *Channa punctatus* (N) *Notopterus notopterus* (N) *Chela laubuca* (N) *Oxygaster phulo* (N) *Oxygaster bacaila* (N) *Mastacembelus puncalus* (N)	*Centrocestus* type metacercaria	Hamster (E) Rhesus monkey (E)	Fande and Shukla, 1972
Cirrhina reba (N) *Amblypharyagodon mola* (N) *Labeo bata* (N) *Puntius sarana* (N)	Metacercaria, *Centrocestus* Type	Chick (E) Pigeon (E) Domestic duck (E) Albino rat (E) Guinea pigs (E) Pup (E)	Dharmendra Nath, 1974

Intermediate Host	Larva	Final Host	Reference
Fishes (N) *Channa (Ophiocephalus) punctatus* (N) *Cirrhina reba* (N) *Mystus vittatus* (N) *Ompok bimaculatus* (N) *Chela laubuca* (N) *Nandus nandus* (N) *Puntius chola* (N) *Colisa fatius* (N)	Metacercaria	Chicks, WLH (E)	Premvati and Pande, 1974
Fish: *Aplocheilus panchax* (N) *Aplocheilus megastigma* (N)	Metacercariae		Madhavi, 1980
Fish, Cyprid: *Aplocheilus panchax* (N)	Metacercariae		Madhavi, 1986
Fish: *Aplocheilus panchax* (N)	Metacercariae		Madhavi and Rukmini, 1991
Fish (N)	Metacercariae	—	Dhanumkumari et al., 1993
1st IH: *Thiara tuberculata* (N) 2nd IH: Fishes *Channa* spp. *Gambusia* sp. *Puntius* sp. *Oryzias* sp.	Cercariae Metacercariae		Madhavi et al., 1997
Thiara tuberculata (N)	Cercariae		Swarnakumari, 2001

Intermediate Host	Larva	Final Host	Reference
		Haplorchis **spp.**	
Freshwater fish: *Channa punctatus* *Channa striatus*	Metacercariae	—	Chakrabarti, 1974
Puntius sophore (N)	Metacercaria, *Haplorchis* type 3	—	Dharmendra Nath, 1974
Snails (N)	Cercariae	—	Rai, 1976
2nd IH: Fishes *Channa* spp. *Gambusia* sp. *Puntius* sp. *Oryzias* sp.	Metacercariae		Madhavi *et al.*, 1997
		Haplorchis pumilio **(Looss, 1896)**	
Fishes (adults): *Mystus vittatus* (N) *Channa* (*Ophiocephalus*) *punctatus* (N) *Cirrhinus reba* (N) *Labeo ariza* (N) *Amblypharyagodon mola* (N) *Labeo bata* (N) *Nandus nandus* (N) *Puntius chola* (N) *Puntius sarana* (N) *Puntius suphore* (N)	Metacercaria	Albino rats (E) Hamsters (E) Rhesus monkey (E)	Pande and Shukla, 1972

Intermediate Host	Larva	Final Host	Reference
Ompok bimaculatus (N) *Glossogobius giurus* (N) Fishes (fingerlings): *Nandus nandus* (N) *Puntius saphore* (N) *Cirrhinus reba* (N) *Colisa lalius* (N) *Mystus vittatus* (N)			
Channa punctatus (N) *Chela laubuca* (N) *Osteobrame cotio* (N) *Nandus nandus* (N)	Metacercariae	Hamster (E)	Pande and Shukla, 1973
Fishes	Metacercariae	Chicks (E)	Premavati and Vibha Pande, 1973
Fishes: *Puntius sophore* (N) *Chela laubuca* (N) *Oxygaster bacaila* (N) *Oxygaster phulo* (N) *Mystus vittatus* (N) *Nandus nandus* (N) *Channa punctatus* (N) *Glossogobius giuris* (N) *Osteobrama cotio* (N)	Metacercarial cysts	White leghorn chicks (E)	Vibha Pande and Premavati, 1977
Thiara tuberculata	Cercariae		Dhanumkumari et al., 1991

Intermediate Host	Larva	Final Host	Reference
1st IH: *Thiara tuberculata* (N)	Cercariae		Madhavi et al., 1997
1st IH: Snail: *Thiara tuberculata* (N) 2nd IH: Fishes *Channa* spp. *Channa punctatus* (N) *Channa orientalis* (N) *Puntius sophore* (N) *Gambusia affinis* (N) Fingerlings of: *Cyprinus carpio* (N) *Liza macrolepis* (N) Fishes: *Therapon jarbua* (E) *Esomus danricus* (E) *Oreochromis mossabica* (E)	Cercariae Metacercariae	Birds (Piscivorous): *Ardeola grayii* (N) *Bubulcus ibis* (N) Chicks, day old (E)	Urr a Devi and Madhavi, 2006
Freshwater snail: *Thiara tuberculata* (N) Fish: *Channa* spp. (N) *Gambusia affinis* (N)	Cercariae Metacercariae	*Ardeola grayii* (N)	Madhavi, 2009

Intermediate Host	Larva	Final Host	Reference
		Haplorchis taichui (Nishigori, 1924)	
Cirrhinus reba (N) *Amblypharyagodon mola* (N)	*Haplorchis* type Metacercaria	Pups (E) Pigeon (E) Albino rats (E)	Dharmendra Nath and Pande, 1970
Cirrhinus reba (N) *Amblypharyagodon mola* (N) *Labeo bata* (N) *Puntius sarana* (N)	Metacercaria	Pups (E) Pigeons (E) Albino rats (E)	Dharmendra Nath, 1972c
Oxygaster hacaila (N) *Cirrhinus reba* (N) *Chela laubuca* (N) *Puntius sophore* (N)	Metacercaria	Hamster (E)	Pande and Shukla, 1972
Fishes (E)	Metacercaria, *Haplorchis taichui*	Pigeon (E)	Dharmendra Nath, 1973a
Amblypharyagodon mola (N) *Labeo bata* (N) *Puntius sarana* (N)	Metacercaria, *Haplorchis* type 2	Pups (E) Pigeons (E)	Dharmendra Nath, 1974
Cirrhinus reba (N) *Punitius sophore* (N)	Metacercaria	White leghorn chicks	Vibha Pande and Premvati, 1977
Aplocheilus megastigma (N)	Metacercariae	—	Madhavi, 1980
Thiara tuberculata	Cercariae		Dhanumkumari *et al.*, 1991
1st IH: *Thiara tuberculata* (N)	Cercariae		Madhavi *et al.*, 1997

Intermediate Host	Larva	Final Host	Reference
		Haplorchis yokogawai (Katsuta, 1932)	
Freshwater fish: *Puntius ticto* (N) *Mystus vittatus* (N)	Metacercariae	*Corvus splendens* (N) *Corvus macrorhynchos* (N)	Pandey, 1966
Cirrhinus reba (N) *Amblypharyagodon mola* (N)	Metacercariae	Pups (E) Pigeon (E) Albino rats (E)	Dharmendra Nath and Pande, 1970
Puntius sp. (N) *Puntius chola* (N) *Puntius tincto* (N) *Mystus vittatus* (N) *Cirrhinus reba* (N) *Amblypharyagodon mola* (N) *Labeo bata* (N) *Nandus nandus* (N)	Metacercariae	Rhesus monkey (E)	Pande and Shukla, 1970
Cirrhinus reba (N) *Amblypharyagodon mola* (N) *Labeo bata* (N) *Puntius sarana* (N)	Metacercariae	Pups (E) Pigeons (E) Albino rats (E)	Dharmendra Nath, 1973b
Nandus nandus (N)	Metacercariae	Hamsters (E)	Pande and Shukla, 1973
Cirrhinus reba (N) *Amblypharyagodon mola* (N) *Labeo bata* (N) *Puntius sarana* (N)	Metacercariae, Haplorchid type 1	Pup (E) Pigeon (E), Albino rat (E)	Dharmendra Nath, 1974

Intermediate Host	Larva	Final Host	Reference
Puntius sophore (N) *Mystus vittatus* (N) *Glossogobius giuris* (N) *Ompok bimaculatus* (N) *Nandus nandus* (N)	Metacercariae	White leghorn chicks	Vibha Pande and Premvati, 1977
Thiara tuberculata (N)	Cercariae		Swarnakumari, 2001
Procerovum varium Oryi and Nishio, 1916			
1st IH: *Thiara tuberculata* (N) 2nd IH: Fishes *Channa* spp. *Gambusia* sp. *Puntius* sp. *Oryzias* sp.	Cercariae Metacercariae		Madhavi *et al.*, 1997
Snail: 1st IH *Thiara tuberculata* (N) Fish: 2nd IH *Oryzias melastigma* (N)	Cercariae	Pond heron: *Ardeola grayii* (N; E) Chicks (E) Ducks (E) Mice (N)	Umadevi and Madhavi, 2000
Freshwater snail: *Thiara tuberculata* (N) Fish: *Oryzias melastigma* (N)	Cercariae Metacercariae	*Ardeola grayii*	Madhavi, 2009

1.6 FAMILY: PLAGIORCHIDAE

Prosthogonimus sp.

Intermediate Host	Larva	Final Host	Reference
Lymnaea luteola (N)	Xiphidio-cercariae	—	Manohar and Rao, 1976

Prosthogonimus ovatus (Rud, 1803) Luhe, 1899

Intermediate Host	Larva	Final Host	Reference
Dragonflies 2ⁿᵈ IH: *Brachythemis contaminata* (N) *Crocothemis servilia servilia* (N) *Orthetrum sabina* (N) *Pantala flavescens* (N) *Trames basilaris burmeisteri* (N) *Trithemis pallidinervis* (N)	Metacercariae	Chicks, WLH (E)	Muraleedharan and Pande, 1968
Brachythemis contaminata (N) *Orthetrum sabina* (N)	Metacercariae	Pigeons (E)	Pratap Singh and Pande, 1968
Brachythemis contaminata (N) *Crocothemis servilia servilia* (N) *Orthetrum sabina* (N) *Pantala flavescens* (N) *Trames basilaris burmeisteri* (N) *Trithemis pallidinervis* (N)	Metacercariae	Hens (E)	Prakash and Pande, 1969
4 known species of dragonflies	Metacercariae	Pullets, WLH (E)	Pande *et al.*, 1970
Brachythemis contaminata (N) *Orthopterum sabina* (N) *Pantala flavescens* (N) *Trithemis pallidinervis* (N)	Metacercariae	Pullets (E)	Pratap Singh and Pande, 1971

Intermediate Host	Larva	Final Host	Reference
Brachythemis contaminata (N) *Crocothemis servilia servilia* (N) *Orthetrum sabina* (N) *Pantala flavescens* (N) *Sympetrum decolaratum* (N) *Trames basilaris burmeisteri* (N) *Trithemis pallidinervis* (N) Naiads of dragonfly (N)	Metacercariae	Common quails (*Coturnix coturnix coturnix*) (E) Grey patridges (*Fancolirus pondicerianus*) (E) Guinea fowls (*Numida meleagris*) (E)	Dharmendra Nath, 1973
Species of dragonflies	Metacercariae	Pullets, WLH (E)	Chaudhary and Ahluwalia, 1973
Snails: *Alocinma orcula* 1st IH (N)	Cercariae	—	Chaudhary *et al.*, 1973
Alocinma orcula 1st IH (N)	Cercariae	—	Chaudhary, 1974
Dragonfly naiads (N)	Metacercariae	Pullets (E)	Chaudhary, 1974
Species of dragonflies	Metacercariae	—	Dharmendra Nath, 1975

Prosthogonimus putschkowskii Skrajabin, 1912
(Syn: *Prosthogonimus ovatus*)

Intermediate Host	Larva	Final Host	Reference
Sympetrum decolaratum (N)	Metacercariae	Chicks (E) Hens, WLH (E)	Mishra and Pande, 1967

1.7 FAMILY: CYCLOCOELIDAE
Typhlocoelium cymbium (Diesing, 1950) Kossack, 1911

Intermediate Host	Larva	Final Host	Reference
Indoplanorbis exustus (N)	Cercariae and metacercaria	Duck (E)	Sreekumaran and Peter, 1973 a
Indoplanorbis exustus (N)	Rediae	—	Sreekumaran and Peter, 1973 b

Intermediate Host	Larva	Final Host	Reference
Typhlocoelium indicum Sreekumar and Peter, 1973			
Indoplanorbis exustus (E)	—	—	Sreekumaran, 1966

1.8 FAMILY: PHILOPHTHALMIDAE

Philophthalmus sp.

Intermediate Host	Larva	Final Host	Reference
2nd IH: *Melanoides tuberculatus* (N)-	*Cercariae indicae* IV Sewell, 1922 Metacercariae	Guinea pigs (E) Water hen (E) Rats (E) Sea gull (E)	Satyanarayana Murty, 1966
Philophthalmus anatinus Sugimoto, 1928			
Melanoides tuberculatus (E)	Redia Daughter redia Cercaria Metacercaria	Duck (N)	Bhaskar Rao and Rao, 1981
Philophthalmus (Philophthalmus) cochiensis Ismail Koya, 1988			
Fish	Metacercariae	Domestic chicken (E)	Syed Ismail Koya, 1988
Philophthalmus gralli Mathias and Leger, 1910			
Thiara (Melanoides) tuberculata (N)	Cercariae Metacercariae	Chick (E)	Rezul Karim, 1981
Thiara (Melanoides) tuberculata (N)	Cercariae Metacercariae	Chick (E)	Rezual Karim et al., 1982
Snails (N)		Chick (E)	Bhatia et al., 1985

Intermediate Host	Larva	Final Host	Reference
	Philophthalmus lucknowensis, Baugh, 1962		
Melanoides tuberculata (N) (E)	Miracidium		Saxena, 1981
Melanoides tuberculata (N) (E)	Mother rediae Daughter rediae Grand-daughter rediae Cercariae		Saxena, 1984
	Cercariae Metacercariae	Chicks (E)	Saxena, 1985
	Philophthalmus notornus Looss, 1907		
Melanoides tuberculatua (N)- 2nd IH	Metacercariae	Chick (E) White rats (E) Guinea pigs (E)	Satyanarayana Murthy, 1966
Thiara tuberculata (N)- 2nd IH	Metacercariae	White lehhorn chick (E)	Swarnakumari and Madhavi, 1992a
Thiara tuberculata (N)- 2nd IH	Metacercariae	White lehhorn chick (E)	Swarnakumari and Madhavi, 1992b
1st IH: *Thiara tuberculata* (N)	Cercariae		Madhavi et al., 1997

1.9 FAMILY: LECITHODENDRIDAE

Eumegacetes sp.

Intermediate Host	Larva	Final Host	Reference
Dragonfly naiads	Metacercariae		Hanumantha Rao and Madhavi, 1961

Intermediate Host	Larva	Final Host	Reference
Dragonfly: *Brachythemis contaminata* (N)	Metacercaria type I		Ram Prakash and Pande, 1970
Thiara tuberculata (N)	Cercariae		Swarnakumari, 2001

Eumegacetes artamii Mehra, 1935

Intermediate Host	Larva	Final Host	Reference
Dragon flies 2ⁿᵈ IH: *Brachythemis contaminata* (N) *Orthopterum sabina* (N)	Metacercariae	White Leghorn chicks (E)	Singh and Pande, 1968
F. W. snail 1ˢᵗ IH: *Thiara tuberculata* (N)	Cercariae		Swarna Kumari and Madhavi, 1994
Thiara tuberculata (N)	Cercariae		Swarnakumari, 2001

Eumegacetes singhi Jaiswal, 1957

Intermediate Host	Larva	Final Host	Reference
Dragon flies 2ⁿᵈ IH: *Sympetrum decolaratum* (N)	Metacercaria	Common quail (*Coturnix coturnix coturnix*) (E)	Dharmendra Nath, 1971

1.10 FAMILY: PARAGONIMIDAE

Paragonimus sp.

Intermediate Host	Larva	Final Host	Reference
Freshwater crabs, species unspecified (N) Freshwater crab: *Potamiscus manipurensis* (N)	Metacercariae, several types Metacercariae	Laboratory animals (E)	Singh, 1996

Intermediate Host	Larva	Final Host	Reference
Crab: *Potamiscus manipurensis* (N)	Metacercariae (easily excysted thin cyst wall)	—	Singh and Singh, 1997
Freshwater crabs: *Barytelpusa lugubris* (N) *Potamiscus manipurensis* (N)	Metacercariae		Mahajan, 2005
Freshwater crabs	Metacercariae	Puppies (E) Albino rats (E)	Singh, 2005
Edible crabs (N)	Metacercariae	—	Veena Tandon, *et al.*, 2007
Freshwater crabs: *Potamiscus manipurensis* (N) (Syn: *Barytelphusa lugubris* (N)	Metacercariae		Takeda *et al.*, 2012
Paragonimus heterotremus Chen and Hsia, 1964			
Crab: *Potamiscus manipurensis* (N)	Metacercariae	—	Singh and Singh, 1997
Freshwater crab: *Potamiscus manipurensis* (N)	Metacercariae	Puppies (E) Albino rats (E) Swiss mice (E)	Singh, 2003
Freshwater crab: *Barytelphusa lugubris* (N)	Metacercariae	Wister rats (E)	Narain *et al.*, 2003a
Freshwater crab: *Barytelphusa lugubris* (N)	Metacercariae	Wister rats, *Rattus norvegicus* (E)	Narain *et al.*, 2003b
Freshwater crabs	Metacercariae		Singh, 2005

Intermediate Host	Larva	Final Host	Reference
Freshwater crab: *Potamiscus manipurensis* (N)	Metacercariae	Puppies (E) Albino rats (E)	Singh *et al.*, 2007
Freshwater crab: *Potamiscus manipurensis* (N)	Metacercariae		Singh *et al.*, 2009
Mountain crab: *Indochinamon manipurensis* (N)	Metacercariae	Puppies (E) (Adults & Immature flukes) Albino rats (E), juveniles	Singh *et al.*, 2010
Mountain crab: *Indochinamon manipurensis* (N)	Metacercariae	Puppies (E) Albino rats (E)	Singh *et al.*, 2011
Crab: *Alcomon superciliosum* (N)	Metacercariae		Singh *et al.*, 2012
Freshwater crabs: *Maydelliatelphusa lugubris* (N) *Sartoriana spinigera* (N)	Metacercariae, Type 1 & Type 2 Metacercariae, Type 3		Rekha Devi *et al.*, 2013
Freshwater crab: *Potamiscus manipurensis* (N)	Metacercariae	Puppies (E) Albino rats (E)	Singh *et al.*, 2007
Mountain crab: *Indochinamon manipurensis* (N)	Metacercariae	Puppies (E) Albino Rats (E)	Singh *et al.*, 2011
Crab: *Alcomon superciliosum* (N)	Metacercariae		Singh *et al.*, 2012

Paragonimus hueit'ungensis **Lan, Piao, Yin, Chih, Lan, Hsi, Teh, Han, Ping, Kuang, Chung and Sheng, 1977**

***Paragonimus skrjabini* Chen, 1959**

***Paragonimus westermanii* (Kerbert, 1878) Braun, 1899**

Intermediate Host	Larva	Final Host	Reference
Paragonimus hueit'ungensis Lan, Piao, Yin, Chih, Lan, Hsi, Teh, Han, Ping, Kuang, Chung and Sheng, 1977			
Freshwater crab: *Potamiscus manipurensis* (N)	Metacercariae	Pup (E)	Singh, 2002
Paragonimus skrjabini Chen, 1959			
Freshwater crab: *Potamiscus manipurensis* (N)	Metacercariae	Pup (E)	Singh *et al.*, 2006
Paragonimus westermanii (Kerbert, 1878) Braun, 1899			
Crab: *Potamiscus manipurensis* (N)	Metacercariae (large and small types)	—	Singh and Singh, 1997
Crab: *Potamons duhani* (N)	Metacercariae	—	Bhatia *et al.*, 2004
Freshwater crabs	Metacercariae (large type and small type)	—	Singh, 2005
Crab, edible (N)	Metacercariae	—	Veena Tandon *et al.*, 2007
Edible crab: *Barytelphusa lugubris* (N)	Metacercariae	—	Prasad *et al.*, 2009
Crab (N): *Barytelpusa lugubris* (now known as *Maydelliathelphusa lugubris*)	Metacercariae	Rats, Wistar (E)	Rekha Devi *et al.*, 2010
Crustacean (N)	Metacercariae	—	Prasad *et al.*, 2011

Intermediate Host	Larva	Final Host	Reference
Crab: *Alcomon superciliosum* (N)	Metacercariae		Singh *et al.*, 2012
1.11 FAMILY: PARAMPHISTOMATIDAE			
Paramphistomes (unidentified)			
Lymnaea auricularia (N) *Indoplanorbis exustus* (N) *Gyraulus convexiusculus* (N) *Bithynia pulchella* (N)	Cercariae	—	Sahai, 1969-70
Indoplanorbis exustus (N)	Cercariae	—	Jain, 1970
Indoplanorbis exustus (N)	Cercariae	—	Muraleedharan, 1972
Lymnaea luteola (N) *Indoplanorbis exustus* (N)	Cercariae	—	Mohandas, 1974a
Indoplanorbis exustus	Cercariae	—	Jain, 1976
Lymnaea (Pseudosuccinea) acuminata (N)	Cercariae	—	Pokhriyal *et al.*, 1996
Thiara (M.) tuberculata (N)	Cercariae	—	Pokhriyal *et al.*, 1998
Indoplanorbis exustus (N) *Lymnaea* spp. (*Lymnaea acuminata* and *Lymnaea luteola*) (N)	Cercariae	—	Muraleedharan *et al.*, 1977
Gyraulus convexiusculus (N) *Lymnaea luteola* (N) *Indoplanorbis exustus* (N)	Larval stages	—	Borkakoty and Das, 1980

Intermediate Host	Larva	Final Host	Reference
Gyraulus convexiusculus (N) *Lymnaea luteola* (N)	Cercariae	—	Dutt and Bali, 1980
Lymnaea luteola f. typica (N) *Indoplanorbis exustus* (N) *Gyraulus convexiusculus* (N)	Cercariae	Lambs (E)	Chaudhri *et al.*, 1982
Indoplanorbis exustus (N)	Cercariae		Sardey, 1982
Lymnaea luteola f. typica (N) *Indoplanorbis exustus* (N)	Cercariae Metacercariae	—	Chaudhri and Gupta, 1985
Gyraulus convexiusculus (E)	Cercariae		Upadhyaya and Sahai, 1986
Indoplanorbis exustus (N) *Gyraulus convexiusculus* (N)	Cercariae of 5 species of paramphistomes		Sahay, 1987
Indoplanorbis exustus (N) *Gyraulus convexiusculus* (N) *Bithynia pulchella* (N) *Lymnaea luteola* (N) *Lymnaea acuminata* (N) *Melanoides tuberculata* (N) *Melanoides lineata* (N) *Vivipara bengalensis* (N)	Cercariae (*Cercaria pigmentata*)		Kumar *et al.*, 1992
Indoplanorbis exustus (N) *Lymnaea luteola* (N)	Cercariae		Agrawal, 1994a
Indoplanorbis exustus (N)	Cercariae		Agrawal, 1994b

Intermediate Host	Larva	Final Host	Reference
Indoplanorbis exustus (N)	Cercariae	—	Shrivastav and Sengar Kanraljeet, 1995
Lymnaea luteola (Pesudosuccinea) acuminata	Cercariae	—	Pokhriyal et al., 1996
Indoplanorbis exustus (N) Gyraulus convexiusculus (N)	Cercariae	—	Choubisa, 1997
Thiara (M.) tuberculata (N)	Cercariae	—	Pokhriyal et al., 1998
Indoplanorbis exustus (N)	Cercariae Metacercariae	Lambs (E) Kids (E)	Urvasi et al., 1999
Lymnaea luteola (N)	Cercariae	—	Agrawal et al., 2000
Indoplanorbis exustus (N)	Cercariae	—	Muraleedharan, 2000
Indoplanorbis exustus (N)	Cercariae Metacercariae	Lambs (E) Kids (E)	Urvashi et al., 2000
Indoplanorbis exustus (N) Gyraulus convexiusculus (N)	Cercariae	—	Choubisa, 2002
Lymnaea (Psuedosucinea) luteola f. ovalis (N) Indoplanorbis exustus (N) Gyraulus convexiusculus (N)		—	Pallavi Devi et al., 2003
Indoplanorbis exustus (N)	Cercaria pigmentata		Vohra et al., 2006

Intermediate Host	Larva	Final Host	Reference
Lymnaea luteola f. australis (N) *Lymnaea luteola f. typical* (N) *Lymnaea luteola f. impure* (N) *Indoplanorbis exustus* (N) *Gyraulus convexiusculus* (N)	Cercariae	—	Choubisa, 2008
The various species of snails claimed to have been positive for cercariae of liver fluke and amphisomes: *Lymnaea acuminata* (N) *Lymnaea luteola* (N) *Thiara (Melanoides) tuberculata* (N) *Indoplanorbis exustus* (N) *Paludomus obesus* (N) *Ferrissia tenuis* (N) *Bellamyia bengalensis* (N) *Bellamyia dissimilis* (N)	Cercariae	—	Pemola Devi and Jauhari, 2008
Indoplanorbis exustus (N) *Gyraulus convexiusculus* (N)	—	—	Singh *et al.*, 2009
Indoplanorbis exustus (N)	Metacercariae	—	Syed Sahib Hussain *et al.*, 2009a
Indoplanorbis exustus (N)	Metacercariae	—	Syed Sahib Hussain *et al.*, 2009 b
Indoplanorbis exustus (N)	Larval parasitic forms	—	Johnpaul *et al.*, 2010
Indoplanorbis exustus (N)	—	—	Gupta, 2012

Ceylonocotyle dicranocoelium (Fischoeder, 1904) Nasmark, 1937
Syn: *Orthocoelium dicranocoelium* (Fischoeder, 1901) Stiles and Goldberger, 1910)
Vide infra

Ceylonocotyle scoliocoelium (Fischoeder, 1904) Nasmark, 1937
(Syn: *Orthocoelium* Stiles and Goldberger, 1910)
Vide infra

Cotylophoron cotylophorum (Fischoeder, 1901) Stiles and Goldberger, 1910

Intermediate Host	Larva	Final Host	Reference
Indoplanorbis exustus (N)	Cercaria *Cotylophoron cotylophorum*		Chatterjee, 1931
Indoplanorbis exustus (E)	Cercariae Metacercariae	Lambs (E)	Srivatsava, 1938
Indoplanorbis exustus (N)	Cercariae	—	Mudaliyar, 1944
Indoplanorbis exustus (N)	Cercariae		Bhalerao, 1945
Indoplanorbis exustus (N)	Cercariae	—	Mudaliyar, 1945
Indoplanorbis exustus (N)	Cercariae	—	Mudaliyar, 1944
Indoplanorbis exustus (N)	Cercariae	—	Mudaliyar and Alwar, 1947
Indoplanorbis exustus (N)	Cercariae Metacercariae	Buffalo calf (E)	Anantaraman and Balasubramaniam, 1949
Indoplanorbis exustus (E)	Metacercariae	Cattle (E) Goats (E) Sheep (E) Kids (E)	Sinha, 1950

Intermediate Host	Larva	Final Host	Reference
Indoplanorbis exustus (N)			Peter and Srivastava, 1960b
Indoplanorbis exustus (N)	Sporocysts Rediae Metacercariae	Lambs (E)	Varma, 1961
Indoplanorbis exustus (N)	Cercariae Metacercariae	Lambs (E)	Prasad, 1973
	Metacercariae	Goats (E)	Prasad *et al.*, 1974
Indoplanorbis exustus (N)	Metacercariae	Lambs (E)	Chaudhri, *et al.*, 1982
Cotylophoron indicum Stiles and Goldberger, 1910			
Indoplanorbis exustus (N)	*Cercaria indicae* XXVI Sewell , 1922	—	Mukherjee, 1962
Indoplanorbis exustus (N) (E)	Cercaria Metacercariae	Goats (E)	Mukherjee, 1966a
Indoplanorbis exustus (N)	Cercariae	—	Mukherjee, 1968
Indoplanorbis exustus (N)	Cercariae	—	Hafeez Md. and Rao, 1979
Indoplanorbis exustus (N)	Cercariae	Lambs (E)	Chaudhri *et al.*, 1982
Fischoederius elongatus (Poirier, 1883) Stiles and Goldberger, 1910			
Lymnaea acuminata (N) *Lymnaea succinea* (N) *Gyraulus euphraticus* (N)	*Cercariae indicae* XXIX Sewell, 1922	—	Sewell, 1922

Intermediate Host	Larva	Final Host	Reference
Lymnaea luteola (N)	*Cercariae indicae* XXIX Sewell, 1922	Calf (E)	Rao, 1932
Lymnaea luteola (N)	*Cercariae indicae* XXIX Sewell, 1922	—	Rao and Ayyar, 1932
Lymnaea luteola (N)	*Cercariae indicae* XXIX Sewell, 1922	—	Bhalerao, 1943
Lymnaea luteola (N)	*Cercariae indicae* XXIX Sewell, 1922	Calf (E)	Vaidyanathan, 1941
Lymnaea luteola (N)	*Cercariae indicae* XXIX Sewell, 1922	—	Mudaliyar and Alwar, 1947
Lymnaea succinea (N)	*Cercariae indicae* XXIX Sewell, 1922	—	Peter. 1956
Lymnaea luteola (N)	*Cercariae indicae* XXIX Sewell, 1922	—	Peter and Srivastava, 1955
Lymnaea luteola (N)	*Cercariae indicae* XXIX Sewell, 1922	—	Tandon, 1958
Lymnaea luteola f. succinia (N)	*Cercariae indicae* XXIX Sewell, 1922	—	Peter and Srivastava, 1960b
Lymnaea luteola f. succinia	*Cercariae indicae* XXIX Sewell, 1922	Cow	Mukherjee, 1966b
Lymnaea luteola (N)	*Cercariae indicae* XXIX Sewell, 1922	—	Jain, 1970

Intermediate Host	Larva	Final Host	Reference
Lymnaea luteola f. *australis*	*Cercariae indicae* XXIX Sewell, 1922	Lamb (E)	Agrawal, 1971
Lymnaea luteola f. *australis*	*Cercariae indicae* XXIX Sewell, 1922	Lamb (E) Kids (E) Guinea pig (E)	Agrawal and Pande, 1971
Lymnaea luteola	*Cercariae indicae* XXIX Sewell, 1922 (Cercariae of *Fischoederius elongatus*)	—	Jain, 1976
Lymnaea luteola f. *typica* (N)	*Cercariae indicae* XXIX Sewell, 1922 Metacercariae Mother rediae Daughter rediae		Mohandas, 1976
Lymnaea luteola (N) (E)			Prasad *et al.*, 1989
Lymnaea luteola (N)			Varma and Prasad, 2000
Lymnaea luteola f. *typica* (N)	Cercariae Metacercaria		Chaudhri, *et al.*, 1982
***Gastrodiscus secundus* Looss, 1907**			
Planorbis exustus (E) (N)	Cercaria *Gastrodisci secondi* Peter and Srivastava, 1948 Metacercariae	Donkey foal (E)	Peter and Srivastava, 1948
Indoplanorbis exustus (N)	Metacercariae	—	Ramanujachari and Alwar, 1954

Intermediate Host	Larva	Final Host	Reference
Indoplanorbis (N)	Metacercariae	—	Peter, 1954
Indoplanorbis exustus (N)	Metacercariae	—	Peter, 1956
Indoplanorbis exustus (N) (E)	Metacercariae	Ass foal (E)	Peter, 1960
Gastrodiscoides hominis (Lewis and McConnel, 1876) Leiper, 1913			
Helicorbis coenosus (E)	Cercaria fraseri		Buckley, 1939
Helicorbis coenosus (E)	Cercariae Metacercariae	Pig (E)	Dutt and Srivastava, 1966
Helicorbis coenosus (E)	Rediae Cercariae Metacercariae	Pig (E)	Dutt and Srivastava, 1972
Gastrothylax crumenifer (Creplin, 1847) Otto, 1896			
Indoplanorbis exustus (E) Gyraulus convexiusculus (E)	Cercariae	—	Srivastava, 1944b
Indoplanorbis exustus (E) Gyraulus convexiusculus (E)	Cercariae	—	Srivastava, 1947
Gyraulus euphraticus (N	Cercaria chungathi	—	Peter and Srivastava, 1955
Gyraulus euphraticus	Cercariae Metacercariae	Kid (E)	Tandon, 1957
Gyraulus euphraticus (N)	Cercaria chungathi	—	Pandey and Jain, 1961
	Cercaria chungathi	—	Pandey and Jain, 1971

Intermediate Host	Larva	Final Host	Reference
Gyraulus euphraticus (N)	*Cercaria chungathi*	Buffalo calf (E) Kid (E)	Peter and Srivastava, 1961
Gyraulus euphraticus (N)	—	—	Jain, 1976
Gyraulus convexiusculus (N)	Cercariae Metacercariae	Buffalo calves (E) Goats (E)	Panzoo et al., 1988
Gyraulus convexiusculus (N)	—	—	Panzoo et al., 1989a
Gyraulus convexiusculus (N)	—	—	Panzoo et al., 1989b
Indoplanorbis exustus (E)	—	—	Kumar et al., 1992
Gyraulus convexiusculus (N)	Larvae	—	Saifullah et al., 2000
Gyraulus convexiusculus (N)	—	—	Varma and Prasad, 2000
Indoplanorbis exustus (N)	Cercariae Metacercariae	Sheep (E)	Kaur et al., 2009
	Metacercariae	Sheep (E)	Syed Shabih Hussain et al., 2009b

Gigantocotyle bathycotyle (Fischoeder, 1901) Nasmark, 1937

Intermediate Host	Larva	Final Host	Reference
Snail (N)	Cercariae		Jain, 1969
Bulimus pulchellus (N)	Cercariae		Jain, 1978

Gigantocotyle explanatum (Creplin, 1847) Nasmark, 1937*
(Syn: *Paramphistomum explanatum*)

Intermediate Host	Larva	Final Host	Reference
Gyraulus convexiusculus (N)	Cercariae Metacercariae	Kid (E)	Singh, 1958

Intermediate Host	Larva	Final Host	Reference
	Cercaria pigmentata	—	Mukherjee and Srivatsava, 1960
	Cercaria pigmentata	Buffaloes (N)	Mukherjee and Srivatsava, 1960
	Cercaria pigmentata	—	Mukherjee, 1962
Gyraulus convexiusculus (N)	Cercariae	—	Mukherjee, 1966a
Gyraulus convexiusculus (N)	Cercariae	—	Ravi Shankar and Singh, 1978
Gyraulus convexiusculus (E)	Cercariae	—	Upadhyay and Sahai, 1986
Gyraulus convexiusculus (N) Indoplanorbis exustus (N)	Cercaria Metacercariae	Buffaloes (N)	Upadhyay, 1987
Gyraulus convexiusculus (N)	Cercaria	—	Pati and Bali, 1988
Gyraulus convexiusculus (N)	Larvae	—	Saifulah et al., 2000
Gyraulus convexiusculus (N)	—	—	Varma and Prasad, 2000
	Metacercariae	Antigen raised in rabbits	Velusamy, et al., 2006

See also *Paramphistomum explanatum*

Orthocoelium dicranocoelium (Fischoeder, 1901) Stiles and Goldberger, 1910)
Syn: Ceylonocotyle dicranocoelium (Fischoeder, 1904) Nasmark, 1937

Intermediate Host	Larva	Final Host	Reference
Bulimus pulchellus (N)	Cercariae Metacercariae	Kids (E)	Jain, 1976
Bulimus pulchellus (N) (E)	Cercariae Metacercariae	Kids (E)	Jain, 1977

Intermediate Host	Larva	Final Host	Reference
Orthocoelium scoliocoelium (Fischoeder, 1904) Yamaguti, 1971 Syn: *Ceylonocotyle scoliocoelium* (Fischoeder, 1904) Nasmark, 1937			
Bulimus pulchellus (N)	*Cercaria bulimusi*		Peter and Srivastava, 1955
Snails (N)	Cercariae		Mukherjee, 1960
Bulimus pulchellus (N)	*Cercaria bulimusi* Metacercariae	Goats (E)	Mukherjee, 1962
Bulimus pulchellus (N)	Sporocysts Rediae Cercariae (*Cercaria bulimusi*) Metacercariae	Kids (E)	Jain and Srivatsava, 1969b
Bithynia pulchellus (N)		—	Mukherjee, 1966a
Bithynia pulchellus (N)		Goat (E)	Mukherjee, 1975
Bulimus pulchellus (N)		—	Jain, 1976
Orthocoelium streptocoelium (Fischoeder, 1901) Yamaguti, 1971 Syn: *Paramphistomum streptocoelium* Fischoeder, 1901			
Bulimus tentaculata var. *kashmirensis* (N)	Cercariae, *Orthocoelium streptocoelium*	—	Dutt and Bali, 1980
Palamphistomum lobatum Srivastava and Tripathi, 1980*			
Gyraulus convexiusculus (E)	Cercariae	Sheep (N) Goats (N) Buffaloes (N)	Srivastava and Malaviya, 1987

Intermediate Host	Larva	Final Host	Reference
Gyraulus convexiusculus (E)	Cercariae	—	Malaviya et al., 1989
Gyraulus convexiusculus (E)	Cercariae Metacercariae	Buffaloes (N)	Malaviya et al., 1992
*Considered as synonym of Orthocoelium dicranocoelium by Jones, Bray and Gibson, 2001			
Paramphistomum cervi (Zeder, 1790) Fischoeder, 1901			
Lymnaea luteola (N) Planorbis exustus (Indoplanorbis exustus) (N)	Cercariae indicae XXVI Sewell, 1922 Metacercariae	Duck (E) —	Chatterji, 1931
Planorbis exustus (Indoplanorbis exustus) (N)	—	Calf (E)	Rao, 1932
Planorbis exustus (Indoplanorbis exustus) (N)	Cercariae indicae XXVI Sewell, 1922 (Cercaria No. 2, Madras) Metacercariae	Calf (E)	Rao and Ayyar, 1932
Indoplanorbis exustus (N)	Cercariae indicae XXVI Sewell, 1922	—	Jain, 1976
Indoplanorbis exustus (N)	Mother rediae Daughter rediae Cercariae indicae XXVI Sewell, 1922		Mohandas, 1976
Indoplanorbis exustus (E)	Cercariae indicae XXVI Sewell, 1922	—	Dutt and Gill, 1978

Intermediate Host	Larva	Final Host	Reference
Indoplanorbis exustus (N)	Metacercariae, normal and irradiated	Buffalo calves (E)	Patel and Dutt, 1978
Indoplanorbis exustus (N)	*Cercariae indicae* XXVI	—	Hafeez and Rao, 1979
Indoplanorbis exustus (N)	Cercariae	—	Dutt and Bali, 1980
		Goats, kids (E)	Singh, 1983
	Metacercariae laboratory maintained	Goats, kids (E)	Singh *et al.*, 1983a
	Metacercariae laboratory maintained	Goats, kids (E)	Singh *et al.*, 1983b
	Metacercariae laboratory maintained	Goats, kids (E)	Singh *et al.*, 1984a
	Metacercariae laboratory maintained	Goats, kids (E)	Singh *et al.*, 1984b
	Metacercariae laboratory maintained	Goats, kids (E)	Sahai *et al.*, 1985
	Metacercariae	Buffalo calves (E)	Gill and Bali, 1987a
	Metacercariae	Buffalo calves (E)	Gill and Bali, 1987b
Indoplanorbis exustus (N, E)	Cercariae Metacercariae		Gill, 1987
	Metacercariae	Sheep (E)	Gill and Bali, 1988
Indoplanorbis exustus (N)	Metacercaria	Lamb (E)	Malaviya *et al.*, 1989

Intermediate Host	Larva	Final Host	Reference
Indoplanorbis exustus (N, E)	Cercariae Metacercariae		Prasad *et al.*, 1989
Gyraulus convexiusculus	Cercariae	—	Panzoo *et al.*, 1993
Indoplanorbis exustus (N)	Cercariae Metacercariae	New Zealand white rabbits (E)	Singla *et al.*, 1998
Indoplanorbis exustus (N) *Gyraulus convexiusculus* (N)	Cercariae	—	Kurrar, 2004
Indoplaorbis exustus (N)	Cercariae (Cercaria pigmentata) Metacercariae	Lambs (E)	Syed Shabih Hussain *et al.*, 2009a
Lymnaea luteola (N)	Cercaria pigmentata		Vohra *et al.*, 2009
***Paramphistomum epiclitum* (Fischoeder, 1904)**			
Snail (N)	*Cercariae indicae* XXVI Sewell, 1922		Mohandas, 1974a
	Metacercaria of *Cercariae indicae* XXVI Sewell, 1922	Kid (E) Lambs (E)	Hafeez, 1975
	Metacercaria of *Cercariae indicae* XXVI Sewell, 1922	Kid (E) Lambs (E)	Hafeez, 1981
Indoplanorbis exustus (N)	*Cercariae indicae* XXVI Sewell, 1922 Metacercaria		Hafeez and Rao, 1979a

Intermediate Host	Larva	Final Host	Reference
Indoplanorbis exustus (N)	*Cercariae indicae* XXVI Metacercaria	Lambs (E)	Hafeez and Rao, 1979b
Indoplanorbis exustus (N)	*Cercariae indicae* XXVI Sewell, 1922 Metacercariae	—	Hafeez and Rao, 1981a
Indoplanorbis exustus (N)	Metacercaria of *Cercariae indicae* XXVI Sewell, 1922	Kid (E) Lambs (E)	Hafeez and Rao, 1981b
	Metacercaria of *Cercariae indicae* XXVI	Sheep and Goats (E)	Hafeez and Rao, 1983a
	Metacercaria of *Cercariae indicae* XXVI Sewell, 1922	—	Hafeez and Rao, 1983b
	Metacercaria of *Cercariae indicae* XXVI Sewell, 1922	Kid (E) Lambs (E)	Hafeez and Rao, 1984
	Metacercaria of *Cercariae indicae* XXVI Sewell, 1922	Lambs (E)	Hafeez *et al.*, 1984
Indoplanorbis exustus (N) (E)	Cercaria Metacercaria	Sheep (E)	Gill, 1985
	Metacercaria of *Cercariae indicae* XXVI Sewell, 1922	Lambs (E)	Hafeez *et al.*, 1985

Intermediate Host	Larva	Final Host	Reference
Indoplanorbis exustus (N) (E)	Cercaria Metacercaria	Sheep (E)	Gill, 1987
	Metacercaria of *Cercariae indicae* XXVI	Lambs (E)	Hafeez and Rao, 1987
	Metacercaria of *Cercariae indicae* XXVI	Lambs (E)	Hafeez et al., 1987
	Cercaria (*Cercariae indicae* XXVI) Metacercaria	—	Hafeez and Rao, 1988a
	Cercaria (*Cercariae indicae* XXVI) Metacercaria	—	Hafeez and Rao, 1988b
	Cercaria (*Cercariae indicae* XXVI) Metacercaria	—	Hafeez and Rao, 1989
	Metacercariae	Lambs (E)	Dwivedi et al., 1989
Indoplanorbis exustus (N) (E)	Metacercariae	Lambs (E)	Malaviya et al., 1993
Indoplanorbis exustus (N) (E)	Metacercariae	Lambs (E)	Varma et al., 1993
Indoplanorbis exustus (E)	Metacercariae	Kid (E) Lambs (E)	Malaviya et al., 1994a
Indoplanorbis exustus (E)	Metacercariae	Kid (E) Lambs (E)	Malaviya et al., 1994b

Intermediate Host	Larva	Final Host	Reference
Indoplanorbis exustus (E)	Metacercariae	Lambs (E)	Malaviya et al., 1994c
Indoplanorbis exustus (N)	Cercariae	—	Prasad et al., 1994a
Indoplanorbis exustus (N)	Cercariae, Note: Differ from Cercariae indicae XXVI	—	Prasad et al., 1994b
—	Metacercariae	Kid (E) Lambs (E)	Dwivedi et al., 1996
Gyraulus convexiusculus (N)	Larvae	—	Saifullah et al., 2000
Indoplanorbis exustus (N)	—	—	Varma and Prasad, 2000
	Metacercariae	Kids (E)	Jyoti, 2001
—	Metacercariae	Buffalo calves Goats	Dixit et al., 2003
	Metacercariae	Goats (E)	Ghosh, 2005a
	Metacercariae	Goats (E)	Ghosh, 2005b
	Metacercariae	Goats (E)	Yokananth et al., 2005
	Metacercariae	Buffalo calves (E)	Velusamy et al., 2006
Indoplanorbis exustus (N)	Metacercariae	Lambs (E)	Ritu Arora et al., 2007
Indoplanorbis exustus (N)	Cercariae Metacercariae	Sheep, Corridale (E)	Shivjot Kaur et al., 2008
Indoplanorbis exustus (N)	Cercariae Metacercariae	Sheep (E)	Shivjot Kaur et al., 2009a

Intermediate Host	Larva	Final Host	Reference
Indoplanorbis exustus (N)	Metacercariae	Sheep (E)	Syed Shabih Hussain et al., 2009b
	Cercariae Metacercariae	Soviet Chinchilla rabbits (E)	Ritu Arora et al., 2010
Paramphistomum explanatum (Creplin, 1847) Fischoeder, 1904* **(Syn: *Gigantocotyle explanatum*)**			
Indoplanorbis exustus (N)		—	Srivatsava, 1944a
Lymnaea luteola f. australis (N)	Cercaria Metacercariae Lambs (E)	Kid (E) Pig (E)	Agrawal and Pande, 1971
Indoplanorbis exustus (N)	Metacercariae	Guinea pigs (E) Rabbit (E) Lambs (E)	Singh and Pande, 1972
See also *Giganotcotyle explanatum*			
Psuedodiscus collinsi (Cobbold, 1875) Sonsino, 1895			
Indoplanorbis exustus (N)	Cercariae Metacercariae	Donkey foal (E)	Peter and Srivastava, 1954
Indoplanorbis exustus (E)	Cercariae Metacercariae	Elephant (N)	Peter, 1955
Indoplanorbis exustus (N)	Cercariae Metacercariae	Donkey foal (E)	Peter and Srivastava, 1955
Indoplanorbis exustus (N)			Peter and Srivastava, 1960a
Indoplanorbis exustus (N)		—	Jain, 1976

Intermediate Host	Larva	Final Host	Reference
		***Oliveria indica* Thapar and Sinha, 1945**	
Gyraulus euphraticus (E)	Cercariae Metacercariae	Buffalo calf (E)	Thapar, 1961
		***Srivastavaia indica* Singh, 1970**	
Indoplanorbis exustus (N)	Cercariae Metacercariae	Lamb (E) Kid (E)	Singh, 1970

1.12 FAMILY STRIGEIDAE

Strigeid

Intermediate Host	Larva	Final Host	Reference
Lymnaea auricularia (N) *Lymnaea luteola* (N) *Indoplanorbis exustus* (N) *Gyraulus convexiusculus* (N) *Vivipara bengalensis* (N) *Melanoides (Tarebia) lineatus* (N) *Melanoides (Tarebia) tuberculatus* (N)	Cercariae	—	Sahai, 1969-70
Indoplanorbis exustus (N) *Lymnaea* spp. (*Lymnaea acuminata* and *Lymnaea luteola*) (N)	Cercariae	—	Muraleedharan et al., 1973
Indoplanorbis exustus (N)	Cercariae	—	Dutta and Bali, 1980

Intermediate Host	Larva	Final Host	Reference
	1.13 FAMILY: ISOPARORCHIDAE		
	Isoparorchis hypselobargri (Billet, 1898) Odhner, 1927		
	Metacercariae		Southwell and Prasad, 1918
	Metacercariae		Bhalaerao, 1932
	Metacercariae		Chauhan, 1947
	Metacercariae		Jaiswal, 1957
Fresh-water fishes (N)	Metacercariae		Rai, 1969
	Metacercariae		Rai and Pande, 1965
Fresh- water fishes: *Mystatus vittatus* *Oxygaster vacaila*	Metacercariae		Pandey, 1969
	1.14 FAMILY: SCHISTOSOMATIDAE		
	Orientobilharzia dattai (Dutt and Srivastava, 1952) Dutt and Srivastava, 1955 (Syn: *Ornithobilharzia dattai* Dutt and Srivastava, 1952)		
Lymnaea luteola (N)	Cercariae	Bull calf (E) Buffalo calf (E) Guinea pigs (E)	Dutt and Srivastava, 1952
Lymnaea luteola (N)	Cercariae	Bull calf (E) Buffalo calf (E) Guinea pigs (E)	Dutt and Srivastava, 1955a

Intermediate Host	Larva	Final Host	Reference
Lymnaea luteola var. *australis* (N) (E) *Lymnaea luteola* (N) (E)	Cercariae	Cattle (E) Buffaloes (E) Sheep (E) Goats (E) Donkey (E) Rabbit (E) Guinea pig (E) White rat (E) White mice (E)	Dutt and Srivastava, 1955b
Lymnaea luteola (N)	Cercariae	—	Dutt and Srivastava, 1955c
Snails (species, not mentioned)	Cercariae	Bull calves (E) Buffalo calf (E) Sheep (E) Goats (E) Donkeys (E) Rabbits (E) Guinea pigs (E) White rat (E) White mice (E)	Dutt and Srivastava, 1961
Lymnaea luteola (N) (E)	Cercariae	—	Dutt and Srivastava, 1962a
Lymnaea luteola (N) (E)	Cercariae	—	Dutt and Srivastava, 1962b
Lymnaea luteola (N) (E)	Cercariae	—	Dutt and Srivastava, 1962c
Lymnaea luteola (N)	Cercariae	Guinea pig (E)	Muraleedharan and Prasanna Kumar, 1974

Intermediate Host	Larva	Final Host	Reference
Lymnaea sp. (N)	Cercariae	—	Muraleedharan *et al.*, 1977
Lymnaea luteola (N)	Cercariae	White mice (E) Guinea pig (E) Rabbit (E)	Singh and Ahluwalia, 1977a
Lymnaea luteola (N)	Cercariae	Sheep (E) Goats (E) White mice (E)	Singh and Ahluwalia, 1977b
Lymnaea luteola (N)	Cercariae	—	Dutt and Bali, 1980
Lymnaea luteola (N)	Cercariae	Rhesus monkeys (E)	Das and Agrawal, 1986
Lymnaea luteola (N)	Cercariae		Agarwal *et al.*, 1991
Lymnaea luteola (N)	Cercariae		Banerjee and Agarwal, 1991
Lymnaea luteola (N)	Cercariae		Kohli and Agarwal, 1994b
Lymnaea luteola (N)	Cercariae		Kohli and Agarwal, 1996
Lymnaea luteola (N)	Cercariae		Agarwal *et al.*, 2000b

Orientobilharzia turkestanicum (Skrjabin, 1952) Dutt and Srivastava, 1964

Intermediate Host	Larva	Final Host	Reference
Lymnaea auricularia (N) (E)	Cercariae	Cattle (E) Sheep (E) Rabbit (E) Guinea pig (E) Albino mice (E)	Dutt and Srivastava, 1964

Intermediate Host	Larva	Final Host	Reference
Lymnaea auricularia sensu stricto (N) (E) / Lymnaea auricularia (E)	Cercariae	Lamb (E)	Vinai Kumar, 1973
Schistosoma incognitum **Chandler, 1926**			
Lymnaea luteola (N)	Cercariae	—	Ramanujachari and Alwar, 1954
Snail (N)	Cercariae	Pig (E), Dog (E), Cat (E), Sheep (E), Goat (E), Cattle (E), Rabbit (E), Guinea pig (E)	Sinha and Srivastava, 1954
Lymnaea luteola var. australis (N) (E)	Cercariae	Pigs (E), Rats (E)	Sinha and Srivastava, 1960
Lymnaea luteola (E)	Cercariae	Dogs (E)	Tewari et al, 1963
Lymnaea luteola (N) (E)	Cercariae	Rhesus monkeys (Macaca mulatta) (E)	Dutt, 1965
Snail (N)	Cercariae		Muralidharan, 1965
Snail (N) (vide Sinha and Srivastava, 1960)	Cercariae	Rabbit (E), White rat (E), Whitemouse (E), Guinea pig (E)	Sinha and Srivastava, 1965

Intermediate Host	Larva	Final Host	Reference
Lymnaea luteola (E)	Cercariae	Squirrel (E) Dog (E) Kitten (E) Goats (E) Sheep (E) Cattle (E)	Tewari et al., 1966
Lymnaea luteola (N)	Cercariae	Dogs (E) Rhesus monkeys (Macaca mulatta) (E)	Dutt, 1967a
Lymnaea luteola (E)	Cercariae	Albino mice (E)	Ahluwalia, 1971
Not mentioned	Cercariae	Pigs (E)	Ahluwalia, 1972a
Not mentioned	Cercariae	Pigs (E)	Ahluwalia, 1972b
Not mentioned	Cercariae	Pigs (E)	Ahluwalia, 1972c
Not mentioned	Cercariae	Pigs (E) Rats (E) Monkeys (E)	Ahluwalia, 1972d
Not mentioned	Cercariae	—	Ahluwalia, 1972e
Not mentioned	Cercariae	Pig (E)	Ahluwalia, 1972f
Not mentioned	Cercariae	Pigs (E)	Ahluwalia and Dutt, 1972
Lymnaea luteola (N)	Cercariae	—	Manohar, et al., 1972
Lymnaea luteola (N)	Cercariae	—	Tewari and Gita Biswas, 1972

Intermediate Host	Larva	Final Host	Reference
Lymnaea luteola (N)	Cercariae	Pigs	Ahluwalia, 1973a
Lymnaea luteola (N)	Cercariae	Pigs	Ahluwalia, 1973b
Lymnaea luteola (N)	Cercariae	Pigs	Subramanyam, *et al.*, 1973
Lymnaea luteola (N)	Cercariae	Pigs (E) Guinea pigs (E) Rabbits (E)	Bhatia and Rai, 1974
Lymnaea luteola australis (E)	Cercariae	Albino mice (E)	Borokakoty, 1975
Lymnaea luteola (N)	Cercariae	—	Gita Biswas, 1975 (Thesis)
Lymnaea luteola (E) (N)	Cercariae	Pigs (E)	Hajela *et al.*, 1975
Lymnaea luteola (E)	Cercariae	Pigs (E)	Hajela *et al.*, 1976
Lymnaea luteola (E)	Cercariae	Pigs (E)	Rai *et al.*, 1975
—	Cercariae	—	Bhatia and Rai, 1976
Lymnaea luteola (E)	Cercariae	Pigglets (E)	Hajela *et al.*, 1976
Lymnaea luteola (E)	Cercariae	Calf	Bhatia *et al.*, 1976
Lymnaea luteola (N)	Cercariae	Albino mice (E)	Tewari and Gita Biswas, 1976
Lymnaea luteola (N)	Cercariae		Agrawal, 1978
Lymnaea luteola (N)	Cercariae	Albino mice (E)	Agrawal *et al.*, 1979
Lymnaea luteola (E)	Cercariae	Albino mice (E)	Tewari and Singh, 1979
Lymnaea luteola (E)	Cercariae	Albino mice (E) Mice (E)	Agrawal, 1981 Shrivastava, *et al.*, 1981a

Intermediate Host	Larva	Final Host	Reference
Lymnaea luteola (N)	Cercariae	Albino mice (E)	Shrivastava et al., 1981b
Lymnaea luteola (N)	Cercariae	—	Agrawal et al., 1982a
Lymnaea luteola (N)	Cercariae	Albino mice (E) Rat (E)	Agrawal et al., 1982b
Lymnaea luteola (N)	Cercariae	Albino mice (E)	Agrawal et al., 1983
Lymnaea luteola (N)	Cercariae	Albino mice (E)	Varma et al., 1983
Lymnaea luteola (N)	Cercariae	Rabbits (E) Pigs (E)	Shrivastava et al., 1983
Lymnaea luteola (N)	—	Albino mice (E) Rats (E)	Agrawal and Sahasrabudhe, 1984
Lymnaea luteola (N)	Cercariae	Rabbits (E) Pups (E)	Agrawal et al., 1984
Lymnaea luteola (N)	Cercariae	Rhesus monkeys (E)	Das, 1984
Lymnaea luteola (N)	Cercariae	White mice (Mus musculus)	Agrawal et al., 1985
Lymnaea luteola (N)	Cercariae	Dogs (pups) (E)	Agrawal et al., 1986
Lymnaea luteola (N)	Cercariae	Rhesus monkeys (Macaca mulatta) (E)	Das and Agrawal, 1986
Lymnaea luteola (N)	—	Rhesus monkeys (Macaca mulatta) (E)	Das et al., 1986
	Cercariae	Albino mice (E)	Panesar and Agrawal, 1986

Intermediate Host	Larva	Final Host	Reference
Lymnaea luteola (N)	Cercariae	Rabbits (E)	Panesar *et al.*, 1986
Lymnaea luteola (N)	Cercariae	Rhesus monkeys (E)	Das *et al.*, 1987a
Lymnaea luteola (N)	Cercariae	Rhesus monkeys (E)	Das *et al.*, 1987b
	Cercariae	Rhesus monkeys (E)	Panesar *et al.*, 1987
	Cercariae	Rabbits (E) (Immunisation)	Avsatthi and Sahasrabudhe, 1988
	Cercariae	Rabbits (E) Albino mice (E)	Agrawal and Sahasrabudhe, 1988
	—	Rhesus monkeys (*Macaca mulatta*) (E)	Das and Joshi, 1988a
Snails (N)	Cercariae	Rhesus monkeys (*Macaca mulatta*) (E)	Das and Joshi, 1988b
Snails (N)	Cercariae	Rhesus monkeys (*Macaca mulatta*) (E)	Das and Joshi, 1988c
Snails (N)	Cercariae	Rhesus monkeys (*Macaca mulatta*) (E)	Das and Joshi, 1988d
Snails (N)	Cercariae	Rhesus monkeys (*Macaca mulatta*) (E)	Das *et al.*, 1988a
Lymnaea luteola (N)	Cercariae	Rhesus monkeys (*Macaca mulatta*) (E)	Das *et al.*, 1988b
	Cercariae	Mice (E)	Gita Biswas and Tewari, 1988

Intermediate Host	Larva	Final Host	Reference
Lymnaea luteola (N) (E)	Cercariae (Gamma irradiation)	—	Bhaskar Rao et al., 1989
Lymnaea luteola (E)	Cercariae	—	Prasad et al., 1989
Lymnaea luteola (E)	Cercariae	—	Agrawal et al., 1991
Lymnaea luteola (E)	Cercariae	Laboratory animals (E) Final hosts (N)	Anjana Misra, 1991
Lymnaea luteola (N)	Cercariae	—	Banerjee and Agarwal, 1990
Lymnaea luteola (N)	Cercariae	—	Kohli, 1991
Lymnaea luteola (E)	Cercariae	Albino rats (E)	Agrawal et al., 1992
Lymnaea luteola (E)	Cercariae	—	Kohli and Agarwal, 1994b
Lymnaea luteola (E)	Cercariae	—	Kohli and Agarwal, 1996
Lymnaea luteola (E)	Cercariae	Albino mice (E)	Agrawal et al., 1999
Snail	Cercariae	Pig (E)	Jain et al., 1999
Snail	Cercariae	Pig (E)	Shames et al., 1999
Lymnaea luteola (E)	Cercariae	Pig (N) Albino mice (E)	Srivastava and Agrawal, 1999a
Lymnaea luteola (E)	Cercariae	Albino mice (E)	Srivastava and Agrawal, 1999b
—	Cercariae	Rabbit (E)	Agrawal and Singh, 2000a
—	Cercariae	Rabbit (E)	Agrawal and Singh, 2000b
		Pig (N)	Agrawal et al., 2000a

Intermediate Host	Larva	Final Host	Reference
Lymnaea luteola (N)	Cercariae	—	Agrawal *et al.*, 2000b
		Pig (E)	Jyothi Jain *et al.*, 2000
Lymnaea luteola (N)	Cercariae	—	Varma and Prasad, 2000
	Cercariae	Piglets (E)	Agrawal *et al.*, 2001
	Cercariae	Piglet (E)	Jyothi Jain *et al.*, 2001
Lymnaea luteola (N)	Cercariae	—	Shameem and Radhika, 2001
	Cercariae	Pig (E)	Shames *et al.*, 2000a
	Cercariae	Pig (E)	Shames *et al.*, 2000b
Lymnaea luteola (N)	Cercariae	Kids (E)	Gupta and Agrawal, 2005
Lymnaea luteola (N)	Cercariae	Barbari goats (E)	Gupta *et al.*, 2006a
Lymnaea luteola (N)	Cercariae	Barbari goats (E)	Gupta *et al.*, 2006b
Lymnaea luteola (N)	Cercariae	Goats (E)	Vohra *et al.*, 2006
	Cercariae	Barberi goats (E)	Gupta and Agrawal, 2007
Lymnaea luteola (N)	Cercariae	Sheep (N)	Vohra *et al.*, 2009
Schistosoma indicum Montgomery, 1906			
Indoplanorbis exustus (N)	Cercariae	Kids (E) Lamb (E) Guinea pig (E)	Srivastava and Dutt, 1951
Indoplanorbis exustus (N)	Cercariae		Ramanujachari and Alwar, 1954

Intermediate Host	Larva	Final Host	Reference
Snail (N) (E)	Cercariae		Srivastava and Dutt, 1955c
Indoplanorbis exustus (N) (E)	Cercariae	Goats (E) Sheep (E) Donkey (E) Rabbit (E) Guinea pig (E) White rat (E) White mouse (E) Grey mouse (E) Buffalo (E) (N)	Srivastava and Dutt, 1955d
Indoplanorbis exustus (N)	Cercariae		Singh, 1959
Indoplanorbis exustus (N) (E)	Cercariae	Cattle (E) Buffalo (E) Goats (E) Sheep (E) (N) Donkey (E) Rabbit (E) Guinea pig (E) White rat (E) White mouse (E) Grey mouse (E) Buffalo (E) (N)	Srivastava and Dutt, 1962a
Indoplanorbis exustus (N) (E)	Cercariae	—	Srivastava and Dutt, 1962b
Indoplanorbis exustus (N) (E)	Cercariae	—	Srivastava and Dutt, 1962c
Indoplanorbis exustus (N)	Cercariae	Sheep (E)	Srivastava et al., 1963

Intermediate Host	Larva	Final Host	Reference
Indoplanorbis exustus (N)	Cercariae	Sheep (E)	Srivastava et al., 1964
Snail (N)	Cercariae		Mohandas, 1974a
Indoplanorbis exustus (N)	Cercariae	Rabbit (E)	Avsathi, 1976
Indoplanorbis exustus (N)	Cercariae	Albino mice (E)	Agrawal et al., 1979
Indoplanorbis exustus (N)	Cercariae		Dutt and Bali, 1980
Indoplanorbis exustus (N) (E)	Cercariae		Prasad et al., 1989
Indoplanorbis exustus	Cercariae		Rajkhowa, 1989
Indoplanorbis exustus (N)	Cercariae		Banerjee and Agrawal, 1990
Indoplanorbis exustus (N)	Cercariae		Agrawal et al., 1991
Indoplanorbis exustus (E)	Cercariae	Laboratory animals (E)	Anjana Mishra, 1991
Indoplanorbis exustus	Cercariae	Bull calves (E)	Rajkhowa et al., 1992
Indoplanorbis exustus (E)	Cercariae		Kohli and Agarwal, 1994b
Indoplanorbis exustus (E)	Cercariae		Kohli and Agarwal, 1996
Indoplanorbis exustus	Cercariae	Cattle (E)	Rajkhowa et al., 1996
Indoplanorbis exustus	Cercariae	Bull calves (E)	Rajkhowa et al., 1997a
Indoplanorbis exustus	Cercariae	Bull calves (E) Mice (E)	Rajkhowa et al., 1997b
Indoplanorbis exustus (N)	Cercariae		Agrawal et al., 2000b
Indoplanorbis exustus (N)	Cercariae		Shameem and Radhika, 2001

Intermediate Host	Larva	Final Host	Reference
Indoplanorbis exustus (N)	Cercariae	Barberi goats (E)	Gupta et al., 2006
Indoplanorbis exustus (N)	Cercariae		Singh et al., 2009
Examined different species of snails including Indoplanorbis exustus (N)	Cercariae	Cattle (N) Sheep (N) Goats (N) Buffaloes (N)	Syed Shabih Hussain and Juyal, 2009
Indoplanorbis exustus (N)	Cercariae		Johnpaul et al., 2010
Schistosoma nasale Rao, 1933			
Planorbis exustus (Indoplanorbis exustus) (N) Limnea amydalum (Lymnaea acuminata) (N)	Cercaria indica, XXX		Sewell, 1922
Lymnaea sp.	Cercariae	Cattle (N)	Mahajan, 1933
Planorbis exustus (Indoplanorbis exutus) (N) Limnea luteola (Lymnaea luteola) (N)	Cercariae, Cercariae indicae, Sewell, 1922	Calves (E)	Rao, 1933
Planorbis exustus (Indoplanorbis exutus) (N) Limnea luteola (Lymnaea luteola) (N)	Cercariae	Calves (E) Guinea pig (E)	Rao, 1934
—	Cercariae	Calves (E)	Rao, 1935
Lymnaea luteola (N)	Cercariae	—	Mudaliyar and Alwar, 1947
Lymnaea luteola (N) (Probable)	Cercariae	Cattle (E) Buffaloes (E)	Varma, 1954

Intermediate Host	Larva	Final Host	Reference
Indoplanorbis exustus (N) *Lymnaea succinea* (N)	Cercariae	—	Peter, 1955
Indoplanorbis exustus (N)	Mother sporocysts, Daughter sporocysts Cercariae	Cattle (E)	Dutt, 1967
Indoplanorbis exustus (N) (E)	Cercariae	Cattle (E) Buffaloes (E) Goats (E) Mouse (E)	Dutt and Srivastava, 1968
Indoplanorbis sp. (N) (E)	Cercariae	Buffalo calf (E) Bull calves (E)	Rajamohanan, 1972
Indoplanorbis sp. (N) (E)	Cercariae	Cattle calves (E) Buffalo calves (E)	Rajamohanan and Peter, 1972
Indoplanorbis exustus (E)	Cercariae	Buffalo calf (E)	Rajamohanan *et al.*, 1972a
Indoplanorbis exustus (E)	Cercariae	Buffalo calf (E)	Rajamohanan *et al.*, 1972b
Indoplanorbis exustus (N)	Cercariae		Muraleedharan, 1973
Indoplanorbis exustus (N) (E)	Cercariae	Buffalo calves (E)	Muraleedharan and Prasanna Kumar, 1974
Indoplanorbis exustus (E)	Cercariae	Cattle (E) Buffaloes (E)	Gita Biswas, 1975
Indoplanorbis exustus (N) *Lymnaea luteola* (N)	Cercariae	Cattle (E) Buffaloes (E)	Koshy *et al.*, 1975

Intermediate Host	Larva	Final Host	Reference
Indoplanorbis exustus (E)	Cercariae	—	Muraleedharan et al., 1975a
Indoplanorbis exustus (E)	Miracidia Cercariae	—	Muraleedharan et al., 1975b
Indoplanorbis exustus (E)	Miracidiae Cercariae	—	Muraleedharan et al., 1975c
Indoplanorbis exustus (N) (E)	Cercariae	—	Muraleedharan et al., 1976a
Indoplanorbis exustus (N)	Cercariae	Buffalo calves (E)	Muraleedharan et al., 1976b
Indoplanorbis exustus (N)	Cercariae	Mice (E) Rabbits (E) Guinea pigs (E) Kids (E) Lambs (E)	Sahay and Sahai, 1976
Indoplanorbis exustus (N)	Cercariae	—	Muraleedharan et al., 1977
Indoplanorbis exustus (N)	Cercariae	Mice (E) Rabbits (E) Guinea pig (E) Kids (E) Lambs (E)	Sahay et al., 1977
Indoplanorbis exustus (N)	Cercariae	—	Bhilegoankar et al., 1978a
Lymnaea luteola (N)	Cercariae	—	Bhilegoankar et al., 1978b
Indoplanorbis exustus (N)	Cercariae	—	Gita Biswas and Subramaiam, 1978a

Intermediate Host	Larva	Final Host	Reference
Indoplanorbis exustus (N)	Cercariae	Calf	Gita Biswas and Subramaniam, 1978b
Indoplanorbis exustus (N)	Cercariae	Mice (E) Rabbits (E) Guinea pig (E) Kids (E) Lambs (E)	Sahay and Sahai, 1978
Indoplanorbis exustus (N)	Cercariae		Dutt and Bali, 1980
Indoplanorbis exustus (N)	Cercariae	Cattle (E) Buffaloes (E)	Gita Biswas and Subramaniam, 1980
Indoplanorbis exustus (N)	Cercariae	Cattle (E) Buffaloes (E) Sheep (E) Goat (E) Mice (E) Rabbits (E) Guinea pig (E) Man, dermatitis (N)	Anandan , 1985
Indoplanorbis exustus (N)	Cercariae	Cattle (N) Buffaloes (N) Mice (E) Rats (E)	Agrawal et al., 1988
Indoplanorbis exustus (N)	Cercariae		Anandan and Ebenezer Raja, 1988a

Intermediate Host	Larva	Final Host	Reference
Indoplanorbis exustus (E) (Exposed to miracidia of cattle, buffalo, and goat origin)	Cercariae	Cattle (N) Buffaloes (N) Goats (N)	Anandan and Ebenezer Raja, 1988b
Indoplanorbis exustus (N)	Cercariae	Calves (E)	Anandan and Ebenezer Raja, 1990
Indoplanorbis exustus (N)	Cercariae		Banerjee and Agrawal, 1990
Indoplanorbis exustus (N)	Cercariae	Mice (E) Rabbits (E) Guinea pig (E)	Gita Biswas and Subramaiam, 1990
Indoplanorbis exustus (N)	Cercariae		Agrawal et al., 1991
Indoplanorbis exustus (E)	Cercariae		Kohli and Agarwal, 1994b
Indoplanorbis exustus (N)	Cercariae	Rabbits (E)	Agrawal, 1996a
Indoplanorbis exustus (N)	Cercariae	Rabbits (E) Lamb (E)	Agrawal, 1996b
Indoplanorbis exustus (N)	Cercariae		Agrawal et al., 1998
Indoplanorbis exustus (N)	Cercariae		Thakre and Bhilegoankar, 1998
Indoplanorbis exustus (N)	Cercariae		Deka, 1999
	Cercariae	Rabbits (E)	Agrawal and Singh, 2000b
	Cercariae	Rabbits (E)	Agrawal and Singh, 2000c
Indoplanorbis exustus (N)	Cercariae		Agrawal et al., 2000b

Intermediate Host	Larva	Final Host	Reference
Indoplanorbis exustus (N)	Cercariae	Man, dermatitis (N) Probable	Muraleedharan, 2000
Indoplanorbis exustus (N)	Cercariae	Golden hamsters (E)	Sapate *et al.*, 2000
Indoplanorbis exustus (N)	Cercariae	Syrian golden Hamsters (E) Mastomyces (*Mastomyces coucha*)	Sapate *et al.*, 2001
Lymnaea sp. (N)	Cercariae	Cattle (N) Buffaloes (N)	Sivaseelan *et al.*, 2000
Snails	Cercariae	Syrian golden Hamsters (E) Mastomyces (*Mastomyces coucha*) (E)	Sunil Wasudeo Kolte, 2002
Indoplanorbis exustus (N)	Cercariae		Johnpaul *et al.*, 2010
Indoplanorbis exustus (N)	Cercariae		Kolte *et al.*, 2012
Schistosoma spindale Montgomery, 1906			
Planorbis exustus (*Indoplanorbis exustus*) (N)	Cercariae	Goats (E) Guinea pig (E)	Liston and Soparker, 1918
Planorbis exustus (N) *Lymnaea acuminata* (E)	Cercariae		Soparker, 1921
Planorbis exustus (N)	Cercariae	Monkey (*Macacus sinicus*)	Fairley, 1927

Intermediate Host	Larva	Final Host	Reference
Planorbis exustus (N)	Cerceriae	Buffaloes (E)	Fairley and Jesudasan, 1927
Planorbis exustus (N)	Cercariae	—	Fairley and Jesudasan, 1930a
Planorbis exustus (N)	Cercariae	Goats (E)	Fairley and Jesudasan, 1930b
Planorbis exustus (N)	Cercariae	—	Fairley and Jesudasan, 1930c
Planorbis exustus (N)	Cercariae	Goats (E)	Fairley and Mackie, 1930
Planorbis exustus (N)	Cercariae	Monkey (*Macacus sinicus*) (E)	Fairley *et al.*, 1930a
Planorbis exustus (N)	Cercariae	Guinea pig (E) Rabbit (E)	Fairley *et al.*, 1930b
Planorbis exustus (N)	Cercariae	—	Rao, 1932
Lymnaea luteola (*Lymnaea luteola*) (N)			Rao, 1933
Planorbis exustus (N) *Lymnaea luteola* (N)	Cercariae	Calves (E) Guinea pigs (E)	Rao, 1934
Planorbis exustus (N)	Cercariae	Calves (E)	Rao, 1935
Indoplanorbis exustus (N)	Cercariae	Guinea pig (E)	Bhalerao, 1938
Indoplanorbis exustus (N)	Cercariae		Mudaliyar and Alwar, 1947
Indoplanorbis (N)	Cercariae	—	Peter, 1954
Indoplanorbis exustus (N)	Cercariae	—	Peter, 1955a
Indoplanorbis exustus (N)	Cercariae	Guinea pig (E)	Dutt, 1957a
Indoplanorbis exustus (N)	Cercariae	Foal (E)	Dutt, 1957b

Intermediate Host	Larva	Final Host	Reference
Indoplanorbis exustus (N)	Cercariae	Man (N) (E), Dermatitis	Anantaraman, 1958
Indoplanorbis exustus (N)	Cercariae		Singh, 1959
Indoplanorbis exustus (N)	Cercariae	Guinea pig (E)	Dutt, 1962
		White mice (E)	Reddy and Rao, 1964
	Worms, males and females	*Bandicota bengalensis* (N) (RH)	Niphadkar and Rao, 1967
Indoplanorbis exustus (N)	Cercariae	—	Dutt and Srivastava, 1968
Indoplanorbis exustus (N)	Cercariae	Man (N), Dermatitis	Rao and Murty, 1968
Indoplanorbis exustus (N)	Cercariae	Goats (E)	Dutt, 1972
Indoplanorbis exustus (N)	Cercariae	—	Muraleedharan et al., 1977
Indoplanorbis exustus (N)	Cercariae	—	Madhavi, 1977
Indoplanorbis exustus (N)	Cercariae	Rabbits (E), Albino mice (E)	Agrawal, 1978
Indoplanorbis exustus (N)	Cercariae	Albino mice (E)	Dutt and Bali, 1980
Indoplanorbis exustus (N)	Cercariae	Albino mice (E)	Agrawal., 1981
Indoplanorbis exustus (N)	Cercariae	Albino mice (E) Rats (E)	Agrawal et al., 1982b
Indoplanorbis exustus (N)	Cercariae	Albino mice (E)	Agrawal et al., 1983
Indoplanorbis exustus (N)		Albino mice (E)	Varma et al., 1983

Intermediate Host	Larva	Final Host	Reference
Indoplanorbis exustus (N)		Albino mice (E) Rats (E)	Agrawal and Sahasrabudhe, 1984
Indoplanorbis exustus (N)	Cercariae of *Schistosoma spindale*	Man (N)	Anandan, 1985
Indoplanorbis exustus (N)	Cercariae	Pups (E)	Agrawal et al., 1986 (Vide Agrawal, 1978)
Indoplanorbis exustus (N)	Cercariae	Rabbits (E)	Panesar et al., 1986 (vide Agrawal and Sahasrabudhe, 1984)
		Mice (E)	Panesar and Agrawal, 1986
	Cercariae	Rabbits (E) Albino mice (E)	Agrawal and Sahasrabudhe, 1988
	Cercariae (gamma radiated)		Bhaskar Rao et al., 1989
	Cercariae	Rabbit (E)	Banerjee et al., 1990
Indoplanorbis exustus (E)	Cercariae	Rabbits (E) Guinea pigs (E) Rats (E) Mice (E)	Anjana Mishra, 1991
Indoplanorbis exustus (N)	Cercariae	—	Agrawal et al., 1991
Indoplanorbis exustus (N)	Cercariae	—	Banerjee and Agrawal, 1991
Indoplanorbis exustus (N)	Cercariae	Rat (E) Rabbit (E) Albino mice (E)	Anjana Mishra and Agrawal, 1993

Intermediate Host	Larva	Final Host	Reference
Indoplanorbis exustus (N)	Cercariae	Guinea pig (E) Rabbits (E) Mice (E) Rats (E)	Agrawal and Mishra, 1994
Indoplanorbis exustus (N)	Cercariae	Albino rats (E)	Anjana Mishra and Agrawal, 1994a
Indoplanorbis exustus (N)	Cercariae	Rabbits (E)	Anjana Mishra and Agrawal, 1994b
Indoplanorbis exustus (N)	Cercariae	Man (N) Human volunteers (E)	Narain *et al.*, 1994
Indoplanorbis exustus (E)	Cercariae		Kohli and Agarwal, 1994b
Indoplanorbis exustus (N)	Cercariae	Guinea pig (E) Rat (E) Mice (E)	Agrawal and Mishra, 1996
Indoplanorbis exustus (E)	Cercariae		Kohli and Agarwal, 1996
Indoplanorbis exustus (N) (E)	Cercariae		Agrawal, 1998
Indoplanorbis exustus (N)	Cercariae	Man (N), dermatitis Swiss albino mice (E)	Narain *et al.*, 1998
Indoplanorbis exustus (N)	Cercariae	Man (N) Rabbits (E)	Narain and Mahanta, 1999
	Cercariae (freshly shed)	Guinea pigs (E)	Shrivastava and Agrawal, 1999c
Indoplanorbis exustus (N)	Cercariae	Man, dermatitis (N)	Agrawal *et al.*, 2000b

Intermediate Host	Larva	Final Host	Reference
Indoplanorbis exustus (N)	Cercariae	Man, dermatitis (N)	Agrawal et al., 2000c
Indoplanorbis exustus (N)	Cercariae	Man, dermatitis (N)	Muraleedharan, 2000
Snails (N)	Cercariae	—	Shameen and Radhika, 2001
Indoplanorbis exustus (N)	Cercariae	—	Agrawal et al., 2003
Indoplanorbis exustus (N)	Cercariae	Kids (E)	Gupta and Agrawal, 2005
Indoplanorbis exustus (N)	Cercariae	—	Agrawal et al., 2006
Indoplanorbis exustus (N)	Cercariae	Barberi goats (E)	Gupta et al., 2006
Indoplanorbis exustus (N)	Cercariae	Barberi goats (E)	Gupta and Agrawal, 2007
Indoplanorbis exustus (N)	Cercariae	—	Daniel and Joseph, 2010
Indoplanorbis exustus (N)	Cercariae	—	Johnpaul et al., 2010
Indoplanorbis exustus (N)	Cercariae	—	Daniel and Joseph, 2011
Trichobilharzia sp.			
Snail	Cercariae	—	Muralidharam, 1965
Trichobilharzia indicum n. sp.			
Indoplanorbis exustus (E)	Cercariae	—	Sreekumar, 1966 (Thesis)
Trichobilharzia rodhaini Fain 1955			
Lymnaea luteola f. impura (E)	Cercariae	—	Sreekumar, 1966 (Thesis)

Schistosomal cercariae causing human dermatitis and including unspecific ones

Intermediate Host	Larva	Final Host	Reference
Snails (N)	Cercariae (species unspecified)	Domestic animals (N) Man (N), dermatits	Bhalerao, 1948
Indoplanorbis exustus (N)	Cercariae of *Schistosoma spindale* Cercariae of *Schistosoma nasale*	Man (N) (E), dermatitis	Anantaraman, 1958
	Cercariae of *Schistosoma incognitum*	Man (N)	Muralidharam and Ananta-raman 1958 (unpublished) vide Anantaraman, 1959
Lymnaea luteola (= *L. auricularia rufescens*) (N)	*Cercariae srivastavai* Dutt, 1957	Man (N), dermatitis	Dutt, 1957
Snail	Cercariae of *Schistosoma incognitum* Cercariae of *Trichobilharzia*		Muralidharam, 1965
Lymnaea acuminata (N) *Indoplanorbis exustus* (N)	Cercariae	—	Sahai, 1969-70
Lymnaea luteola (= *L. auricularia rufescens*) (N) *Indoplanorbis exustus* (N)	*Cercariae srivastavai* *Cercariae hardayali*	Man (N)	Dutt, 1970
Indoplanorbis exustus (N) *Lymnaea luteola* (N)	Cercariae	—	Jain, 1970
Indoplanorbis exustus (N)	Cercariae, non-ocellate, ocellate	Rat (E), recovered immature worms from liver, species unidentified	Muraleedharan, 1972

Intermediate Host	Larva	Final Host	Reference
Indoplanorbis exustus (N)	Cercariae Ocellate		Muraleedharan, 1972
	Cercariae		Nikhale, 1972
Lymnaea luteola (N) *Indoplanorbis exustus* (N)	Larval stages	—	Borkakoty and Das, 1980
Gyraulus convexiusculus (N)	Cercariae	—	Chaudhuri *et al.*, 1982
Indoplanorbis exustus (N)	Cercariae	—	Sardey, 1982
Indoplanorbis exustus (N)	Cercariae of *Schistosoma spindale* Cercariae of *Schistosoma nasale*	Man (N)	Anandan, 1985
Indoplanorbis exustus (N)	Cercariae of *Schistosoma spindale*	Man (N) Mice (E)	Narain *et al.*, 1988
Planorbid snail (N)	Cercariae of animal schistosome	Man (N) (E) Dermatitis	Narain *et al.*, 1994
Planorbid snail (N)	Cercariae of animal schistosome	Man (N), dermatitis Swiss albino mice (E)	Mahanta *et al.*, 1996
Lymnaea luteola (*Psuedosuccinea*) *acuminata*	Cercariae	—	Pokhriyal *et al.*, 1996
Melanoides (*Thiara*) *tuberculata*	Cercariae	—	Pokhriyal *et al.*, 1998
Indoplanorbis exustus (N)	Cercariae of *Schistosoma spindale*	Man (N) Rabbits (E)	Narain and Mahanta, 1999

Intermediate Host	Larva	Final Host	Reference
Indoplanorbis exustus (N)	Cercariae of *Schistosoma spindale*	Man (N) Mice (E)	Narain *et al.*, 1999
Indoplanorbis exustus (N)	Cercariae of *Schistosoma spindale* Cercariae of *Schistosoma nasale*	Man (N) Swiss albino mice (E) Rabbits (E) Man (N) (circumstantial evidence)	Narain and Mahanta (year not available; source: Google)
	New oculate cercariae	Swiss albino mice (E) (developed in liver) *Praomys* (*Mastomys*) *coucha* rats (E) Swiss albino mice (E), developed to adults of a new schistosome to be identified	Narain and Mahanta (year not specified; source: Google)
Lymnaea luteola (N) *Indoplanorbis exustus* (N)	Cercariae of *Orientobilharzia dattai Schistosoma incognitum Schistosoma indicum S. nasale S. spindale*	Man (N) Dermatitis	Agarwal *et al.*, 2000b
Indoplanorbis exustus (N)	Cercariae of *Schistosoma spindale* and *S.nasale*	Man (N) (one or both species might be responsible for dermatitis)	Muraleedharan, 2000
Thiara (*Melanoides*)*tuberculta* (N)	Cercariae, schistosome	—	Tamloorkar *et al.*, 2001

Intermediate Host	Larva	Final Host	Reference
Indoplanorbis exustus (N) *Lymnaea luteola* (N)	Cercariae of *Schistosoma spindale* Used for Cercarian Hullen Reaction	Man (N) Dermatitis	Agarwal et al., 2006
Freshwater snails (N)	Cercariae Cercarian Hullen Reaction	Man (N) Dermatitis	Agarwal et al., 2007
Freshwater snails (N)	Cercariae	Man (N) (E)	Rao et al., 2007
Indoplanorbis exustus (N) *Lymnaea luteola* (N) (Probable species)	Cercariae	—	Pemola Devi and Jauhari, 2008
Indoplanorbis (N)	Larval parasitic forms		Johnpaul et al., 2010
Melanoides (Thiara) tuberculata	Cercariae	—	Poonam Devi, 2011

2. CLASS: CESTODA

Final hosts include both primary and secondary hosts ; Abbreviations used: E- Experimental; N-Natural

2.1 FAMILY: MESOCESTOIDIDAE

Intermediate Host	Larva	Final Host	Reference
Mesocestoides lineatus (Goeze, 1782)			
Snake: 2nd IH *Tropidonotus platypus* (N) Lizards (N) Wild rats (N)		Jackal pup (E)	Srivasatava, 1939

2.2 FAMILY: ANOPLOCEPHALIDAE

Anoplocephala manubriata (Raillet, Henry and Bauchi, 1914)

Intermediate Host	Larva	Final Host	Reference
Oribatid mites: *Galumna flabienlifera* *Scheloribates latipes* *Scheloribates praeincisus* *Protoribates semindus* *Protoribates triangularis*	Cysticercoids		McAloom, 2004
		Moniezia sp.	
First time vector: *Galumna flabellifera orientalis* (N) *Haplacarus imitator* (N) First time from India: *Scheloribates laevigatus* (N)			Balakrishnan and Haq, 1984

Intermediate Host	Larva	Final Host	Reference
Pergalumna intermedia (N)			Balakrishnan and Haq, 1985
Cryptogalumna grandjeani (N)			(issue published only in 1989)
Galumna flabellifera orientalis (N)			
New oribatid mites (N)			Haq, 1984
Two new species oribatid mites (N)			Adolph and Haq, 1988
Oribatid mites (N)			Haq, 1990
Scheloribates fijiensis			Haq, 1991
Ischeloribates lanceolatus (N), (First time vector)			
Eloribates lecipunctatus (N)			
Eupelops claviger (N)			
Galumna longiporosa (N)			
Oribatid mites: (30 species):		—	Haq, 1999
1. *Fosseremus silensis* (N)	1. Early cysticercoids		
2. *Nesotocepheus hauseri* (N)	2. Spherical larvae		
3. *Scheloribates laevigatus* (N)	3. All stages		
4. *Scheloribates latipes* (N)	4. Spherical larvae		
5. *Scheloribates praeincisus interruptus* (N)	5. All stages		
6. *Scheloribates rectus* (N)	6. Eggs, onchospheres		
7. *Scheloribates madrasensis* (N)	7. Onchospheres		
8. *Scheloribates fimbriatus* (N)	8. Spherical larvae		
9. *Scheloribates fijiensis* (N)	9. Cysticercoids		
10. *Scheloribates striolatus* (N)	10. Spherical larvae		
11. *Scheloribates lanceolatus* (N)	11. All stages		

Intermediate Host	Larva	Final Host	Reference
12. *Zygoribatula lineata* (N)	12. Early cysticercoids		
13. *Peloribates levipunctatus* (N)	13. Eggs,spherical & pyriform larvae		
14. *Peloribates pilosellus* (N)	14. Eggs, oncospheres		
15. *Xylobates semindus* (N)	15. All stages		
16. *Xylobates triangularis* (N)	16. All stages		
17. *Hypozetes imitator* (N)	17. Cysticercoids		
18. *Plakoribates vectus* (N)	18. Early cysticercoids		
19. *Eupelops claviger* (N)	19. Cysticercoids		
20. *Protogalumna erecta* (N)	20. Vermiform larvae		
21. *Cryptogalumna grandjeani* (N)	21. Cysticercoids		
22. *Galumna longipluma* (N)	22. Oncospheres, spherical, pyriform & vermiform larvae		
23. *Galumna obvia* (N)	23. Onchospheres, cysticercoids		
24. *Galumna alata* (N)	24. Eggs, Oncospheres, spherical larvae		
25. *Galumna flabellifera orintalis* (N)	25. All stages		
26. *Galumna triquetra* (N)	26. Mature cysticercoids		
27. *Galumna discifera* (N)	27. Spherical& pyriform larvae		
28. *Pergalumna nervosa* (N)	28. Mature cysticercoids		
29. *Pergalumna intermedia* (N)	29. All stages		
30. *Pergalumna bimeculata* (N)	30. All stages		
Ischeloribates lanceolatus (N)	Cysticercoids		Haq et al., 1999
Scheloribates elegans (N) (First time vector)			
Galumna flabellifera orientalis (N)			

Intermediate Host	Larva	Final Host	Reference

1. *Galumna flabellifera orintalis* (N)
2. *Scheloribates lanceolatus* (N)
 	
3. *Scheloribates praeincisus interruptus* (N)
 	
4. *Pergalumna intermedia* (N)
	1. All stages, see infra
	2. Onchospheres
	   Spherical larvae
	   Pyriform larvae
	   Cysticercoids
	3. Onchosphere
	   Spherical larva
	   Cysticercoids
	4. Spherical larvae
	   Cysticercoids
	   Vermiform larvae
	   Onchospheres
	5. *Pickonbates* sp. nov. (N)
			Haq, 2001

Scheloribates laevigatus (N) *Scheloribates madrasensis* (N) *Scheloribates rectus* (N) *Scheloribates praenicieus* (N) *Galumna* sp. (N)	Cysticercoids		Jayathilakan *et al.*, 2010

Moniezia benedini (Moniez, 1879)

Intermediate Host	Larva	Final Host	Reference
Scheloribates madrasensis (E) *Galumna* sp.	Cysticercoids		Anantaraman, 1951
Oribatid mites	Cysticercoids	Lambs (E)	Mehra and Srivastava, 1955
Protoscheloribates sp. (E)	Cysticercoids		Nadakal, 1960b
Scheloribates laevigatus (E) *Scheloribates fimbriatus* (E)	Cysticercoids	—	Narasapur, 1974b

Moniezia expansa (Rudolphi, 1810)

Intermediate Host	Larva	Final Host	Reference
Scheloribates laevigatus (E) *Scheloribates fimbriatus* (E)	Cysticercoids	—	Narasapur, 1976b
Scheloribates spp. (E)	Cysticercoids	—	Narasapur, 1976c
Zygoribatula tortilis (E)	Cysticercoids	—	Kaur, 1995
Scheloribates curvialatus (E)	Cysticercoids	—	Kaur et al., 1995
Zygoribatula tortilis (E)	Cysticercoids	—	Kaur et al., 1996
Zygoribatula tortilis (E)	Cysticercoids		Kaur et al., 1997
Scheloribates curvialatus (E)	upto onchosphere stage		Kaur et al., 2000
Scheloribates madrasensis Galumna sp.	Cysticercoids	—	Anantaraman, 1951
Mites (unidentified sp.) (N)	Cysticercoids	—	Mehra and Srivastava, 1955
Protoscheloribates sp.	Cysticercoids	—	Nadakal, 1960b
Scheloribates laevigatus (E) *Scheloribates fimbriatus* (E)	Cysticercoids	Sheep (E)	Narasapur, 1974
Oribatid mites	Cysticercoids		Pythal, 1974
Scheloribates laevigatus (E) *Scheloribates fimbriatus* (E)		Sheep (E)	Narasapur, 1976a
Scheloribates laevigatus (E) *Scheloribates fimbriatus* (E)		Sheep (E)	Narasapur, 1976b

Intermediate Host	Larva	Final Host	Reference
Scheloribates sp. (E)		Sheep (E)	Narasapur, 1976c
Hypozetes sp. (N)	Cysticercoids		Deshpande et al., 1980a
Hypozetes sp. (N)	Cysticercoids	Kids (E)	Deshpande et al., 1980b
Scheloribates curvialatus (E) Scheloribates parvus (E) Zygoribatula tortilis (E) Galumna sp. (E) Hypozetes sp. (E)	Cysticercoids		Kaur et al., 1993a
Scheloribates parvus (N)	Cysticercoids		Baruah, 1994
Scheloribates curvialatus (E) Scheloribates parvus (E) Zygoribatula tortilis (E) Galumna sp. (E) Hypozetes sp. (E)	Onchospheres Cysticercoids	Lambs (E)	Kaur, 1995
Scheloribates curvialatus (E) Scheloribates parvus (E) Zygoribatula tortilis (E) Galumna sp. (E) Hypozetes sp. (E)	7 developmental stages Lastly, cysticercoids	Lambs (E)	Kaur and Bali, 1995
Scheloribates curvialatus (E)	Cysticercoids		Kaur et al., 1995
Scheloribates curvialatus (E) Scheloribates parvus (E) Zygoribatula tortilis (E) Galumna sp. (E) Hypozetes sp. (E)	Onchosphere Cysticercoids		Kaur, 1996

Intermediate Host	Larva	Final Host	Reference
Scheloribates curvialatus (E) (N)	Cysticercoids		Kaur *et al.*, 1997
Name of IH, not indicated (E)	Cysticercoids (Lab. raised)	Kids (E)	Baruah *et al.*, 2002
Natural infection: *Scheloribates curvialatus* (N) *Galumna* spp. (N) *Hypozetes* spp. (N) Experimental infection: *Scheloribates curvialatus* (E) *Scheloribates parvus* (E) *Galumna* spp. (E) *Hypozetes* spp. (E) *Zygoribatula torttilis* (E)	Cysticercoids All stages like: onchosphere spherical larva pyriform larva vermiform larvae cysticercoids with cercomeres		Kaur *et al.*, 2007

2.3 FAMILY: THYSANOSOMIDAE

Avitellina centripunctata (Revolta, 1874) Gough, 1911

Intermediate Host	Larva	Final Host	Reference
Protoscheloribates sp. *Trichoribates* sp.	Oncosphere only	—	Nadakal, 1960a

Avitellina lahorea Woodland, 1927

Intermediate Host	Larva	Final Host	Reference
Scheloribates laevigatus	Cysticercoids	—	Nadakal, 1960b
Scheloribates laevigatus (E) *Scheloribates fimbriatus* (E)	Cysticercoids	Sheep (N)	Narasapur, 1967a
Scheloribates laevigatus (E) *Scheloribates fimbriatus* (E)	Cysticercoids	—	Narasapur, 1974

Intermediate Host	Larva	Final Host	Reference
		Avitellina sudanea Woodland, 1927	
Oribatid mites	Cysticercoids	—	Kaur, 1995
Scheloribates curvialatus (E)	Cysticercoids	—	Kaur et al., 1995
Scheloribates curvialatus (E)	Onchospheres	—	Kaur et al., 1997
Scheloribates curvialatus (E)	Onchospheres	—	Kaur et al., 2000
		Stilesia globipunctata (Rivolta, 1874)	
Scheloribates indica Erythraeus sp.	Cysticercoids	—	Tandon, 1963
Oribatid mites	Cysticercoids	—	Kaur, 1995
Scheloribates curvialatus (E)	Cysticercoids	—	Kaur et al., 1995
Scheloribates curvialatus (E)	Onchospheres	—	Kaur et al., 1997
Scheloribates curvialatus (E)	Onchospheres	—	Kaur et al., 2000
		Thysaniezia giardi (Moniez, 1879)	
Scheloribates curvialatus (E)	Cysticercoids	—	Kaur et al., 1993b
Scheloribates curvialatus (E)	Cysticercoids	—	Kaur, 1995
Scheloribates curvialatus (E)	Cysticercoids	—	Kaur, et al., 1995
Scheloribates curvialatus (E)	Cysticercoids	—	Kaur et al., 1996
Scheloribates curvialatus (E)	Cysticercoids	—	Kaur et al., 1997
Scheloribates curvialatus (E)	Cysticercoids	—	Kaur et al., 2000

Intermediate Hosts of Parasites

Intermediate Host	Larva	Final Host	Reference
2.4 FAMILY: DAVAINEIDAE			
Cotugnia digonopora (Pasquale, 1890)			
Ants (N)	Cysticercoids	Chicks (E)	Chand, 1964
Monomorium (Holcomyrmet) scabriceps (N)	Cysticercoids	Chicks (E)	Chand, 1970b
Monomorium gracilimum (N) *Monomorium destructor* (N)	Cysticercoids	Chicken, WLH (E)	Nadakal *et al.*, 1970c
Monomorium gracilimum (N) *Monomorium destructor* (N)	Cysticercoids	Chicken, (E)	Nadakal *et al.*, 1971a
Pheidole sp. (N)	Cysticercoids	Chicks (E)	Joseph *et al.*, 1987
Monomorium sp. (N)	Cysticercoids	—	Chellappa *et al.*, 1993
Monomorium floricola (N)	Cysticercoids	Chicks (E)	Sundar and Chellappa, 1995
Monomorium floricola (N)	Cysticercoids	Chicks (E)	Ponnudurai and Chellappa, 2001a
Monomorium floricola (N)	Cysticercoids	Chicks (E)	Sundar and Chellappa, 2001
Monomorium floricola (N)	Cysticercoids	Chicks (E)	Harikrishnan and Ponnudurai, 2010
Cotugnia meggitti Yamaguti, 1935			
Monomorium destructor (N)	Cysticercoids	—	Malaviya and Dutt, 1971

Cotugnia srivastavai Malaviya and Dutt, 1970

Monomorium destructor (N)	Cysticercoids	Domestic pigeon (N)	Malaviya and Dutt, 1970

Raillietina cesticelus (Molin, 1858)

4 species of beetles (N) 10 species of beetles (E) include *Cincidella* sp.	Cysticercoids	Chicks (E)	Dutt *et al.*, 1950
Opatroides vicinus (E)	Cysticercoids	Chicks (E)	Sinha and Srivastava, 1958
Anthicus confuscii	Cysticercoids	Chicks (E)	Dutt, 1961
Beetles: *Anthicus confuscii* (N) (E) *Opatroides vicinus* (N) (E) *Dermestes maculatus* *Dermestes vulpinus* (N) (E) *Hister orientalis* (N) (E) *Cantharsius pithecius* (N) *Onticellus cinctus* (N) *Onticellus palipes* (N) *Cincindella sexpunctata* (N) *Cincindella vigintiguttata* (N) *Cincindella erudata* (N)	Cysticercoids	Chicks (E)	Dutt *et al.*, 1961
Broconidae: species not known (N) *Carcinops striatus* (N)	Cysticercoids	Chicks (E)	Srivastava and Pande, 1967
Carcinops striatus (N)	Cysticercoids	Chicks (E)	Mathur and Pande, 1967
Opatroides vicinus (E)	Cysticercoids	WLH Chicks (E)	Bhowmik *et al.*, 1982

Intermediate Host	Larva	Final Host	Reference
Trilobium confusum (N)	Cysticercoids	Chicks (E)	Gogoi and Chaudhury, 1982a
Trilobium confusum (E)	Cysticercoids	Chicks (E)	Gogoi and Chaudhury, 1982c
Trilobium confusum (N) *Alphitbius diaperinus* (N) *Onthophagus bifasciatus* (N)	Cysticercoids	Chicks (E)	Gogoi and Chaudhury, 1982d
Trilobium confusum (E)	Cysticercoids	Chicks (E)	Gogoi and Chaudhury, 1982e
Opatroides vicinus (E) *Epierus pulicarius* (E)	Cysticercoids	WLH Chicks (E)	Bhowmik and Sinha, 1983
Opatroides vicinus (E) *Epierus pulicarius* (E)	Cysticercoids	WLH Chicks (E)	Bhowmik *et al.*, 1983
Opatroides vicinus *Epierus pulicarius*	Cysticercoids	WLH Chicks (E)	Bhowmik *et al.*, 1985
Tetramorium sp. (N) *Pheidole* sp. (N)	Cysticercoids		Sinha *et al.*, 1986
Tenebrionid sp.	Cysticercoids		Ponnudurai and Chellappa, 2001b
Raillietina echinobothridia (Megnin, 1880)			
Ant: *Monomorium salmonis indicum* (N)	Cysticercoids	Chicks (E)	Chand, 1961
Monomorium salmonis indicum *Pheidole fossulata*	Cysticercoids	Chicks (E)	Chand, 1964a

Intermediate Host	Larva	Final Host	Reference
Tetramorium simillimum (N)	Cysticercoids	Chicks (E)	Chand, 1970b
Tetramorium sp.1 (N) *Tetramorium* sp.2 (N) *Pheidologeton* sp. (N)	Cysticercoids	Chicken (E)	Nadakal *et al.*, 1971a
Tetramorium sp. (N) *Tetramorium simillimum* (N) *Pheidole* sp. (N) *Triglypholthrix striatus* (N) *Xiphomyrmex* sp. (N)	Cysticercoids	Chicken, White Rocks (E)	Nadakal *et al.*, 1973f
Ants	Cysticercoids	Chicks (E)	Gogoi and Hazarika, 1975a
Tetramorium tortosum (N)	Cysticercoids	Chicks (E)	Gogoi and Hazarika, 1975b
Ants	Cysticercoids	WLH, pullets (E)	Gogoi and Hazarika, 1975c
Pheidole sp. (N)	Cysticercoids (abnormal) with 6 hooks		Rajendran *et al.*, 1981
Tetramorium simillimum (E)	Cysticercoids	WLH chicks (E)	Gogoi and Chaudhury, 1982a
Ants (E)	Cysticercoids	Chicks (E)	Gogoi and Chaudhury, 1982b
Tetramorium simillimum (E) *Pheidole* sp. (E)	Cysticercoids	Chicks (E)	Gogoi and Chaudhury, 1982c
Tetramorium tortosum (N) *Tetramorium trothneyi* (N) *Tetramorium simillimum* (N) *Pheidole* sp. (N)	Cysticercoids	Chicks (E)	Gogoi and Chaudhury, 1982d

Intermediate Host	Larva	Final Host	Reference
Tetramorium simillimum (E) Pheidole sp. (E)	Cysticercoids	Chicks (E)	Gogoi and Chaudhury, 1982e
Raillietina mehrai Malaviya and Dutt, 1971			
Pheidole rhombinoda (N)	Cysticercoids	Pigeon (N) (E)	Malaviya and Dutt, 1971
Raillietina nagpurensis Moghe, 1925			
Pheidole rhombinoda (N)	Cysticercoids		Malaviya and Dutt, 1971
Raillietina singhi Malaviya and Dutt, 1971			
Pheidole rhombinoda (N)	Cysticercoids		Malaviya and Dutt, 1971
Raillietina tetragona (Molin, 1858)			
Monomorium salmonis indicum (N)	Cysticercoids	Chicks (E)	Chand, 1961
Pheidole fossulata (N)	Cysticercoids	Chicks (E)	Chand, 1964a
Monomorium salmonis indicum (N)	Cysticercoids	Chicks (E)	Chand, 1964b
Monomorium sp. (N) Monomorium floricola (N) Pheidole rhombinoda (N)	Cysticercoids	Chicks (E)	Srivastava and Pande, 1967a
Monomorium sp. (N) Monomorium floricola (N) Pheidole rhombinoda (N)	Cysticercoids	Chicks (E)	Mathur and Pande, 1969
Ants (N)	Cysticercoids	Chicks (E)	Chand, 1969
Ants (N)	Cysticercoids	Chicks (E)	Chand, 1970a

Intermediate Host	Larva	Final Host	Reference
Tetramorium simillimum (N)	Cysticercoids	Chicks (E)	Chand, 1970b
Tetramorium simillimum (N)	Cysticercoids	Chicken, WLH (E)	Nadakal *et al.*, 1970a
—	Cysticercoids	Chicken, WLH (E)	Nadakal *et al.*, 1970b
Tetramorium sp. (N) *Tetramorium simillimum* (N) *Pheidole* sp. (N)	Cysticercoids	Chicken (E)	Nadakal *et al.*, 1971a
Tetramorium sp. (N) *Tetramorium simillimum* (N) *Pheidole* sp. (N)	Cysticercoids	Chicken, WLH, White Rock (WR), Desi, WLHx Desi (E)	Nadakal *et al.*, 1971b
Tetramorium sp.2 (N) *Tetramorium simillimum* (N) *Pheidole* sp. (N)	Cysticercoids	Chicks WLH, WR, Desi, WLHx Desi (E)	Nadakal *et al.*, 1972
Tetramorium simillimum (N) *Tetramorium* sp. 2 (N) *Pheidole* sp. (N) *Triglypholthrix striatidens* (N) *Xiphomyrmex* sp. (N)	Cysticercoids	Chicken, WLH, WR, Desi, WLHx Desi (E)	Nadakal *et al.*, 1973a
—	Cysticercoids	Chicken, WLH, WR, Desi, WLHx Desi (E)	Nadakal *et al.*, 1973b
—	Cysticercoids	Chicken, WLH, WR, Desi, WLHx Desi (E)	Nadakal *et al.*, 1973c
—	Cysticercoids	Chicken, WLH, WR, Desi, WLHx Desi (E)	Nadakal *et al.*, 1973d

Intermediate Host	Larva	Final Host	Reference
Tetramorium simillimum (N) *Triglypholthrix striatus* (N) *Xiphomyrmex* sp. (N)	Cysticercoids	Chicken, WLH, WR, Desi,WLHx Desi (E)	Nadakal et al., 1973e
Tetramorium simillimum (N) *Triglypholthrix striatidens* (N) *Xiphomyrmex* sp. (N)	Cysticercoids	Chicken, WLH, WR, Desi, WLHx Desi (E)	Nadakal et al., 1974a
Infected ants (N) (vide 1971a)	Cysticercoids	Chicken, WLH, WR, Desi, WLHx Desi (E)	Nadakal et al., 1974b
Tetramorium tortosum (N)	Cysticercoids	Chicks (E)	Gogoi and Hazarika, 1975
Infected ants (vide Nadakal et al., 1971a)	Cysticercoids	Chicken, WLH, WR, Desi, WLHx Desi (E)	Nadakal et al., 1975a
—	Cysticercoids	Chicken, WLH, WR, Desi, WLHx Desi (E)	Nadakal et al., 1975b
Tetramorium simillimum (N) *Pheidole* sp. (N)	Cysticercoids	Chicken, WLH (E)	Nadakal and Vijayakumaran Nair, 1979
Tetramorium simillimum (N) *Tetramorium caespetum* (N) *Pheidole* sp. (N)	Cysticercoids	Chicks, WLH (E)	Vijayakumaran Nair and Nadakal, 1980
Tetramorium simillimum (N) *Triglypholthrix striatidens* (N) *Pheidole* sp. (N)	Cysticercoids	Chicks, WLH (E)	Vijayakumaran Nair and Nadakal, 1981
Pheidole sp. (E)	Cysticercoids	Chicks (E)	Gogoi and Chaudhury, 1982a

Intermediate Host	Larva	Final Host	Reference
Tetramorium tortosum (N) *Tetrmorium rothneyi* (N) *Tetramorium simillimum* (N) *Pheidole* sp. (N)	Cysticercoids	Chicks (E)	Gogoi and Chaudhury, 1982c
Tetramorium simillimum (E) *Pheidole* sp. (E)	Cysticercoids	Chicks (E)	Gogoi and Chaudhury, 1982d
Tetramorium simillimum (E) *Pheidole* sp. (E)	Cysticercoids	Chicks (E)	Gogoi and Chaudhury, 1982e
Ants (N),different species as mentioned in Nadakal et al., 1971	Cysticercoids	Chicks, WLH (E)	Nadakal and Vijayakumaran Nair, 1982
Tetramorium simillimum (N) *Tetramorium* sp.2 (N) *Triglypholthrix striaiidens* (N) *Pheidole* sp. (N) *Xiphomyrmex* sp. (N)	Cysticercoids	Chicks (E)	Vijayakumaran Nair et al., 1982
Tetramorium simillimum (N) *Pheidole* sp. (N)		Chicks, day old (E)	Rajendran and Nadakal, 1984
Tetramorium sp. (N) *Pheidole* sp. (N)	Cysticercoids		Sinha et al., 1986
Tetramorium simillimum (N) *Tetramorium caespetum* (N) *Pheidole* sp. (N)	Cysticercoids	Chicks (E)	Vijayakumaran Nair and Nadakal, 1986
Tetramorium simillimum (N) *Pheidole* sp. (N)	—	Domestic fowl (E)	Rajendran and Nadakal, 1988

Intermediate Host	Larva	Final Host	Reference
Raillietina torquata (Meggit, 1924) Southwell, 1930			
	Cysticercoids		Malaviya and Dutt, 1971
2.5 FAMILY: DILEPIDIDAE			
Anomotaenia sp.			
Gonocephalum depressum	Cysticercoids		Srivastava and Pande, 1967b
Amoebotaenia sphenoides (Railliet, 1892)			
Earthworm (N)	Cysticercoids	Birds (E)	Sunderam and Radha-krishnan, 1962
Earthworm (N)	Cysticercoids	Chicks (E), (N)	Chandra and Singh, 1972
Amoebotaenia spinosa Yamguti, 1956			
Earthworm: *Megascolex* sp. (N)	Cysticercoids	Fowl (E)	Madhavan Pillai, 1968
Megascolex sp. (N)	Cysticercoids	Fowl (E)	Madhavan Pillai and Peter, 1971
Choanotaenia infundibulum (Bloch, 1779)			
16 species of beetle (E) 15 species of beetle (E) 3 species of beetle (E) 2 species of beetle (N)	Cysticercoids	Chicks (E)	Dutt and Sinha., 1950

Intermediate Host	Larva	Final Host	Reference
Beetle:	Cysticercoids	Chicks (E)	Dutt and Sinha, 1961
Anthicus confuscii (N)			
Opatroides vicinus (N)			
Dermestes maculatus (N)			
Hister orientalis (E)			
Cantharsius pithecius (E)			
Onticellus cinctus (E)			
Onticellus pallipes (E)			
Rhizopertha dominica (E)			
Trilobium confusum (E)			
Latheticus oryzae (E)			
Gonocephalum depressum (E)			
Sphenariopsis tristis (E)			
Cincindella sexpunctata (E)			
Cincindella vigintiguttata (E)			
Cincindella erudata			
Grasshoppers:			
Acrida exaltata (E)			
Oedaleus abruptus (E)			
Acrotylus humbertianus (E)			
Attractomorpha sp. (E)			
Hieroglyphus sp. (E)			
Hieroglyphus banian (E)			
Hieroglyphus nigroreplatus (E)			
Spathosternum parsiniferum (E)			
Cataloipus indicus			
Aiolopus sp. (E)			

Intermediate Host	Larva	Final Host	Reference
Oxya sp. (E)			
Stenohippus mundus (E)			
Grouse locusts			
3 species of family Tetrigidaesa (E)			
Hyparpalus indicus (N) Optiopalpus obesus (N) Conocephalus pallidus (E) Heyacentrus munda (E)	Cysticercoids		Chand, 1970
Dermestes sp. (N) Tenebrionid sp. (N)	Cysticercoids		Ponnudurai and Chellappa, 2001
Musca domestica (N)	Cysticercoids		Ponnudurai et al., 2003

2.6 FAMILY: DIPYLIDIIDAE

Dipylidium caninum (Linnaeus, 1758) Railliet, 1858

Intermediate Host	Larva	Final Host	Reference
Ctenocephalides felis orientis (E)	Onchospheres	—	Joseph, 1974
		Cat (N) toMan (N)	Gadre et al., 1993
Ctenocephalides canis (N) (E) Ctenocephalides felis (N) (E)	Cyticercoids	Dogs (N)	Sangarika Devi et al., 2009
Ctenocephalides felis orientis (E)	Cyticercoids	Dogs	Sangarika Devi, 2010a
Ctenocephalides felis orientis (E)	Cyticercoids		Sangarika Devi, 2010b
Ctenocephalides canis (N) Ctenocephalides felis felis (N)	Cysticercoids	Dogs (N)	Devi et al., 2011
Ctenocephalides felis felis (N)	Cysticercoids		Sangarika Devi et al., 2011

Intermediate Host	Larva	Final Host	Reference
Joyeuxiella pasqualei (Diamara, 1893)			
Hemidactylus flaviviridis	Cysticercoids	Kitten (E)	Agrawal, 1971
Joyeuxiella echinorhynchoides (Sonsino, 1884)			
Uromatix hardwickii (N)	Cysticercoids	Pup (E)	Gupta, 1970
Uromatix hardwickii (N)	Cysticercoids	Kitten (E)Pup (E)	Singh and Pande, 1972
2.7 FAMILY: HYMENOLEPIDAE			
Hymenolepis cantaniana (Polonio, 1860)			
Beetle	Cysticercoids Onthophagus quadridentatus	—	Srivasatava and Pande, 1968
Hymenolepis carioca (Maghalaes, 1898) Ransom, 1902			
Beetle	Cysticercoids	Chicks	Mehra, 1950
Hymenolepis (Microsomacanthus) collaris (Batsch, 1786)			
Cyclops (E)	Cysticercoids	Duckling E	Baruah and Gogoi, 1983
Cyclops gigas (E)	Cysticercoids	Ducking (E)	Baruah and Gogoi, 1985
2.8 FAMILY: TAENIIDAE			
Echinococcus granulosus (Batsch, 1786)			
Bullock (N)	Hydatid		Gangulee, 1922
Bullock (N)	Hydatid		Pillai, 1928

Intermediate Host	Larva	Final Host	Reference
Elephant (N)	Hydatid		Southwell, 1930
	Hydatid		Sami and Khan, 1938
Ram (N)	Hydatid		Maplestone and Bahaduri, 1940
Cow (N)	Hydatid		Venkatachala Iyer, 1941
Giraffe	Hydatid		Ramanujachari and Alwar, 1955
Bovine (N)	Hydatid		Sambamurthy, 1956
Cattle (N) Sheep (N) Goat (N)	Hydatid		Thapar, 1956
Sheep (N)	Hydatid		Rao, 1958
Sheep (N)	Hydatid		Johri, 1959
Sheep (N) Goat (N)	Hydatid		Sachdeva and Talwar, 1960
Cattle (N) Buffalo (N)	Hydatid, unilocular cysts Multilocular cysts (Such cysts were preferred to be included under *E. multilocularis* by Pal and Sinha, 1970)		Sunderam and Natarajan, 1960
Cattle (N) Buffalo (N)	Hydatid		Dwivedi, 1963
Sheep (N) Goat (N)	Hydatid		Sharma and Chitkara, 1963

Intermediate Host	Larva	Final Host	Reference
Cattle (N) Goat (N) Pig (N)	Hydatid		Endrejat, 1964
Cattle (N) Goat (N)	Hydatid		Gogoi, 1964
Pig (N)	Hydatid		Sinha, 1966
Buffalo (N)	Hydatid	Dogs (E) Cats (E)	Gill and Rao, 1967a
Buffalo (N)	Hydatid	Dog (E) Cat (E) Mice (E)	Gill and Rao, 1967b
Buffalo (N)	Hydatid		Gill and Rao, 1967c
Sheep (N) Goat (N)	Hydatid		Satyaprakash et al., 1967
Buffalo (N)	Hydatid		Hardev Singh and Rao, 1967
Buffalo (N)	Hydatid		Gill, 1968
Buffalo (N)	Hydatid		Gill and Rao, 1968
Sheep (N) Goat (N)	Hydatid		Reddy et al., 1968
Pigs (N)	Hydatid		Srivastav and Shaw, 1968
Buffaloes (N)	Hydatid		Gill and Rao, 1969

Intermediate Host	Larva	Final Host	Reference
Goat (N)	Hydatid		Singh and Kuppuswamy, 1969
Bullock (N)	Hydatid		Arora and Dixit, 1970
Bullock (N)	Hydatid		Pal and Sinha, 1970
Buffalo (N)	Hydatid		Rao, 1970
Indian squirrel: *Ratufa indica maxima* (N)	Hydatid		Radhakrishna Reddy and Gowher Ali Khan, 1970
Buffalo (N)	Hydatid		Singh and Singh, 1970
Pig (N)	Hydatid		Sinha, 1970
Bullock (N)	Hydatid		Awachat and Iyer, 1971
Bullock (N)	Hydatid		Bali *et al.*, 1971
Sheep (N) Cattle (N)	Hydatid		Devi *et al.*, 1971
Buffaloes (N)	Hydatid		Sinha *et al.*, 1971
Goat (N)	Hydatid		Pandey 1971a
Goat (N)	Hydatid		Pandey 1971b
Goat (N)	Hydatid		Pandey 1971c
Cattle (N) Buffaloes (N)	Hydatid		Pandey *et al.*, 1971
Animals (N)	Hydatid		Reddy and Suvarnakumari, 1971

Intermediate Host	Larva	Final Host	Reference
Pig (N)	Hydatid		Chacchan and Rao, 1972
Buffalo (N) Goat (N)	Hydatid		Gill, 1972
Sheep (N)	Hydatid		Jain, 1972
Goat (N)	Hydatid	Cat (E)	Pandey, 1972a
Goats (N)	Hydatid	Dog	Pandey, 1972b
Cattle (N)) Buffaloes (N	Hydatid cyst		Pethkar and Hiregaudar, 1972
Giant flying squirrel: *Petaurista petaurista* Spotted deer: *Axis axis* Indian pangolin: *Manis crassicaudata*	Hydatid		Rao *et al.*, 1972
Cattle (N)	Hydatid		Hegde *et al.*, 1974
Cattle (N) Buffaloes (N) Sheep (N) Goats (N)	Hydatid		Pytha⁻, 1974
Bull, crossbred (N)	Hydatid		Ravindran Nair, 1974
Bullock (N)	Hydatid		Rao ar.d Mohiyuddin, 1974
Buffaloes (N)	Hydatid		Gupta and Singh, 1975

Intermediate Host	Larva	Final Host	Reference
Sheep (N)	Hydatid		Bali et al., 1976
Buffaloes (N)	Hydatid		Das et al., 1976
Bovines (N)	Hydatid		Rao and Kohli, 1976
Cattle (N)	Hydatid		Deshpande, 1977a
Pigs (N)	Hydatid		Deshpande, 1977b
Suckling calf (N)	Hydatid		Gaur et al., 1977
Buffaloes (N)	Hydatid		Mandal, 1977
Sheep (N) Goat (N)	Hydatid		Mathur and Khanna, 1977
Buffaloes (N)	Hydatid		Sinha et al., 1977
Buffaloes (N)	Hydatid	Pig (E)	Bali and Chhabra, 1978
Goats (N)	Hydatid		Prasad and Mandal, 1978
Sheep (N) Goats (N) Pigs (N)	Hydatid		Varma, 1978
Cattle (N) Buffaloes (N)	Hydatid		Abraham, 1979
Camel (N)	Hydatid		Gupta, 1979
Cow calf (E) Buffalo calf (N) Kids (E)	Hydatid	Dog (E)	Pythal, 1979

Intermediate Host	Larva	Final Host	Reference
Buffaloes (N)	Hydatid		Prasad and Mandal, 1979
Cattle (N) Buffaloes (N) Sheep (N) Goats (N) Pigs (N)	Hydatid		Abraham *et al.*, 1980a
Cattle (N) Buffaloes (N) Sheep (N) Goats (N) Pigs (N)	Hydatid		Abraham *et al.*, 1980b
Buffaloes (N)	Hydatid		Kosalaraman and Ranganathan, 1980
Sheep and goats (N)	Hydatid		Janardhana Pillai *et al.* 1980
Cattle (N)	Hydatid		Prabhakar *et al.*, 1980
Sheep and goats (N)	Hydatid		Prasad and Prasad, 1980
Pigs (N)	Hydatid		Prasad, 1981
Cattle (N) Buffaloes (N) Sheep (N) Goats (N) Pigs (N) Camel (N)	Hydatid (patho-anatomy)		Pattanayak and Gupta, 1981

Intermediate Host	Larva	Final Host	Reference
Ruminants (N)	Hydatid		Deka, *et al.* 1982
Indian camel	Hydatid		Lodha *et al.* 1982
Sheep (N)	Hydatid		Prasad and Mandal, 1982
Buffaloes (N) Sheep (N)	Hydatid		Singh *et al.* 1982
Cattle (N) Buffaloes (N) Sheep (N) Goats (N)	Hydatid		Deka *et al.*, 1983a
Buffaloes (N)	Hydatid	Dogs (E)	Deka *et al.*, 1983b
Cattle (N) Buffaloes (N) Sheep (N) Goats (N)	Hydatid		Deka *et al.*, 1983c
Cattle (N) Buffaloes (N) Sheep (N) Goats (N)	Hydatid		Deka *et al.*, 1983d
Cattle (N) Buffaloes (N) Sheep (N) Goats (N)			Kulkarni *et al.*, 1983

Intermediate Host	Larva	Final Host	Reference
Cattle (N) Buffaloes (N)	Hydatid cysts		Gopalakrishna Rao et al. 1983
Buffaloes (N)	Hydatid cysts		Chhabra, 1983
Cattle (N) Buffaloes (N) Pig (N) Goats (N) Sheep (N) Asses (N) Camels (N) Black ants (N), TH	Hydatid		Farooqi et al., 1983
Cattle (N) Buffaloes (N)	Hydatid cysts		Sanyal and Sinha, 1983b
Food animals (N)	Hydatid		Deshpande et al., 1984
Buffaloes (N)	Hydatid cysts	Dogs	Koshy, 1984
Buffaloes (N)	Hydatid cysts		Kulkarni et al., 1984
Pig (N)	Hydatid		Baruah et al., 1985
Cattle (N) Buffaloes (N) Pig (N) Goats (N)	Hydatid		Deka et al., 1985

Intermediate Host	Larva	Final Host	Reference
Cattle (N) Buffaloes (N) Pig (N) Goats (N)	Hydatid		Deka et al., 1985a
Domestic animals (N)	Hydatid		Gopalakrishna Rao, 1985
Sheep (N) Goats (N)	Hydatid		Montosh Banerjee et al., 1985
Buffalo (N)	Hydatid cysts, secondary		Chowdhury et al., 1986
Cattle (N) Buffalo (N) Sheep (N) Goats (N)	Hydatid		Kulkarni et al., 1986
Mice (E)	Hydatid		Ganguly et al., 1986
Domestic animals (N)	Hydatid		Rana et al., 1986
Buffaloes (N)	Hydatid		Singh et al., 1986
Goats (N) Pig (N)	Hydatid		Varma and Malaviya, 1986
Buffaloes (N)	Hydatid, protoscolices		Bandyopadhyay and Basu, 1987
Bison, American (*Bison bison*) (N)	Hydatid		Choudary, Ch, et al., 1987
Water buffaloes (N)	Hydatid		Kumar and Parihar, 1987

Intermediate Host	Larva	Final Host	Reference
Buffaloes (N) Lambs (E) Kids (E)	Hydatid		Singh et al., 1987
Mice (E)	Protoscolices		Wangoo et al., 1987
Sheep (N) Goat (N)	Hydatid		Chatapadhaya et al., 1988
Cattle (N) Sheep (N)	Hydatid		Muralidhar and Sastry, 1988a
Sheep (N) Goats (N)	Hydatid		Muralidhar and Sastry, 1988b
Yak (*Bos poephagus*) (N)	Hydatid		Rai and Ansari, 1988
Buffalo (N)	Hydatid	Pups (E)	Singh and Dhar, 1988a
Buffaloes (N) Sheep (N) Goats (N) Dogs (shepherd; street) (N)	Hydatid		Singh and Dhar, 1988b
Sheep (E) Guinea pigs (E) Golden hamsters (E) Mice (E)	Hydatid, protoscolices		Singh et al., 1988c
Buffaloes (N)	Hydatid		Srivastava et al., 1988
Food animals (N)	Hydatid		Varma and Malaviya, 1988

Intermediate Host	Larva	Final Host	Reference
Mice (E) Sheep (E)	Hydatid		Wangoo et al., 1988
Yak (*Bos poephagus*)	Hydatid		Ansari et al., 1989
Cattle (N)	Hydatid		Gatne et al., 1989
Cattle (N) Buffaloes (N) Goats (N) Sheep (N)	Hydatid		Gita Biswas et al., 1989
Cattle (N)	Hydatid		Ghouri et al., 1989
Cattle (N) Buffaloes (N) Goats (N) Sheep (N)	Hydatid		Ghouri and Sahai, 1989
Buffaloes (N)	Hydatid	Puppies (E)	Irshadullah and Nizami, 1989
Buffaloes (N) Goats (N) Sheep (N) Camel (N) Pigs (N)	Hydatid		Irshadullah et al., 1989
Buffaloes (N)	Hydatid		Varma and Malaviya, 1989e
Buffaloes (N)	Hydatid		Deka and Gaur, 1990a
Cattle (N)	Hydatid		Ghorui et al., 1990

Intermediate Host	Larva	Final Host	Reference
Cattle	Hydatid		Gatne et al., 1990
Sheep (N) Goats (N) Pigs (N)	Hydatid		Varma, 1990
Buffaloes (N) Pigs (N) Sheep (N) Goats (N)	Hydatid	Dogs (N)	Varma and Ahluwalia, 1990c
Swamp deer (Cervus dawvauceli) (N) Langur (Presbytis entellus) (N) Giantsquirrel (N)	Hydatid		Arora, 1991
Cattle	Hydatid		Gatne et al., 1991
Domestic animals (N)	Hydatid		Roy and Tandon, 1991
Sheep (N) Goats (N) Pigs (N)	Hydatid		Varma, 1991
Cattle (N) Buffaloes (N)	Hydatid		Dhote et al., 1992a
Cattle, bullock (N)	Hydatid		Dhote et al., 1992b
Cattle	Hydatid		Gatne and Narsapur, 1992
Cattle (N)	Hydatid		Ghorui et al., 1992

Intermediate Host	Larva	Final Host	Reference
Sheep (N) Goat (N) Pig (N) Human (N)	Hydatid		Kanwar et al., 1992
Food animals (N)	Hydatid		Varma and Malviya, 1992
	Hydatid		Prasanna et al., 1992
Buffaloes (N) Sheep (N)	Hydatid		Singh et al., 1992a
Pig (N)	Hydatid		Singh et al., 1992b
Buffalo (N) 1993	Hydatid		Chowdhury and Rajvir Singh, 1993
Pig (N)	Hydatid		Deka and Gaur, 1993b
Pig (N)	Hydatid		Deka and Gaur, 1993c
Pig (N)	Hydatid		Deka and Gaur, 1993d
Buffaloes	Hydatid		Deka and Gaur, 1993f
Sheep (N) Goats (N)	Hydatid		Reddy et al., 1993
Buffaloes (N) Cattle calf (E) Piglets (E)	Hydatid	Pups (E)	Singh, 1993

Intermediate Host	Larva	Final Host	Reference
Cattle (N) Buffaloes (N) Sheep (N)	Hydatid		Vijayasmitha et al., 1993
Buffaloes (N) Goats (N) Sheep (N)	Hydatid		Bancyopadhyay, 1994
Buffaloes (N)	Hydatid		Binjola and Gour, 1994
Cattle (N) Buffaloes (N) Sheep (N) Pig (N)	Hydatid		Hafeez Md. et al., 1994
Sheep (N)	Hydatid		Kanwar and Kanwar, 1994
Buffaloes (N) BALB/C mice (E)	Hydatid (brood capsules containing protoscolices)		Prasad, et al., 1994
Food animals (N)	Hydatid		Sangaran, 1994
Goats (N)	Hydatid cysts		Singh and Pal, 1994
Swamp deer (*Cervus duvauceli duvauceli*)	Hydatid		Verma et al., 1994
Cattle (N) Buffaloes (N) Sheep (N) Pigs (N)			Hafeez et al., 1994
Buffaloes (N)	Hydatid	Pups, Mongrel (E)	Katoch and Singh, 1994a

Intermediate Host	Larva	Final Host	Reference
Cattle (N)	Hydatid		Katoch and Singh, 1994b
Buffaloes (N)	Hydatid, protoscolices	Pups (E)	Srivastava and Singh, 1994
Goats (N) Sheep (N) Bullocks (N) Buffaloes (N)	Hydatid		Dhakshayani, 1995
Buffaloes (N)	Hydatid, protoscolices		Ali Afsar and Nizami, 1995
Yak (N)	Hydatid		Ranga Rao et al., 1994
Cattle (N)	Hydatid		Katoch and Singh, 1995
Cattle (N)	Hydatid		Deka et al., 1995
Cattle (N)	Hydatid, protoscolices	Pups (E)	Dhar, 1995
Cattle (N)	Hydatid		Ghouri et al., 1995
		Man (N)	Nadkarni et al., 1995
Buffaloes (N)	Hydatid		Raina and Singh, 1995
Camel (N)	Hydatid		Ranga Rao and Sharma, 1995
Sheep (N) Buffaloes (N)			Sangaran et al., 1995
Giant squirrel (Ratufa indica) (N)	Hydatid	Pups (E)	Varma et al., 1995
Cattle (N)	Hydatid		Bandyopadhyay and Basu, 1996
Cattle (N)	Hydatid		Dhar and Singh, 1996a

Intermediate Host	Larva	Final Host	Reference
Bovines (N)	Metacestodes		Dhar and Singh, 1996b
Sheep (N) Goats (N)	Hydatid		Jithendran, 1996
Buffaloes (N)	Hydatid		Khan, 1996
Cattle (N) Buffaloes (N)	Hydatid		Konapur, 1996
Cattle (N) Buffaloes (N)	Hydatid		Konapur et al., 1996
Cattle (N)	Hydatid		Bandyopadhyay and Basu, 1997
Buffaloes (N) Swiss albino mice (E) I/P injn.	Hydatid Protoscolices		Bandyopadhyay and Singh, 1997
Lion (N)	Hydatid, cystic stage		Ganorkar et al., 1997
Buffaloes (N)	Hydatid		Rama and Singh, 1997a
Buffaloes (N)	Hydatid		Raina and Singh, 1997b
Sheep (N) Goats (N) Man (N)	Hydatid		Ayub, 1998
Sheep (N) Goats (N) Pig (N)	Hydatid		Deka and Gaur, 1998

Intermediate Host	Larva	Final Host	Reference
Sheep (N) Goats (N)	Hydatid		Das and Sreekrishnan, 1988
Food animals (N)	Hydatid		Vijaya Smitha, 1991
Food animals (N)	Hydatid		Vijaya Smitha et al., 1993
Animals (N)	Hydatid		Sharma et al., 1998
	Hydatid	Pups (E)	Singh and Dhar, 1998a
Buffalo (N) Goats (N) Sheep (N) Goats (N)	Hydatid		Singh and Dhar, 1998b
Sheep (N)	Hydatid Antibodies		Singh and Dhar, 1998c
Pig (N)	Hydatid		Singh et al., 1998a
Animals (N)	Hydatid		Singh et al., 1998b
Buffaloes (N)	Hydatid		Singh et al., 1998c
Cattle (N) Buffaloes (N)	Hydatid		Reddy and Rao, 1998
Animals (N)	Hydatid		Reddy et al., 1998
Buffalo (N)	Metacestode	Foxes (N)	Rao, 1998

Intermediate Host	Larva	Final Host	Reference
Cattle (N) Buffaloes (N) Goats (N) Sheep (N) Pigs (N)	Hydatid cysts		Utpal Das and Arup Kr. Das, 1998
Squirrel, Giant (N)	Hydatid		Varma et al., 1998
Cattle (N) Buffaloes (N)	Hydatid		Bhattacharya et al, 1999
Lion (N)	Hydatid		Charaborthy, 1999
Sheep (N) Human (N)	Hydatid		Kaur, 1999
Cattle (N) Buffaloes (N)	Hydatid		Konapur et al., 1999a
Cattle (N) Buffaloes (N)	Hydatid		Konapur et al., 1999b
Pig (N)	Hydatid		Srinivasa Murthy et al., 1999
Cattle (N) Buffaloes (N)	Hydatid		Bhattacharya et al., 2000
Buffaloes (N)	Hydatid		Bandyopadhyay and Singh, 2000
Cattle (N) Buffaloes (N)	Hydatid		Bandyopadhyay et al., 2000

Intermediate Host	Larva	Final Host	Reference
Cattle (N) Buffaloes (N) Goats (N) Pigs (N)	Hydatid		Dev Sharma *et al.*, 2000
Cattle (N) Buffaloes (N) Goats (N) Sheep (N) Pigs (N)	Hydatid		Jagannath and D' Souza, 2000
Buffaloes (N)	Hydatid		Bhattacharya *et al.*, 2001
Goat, Ladahki (N)	Hydatid		Dhoot and Upadhaye, 2001
Food animals (N)	Metacestode		Gatne, 2001
Sheep (N)	Hydatid		Manisha Mathur *et al.*, 2001
Cattle (N) Pigs (N) Goats (N)	Hydatid		Rajkhowa and Bandopaydhyay, 2001
Buffaloes (N) Sheep (N) Goats (N) Rabbits (E) Mice (E)	Hydatid, Protoscolices		Samantha, 2000 (Ph.D. Thesis, Subhamoy Samanta, 2000)
Buffaloes (N) Sheep (N) Goats (N)	Hydatid, Protoscolices		Samantha, 2001

Intermediate Host	Larva	Final Host	Reference
Buffaloes (N)	Hydatid		Sarkar et al., 2001
Cattle (N) Goats (N) Pigs (N)	Hydatid		Dev Sharma et al., 2002
Lion (N)	Hydatid, cysts		Dhoot and Upadhye, 2002
Goats (N)	Hydatid		Kumari et al., 2002
Cattle (N) Buffaloes (N) Sheep (N) Horse (N)	Hydatid		Debasis Bhattacharya, 2003
Human (N) Cattle (N)	Hydatid		Devi and Parija, 2003
Domestic animals (N)	Hydatid		Kumar et al., 2003
Sheep (N) Goats (N)	Hydatid		Raman and Lalitha John, 2003
Buffaloes (N) Sheep (N) Goats (N)	Hydatid, Protoscolices		Samantha, et al., 2003
Cattle (N) Buffalo (N) Sheep (N) Goat (N)	Hydatid cysts		Pathak et al., 2004
Sheep (N)	Hydatid		Abdul Basith et al., 2005

Intermediate Host	Larva	Final Host	Reference
Buffaloes (N)	Hydatid		Choudhari *et al.*, 2005
Pig (N)	Hydatid		Gaurat and Gatne, 2005
	Hydatid		Yokananth *et al.*, 2005
Goats (N) Sheep (N)	Hydatid		Rashid *et al.*, 2005
	Hydatid		Khan and Purohit, 2006
Cattle (N) Buffaloes (N) Sheep (N)	Hydatid		Bhattacharya *et al.*, 2003
Cattle (N) Buffaloes (N) Sheep (N)	Hydatid		Bhattacharya *et al.*, 2007
Cattle (N) Buffaloes (N) Sheep (N)	Hydatid		Bhattacharya *et al.*, 2008a
Cattle (N) Buffaloes (N) Sheep (N) Goats (N)	Hydatid		Bhattacharya *et al.*, 2008b
Buffaloes (N)	Hydatid		Maity *et al.*, 2007
Cattle (N) Buffaloes (N) Goats (N) Pigs (N)	Hydatid		Deka *et al.*, 2008

Intermediate Host	Larva	Final Host	Reference
Buffaloes (N)	Hydatid		Pan *et al.*, 2008
Buffaloes (N) Mice (E)	Hydatid		Samanta *et al.*, 2008
Sheep (N)	Hydatid		Abdul Basith *et al.*, 2009
Cattle (N) Buffaloes (N)	Hydatid cysts		Bari *et al.*, 2009
Pig (N)	Hydatid		Gaurat and Gatne, 2009
Buffaloes (N)	Hydatid		Gudewar *et al.*, 2008
Buffaloes (N)	Hydatid		Battacharya *et al.*, 2009
Buffaloes (N)	Hydatid		Pan *et al.*, 2009a
Buffaloes (N)	Hydatid		Pan *et al.*, 2009b
Cattle (N) Buffaloes (N) Sheep (N) Pigs (N)	Hydatid		Pednekar *et al.*, 2009a
Cattle (N) Buffaloes (N) Sheep (N) Pigs (N)	Hydatid		Pednekar *et al.*, 2009b
Buffaloes (N)	Hydatid		Samanta *et al.*, 2009
Sheep (N) Goats (N)	Hydatid		Sangaran and Lalitha John, 2009a

Intermediate Host	Larva	Final Host	Reference
Cattle (N) 2009b	Hydatid		Sangaran and Lalitha John,
Jaguar (*Panthera onca*) (N)	Hydatid		Sathaivam *et al.*, 2009
Cattle (N) Buffaloes (N)	Hydatid cysts		Bhangale *et al.*, 2010
Nilgai (*Boelaphus tragocamelus*) (N)	Hydatid		Geetha Devi *et al.*, 2010
Cattle (N) Buffaloes (N) Goats (N) Pigs (N)	Hydatid		Gohain Borua *et al.*, 2010
Buffaloes (N)	Hydatid		Irshadulla and Monica Rani, 2010
Jaguar (*Panthera onca*) (N)	Hydatid		Muraleedharan *et al.*, 2010
Buffaloes (N)	Hydatid		Pan *et al.*, 2010
Buffaloes (N)	Hydatid cysts		Patil *et al.*, 2010
Buffaloes (N)	Hydatid		Sangaran and Lalitha John, 2010
Cattle (N)	Hydatid cysts		Sanjiv Kumar *et al.*, 2010
Rabbit, broiler (N)	Hydatid		Sreekumar *et al*, 2010
Buffaloes (N)	Hydatid		Gupta *et al.*, 2011
Sheep (N)	Hydatid cysts		Jeyathilakan *et al.*, 2011

Intermediate Host	Larva	Final Host	Reference
Cattle (N) Buffalo (N)	Hydatid cysts		Pan et al., 2011
Goat (N)	Hydatid cysts		Qazi et al., 2011
Sheep (N)	Hydatid		Saha et al., 2011
Pig (N)	Hydatid		Sajjan et al., 2011
Buffaloes (N)	Hydatid		Sangaran et al., 2011
Sheep (N)	Hydatid		Shahnawaz et al., 2011
Cattle (N) Buffalo (N) Sheep (N) Pigs (N)	Metacestodes	—	Gatne and Gaurat, 2012
Camels	Hydatid cysts		Ghouri et al., 2012
Goats (N)	Hydatid cysts	—	Sangaran and Lalitha John, 2012
Sheep (N)	Hydatid Antibodies	Human (N)	Mohd Irfan Naik et al., 2013
Echinococcus granulosus ortleppi			
Sheep (N) Buffaloes (N)	Hydatid		Pednekar et al., 2009b
Echinococcus granulosus senso stricto			
Cattle (N)	Hydatid		Pednekar et al., 2009b

Intermediate Host	Larva	Final Host	Reference
	***Echinococcus multilocularis* Leuckart, 1863**		
Cattle (N) Buffalo (N)	Hydatid, Multilocular cysts (such cysts were preferred to be included under *E.multilocularis* by Pal and Sinha, 1970, see below)		Sunderam and Natarajan, 1960
Bullock (N)	Hydatid		Pal and Sinha, 1970
		***Taenia* sp.**	
Goat (N)	Cysticercus		Bhatia, 1936
Sheep (N) Goat (N)	Cysticercus		Gill, 1972
Pig (N)	Cysticercus		Gogoi, 1974
Camel (N)	Cysticercus		Ranga Rao and Sharma, 1995
Sheep (N)	Cysticercus		Abdul Basith *et al.*, 2005
Wild rodents (N)	Metacestode of *Taenia* sp.		Malsawmtluange and Tandon, 2009a
		***Taenia (Multiceps)* sp.**	
Sheep (N) Camels (N)	Coenurus		Dey,1909

Intermediate Host	Larva	Final Host	Reference
Sheep (N) Camels (N)	*Coenurus*		Gaiger, 1915
Domestic animals (N)	*Coenurus*		Southwell, 1930
	Coenurus		Bhalerao, 1939
Ram (N)	*Coenurus*		Rahimuddin, 1941
	Coenurus		Rao and Anantaraman, 1966
	Coenurus		Gill and Rao, 1967, abc
	Coenurus		Singh and Rao, 1968
Gerbil: *Tatera indica* (N)	*Coenurus*		Ebenezer Raja, 1971
Sheep (N)	*Coenurus*		Paliwal and Sigh, 1971
Buffalo (N)	*Coenurus*		Rao and Parihar, 1973
Sheep (N) Goat (N)	*Coenurus*		Dewan Muthu Mohammed and Rajendran, 1974
Rodents (N)	*Coenurus*		Ebenezer Raja, 1974
Goat (N)	*Coenurus*		Kulkarni et al., 1974
Goat (N)	*Coenurus*		Sharma and Tyagi, 1974
Goat (N)	*Coenurus*		Sharma and Tyagi, 1975
Goat (N)	*Coenurus*		Deka et al., 1985
Buffalo (N)	*Coenurus*		Gupta and Chowdhury, 1985

Intermediate Host	Larva	Final Host	Reference
Goat (N)	*Coenurus* sp.		Saikia *et al.*, 1987
Buffalo (N)	*Coenurus*		Bali *et al.*, 1990
Goat (N)	*Coenurus*		Gogoi *et al.*, 1992
	Coenurus cerabralis		Deka and Gaur, 1994abc
Yak, Indian, *Bos (poephagus) grunniens* (N)	*Coenurus*		Ranga Rao *et al.*, 1994
Goat (N)	*Coenurus*		Sudhan and Shahardar, 1999
Sheep (N)	*Coenurus*		Abdul Basith *et al.*, 2005
Goat (N)	*Coenurus*		Ingole *et al.*, 2009
Goat, Barbari (N)	*Coenurus*		Shivasharanappa *et al.*, 2011
		Taenia (Multiceps) gaigeri Hall, 1916	
Goat (N)	*Coenurus gaigeri*		Dey, 1909
Goat (N)	*Coenurus gaigeri*		Bhalerao, 1939
Sheep (N) Goat (N)	*Coenurus gaigeri*		Mudaliyar and Alwar, 1947
Ewe (N)	*Coenurus gaigeri*		Rao *et al.*, 1957
Goat	*Coenurus gaigeri*		Pal and De, 1964
Goat	*Coenurus gaigeri*		Rao and Ananatharaman, 1966
Goat (E)			Singh *et al.*, 1968

Intermediate Host	Larva	Final Host	Reference
Sheep (N)	*Coenurus gaigeri*		Mathur and Dutt, 1969
Goat (N)	*Coenurus gaigeri*		Paliwal *et al.*, 1971a
Goat (N)	*Coenurus gaigeri*		Paliwal *et al.*, 1971b
Goats (N)	*Coenurus gaigeri*		Rao, 1971
Goat (E)	*Coenurus gaigeri*		Singh *et al.*, 1971
Sheep (N) Goat (N)	*Coenurus gaigeri*	Dog (E)	Dewan Muthu Mohammad and Rajendran, 1974
	Coenurus gaigeri	Dog (E)	Pythal, 1974
Goats (N)	*Coenurus gaigeri*		Panisup *et al.*, 1979
Sheep (N)	*Coenurus gaigeri*		Jain, 1982
Sheep (N)	*Coenurus gaigeri*		Jain and Shaw, 1982
Goats (N)	*Coenurus gaigeri*		Sanyal and Sinha, 1983a
		Dog- pup (E)	Kulkarni and Satyanarayana Shetty, 1983a
Goat	*Coenurus gaigeri*		Varma and Sharma Deorani, 1980
Goat (N)	*Coenurus gaigeri*		Shastri *et al.*, 1985
Goat (N)	*Coenurus gaigeri*		Varma and Ahluwalia, 1986
Goat (N)	*Coenurus gaigeri*		Dey *et al.*, 1988
Goat (N) Pig (N)	*Coenurus gaigeri*	Pups (E) Golden hamsters (E)	Varma and Malaviya, 1989a

Intermediate Host	Larva	Final Host	Reference
Sheep (N) Goats (N) Pigs (N)	*Coenurus gaigeri*		Varma and Malaviya, 1989b
Goat (N)	*Coenurus gaigeri*		Maity and Bandopadhyaya, 1991
Sambar (*Cervus unicolor*) (N)	*Coenurus gaigeri*		Varma and Malaviya, 1994
Goat (N)	*Coenurus gaigeri*		Patro *et al.*, 1997
Kid (N)	*Coenurus gaigeri*		Devasena *et al.*, 1998
Goat (N)	*Coenurus gaigeri*		Sharma *et al.*, 1998
Goat (N)	*Coenurus*		Gahlod *et al.*, 1999
Sheep (N)	*Coenurus gaigeri*		Naveen Kumar *et al.*, 2003
Goat (N)	*Coenurus gaigeri*		Tiwari *et al.*, 2004
Goat (N)	*Coenurus gaigeri*		Ghosh *et al.*, 2005
Goat (N)	*Coenurus gaigeri* (Metacestode)		Shivaprakash and Thimma Reddy, 2009
Sheep (N)	*Coenurus gaigeri*		Puttalakshmamma *et al.*, 2011
		Taenia (*Multiceps*) *multiceps* (Leske, 1780)	
Camel (N)	*Coenurus cerebralis*		Gaiger, 1915
Sheep (N)	*Coenurus cerebralis*		Rahimuddin, 1935
Sheep (N)	*Coenurus cerebralis*		Mudaliyar and Alwar, 1947

Intermediate Host	Larva	Final Host	Reference
Goat (N)	Coenurus cerebralis		Thapar, 1956
Goat (N)	Coenurus cerebralis		Bhalla and Negi, 1962
Goat (N)	Coenurus cerebralis		Endrejat, 1964
Cattle (N), male calf	Coenurus cerebralis		Kuppuswamy and Gupta, 1964
Goat (N)	Coenurus cerebralis		Sharma, 1965
Cattle (N)	Coenurus cerebralis		Prasad and Srivastava, 1967
Goat (N)	Coenurus cerebralis		Alam et al., 1971
Sheep (N)	Coenurus cerebralis		Paliwal and Sigh, 1971
Sheep (N)	Coenurus cerebralis	Dog - pup (E)	Bali, 1972
Goat (N)	Coenurus cerebralis		Singh and Singh, 1972
Buffalo (N)	Coenurus cerebralis		Rao and Parihar, 1973
Sheep (N)	Coenurus cerebralis		Saxena and Bhargava, 1973
Goat (N)	Coenurus cerebralis		Verma et al., 1973
Sheep (N)	Coenurus cerebralis		Panisup et al., 1979
Cow (N)	Coenurus cerebralis		Thimmaiah and Jagannath, 1982
		Dog - pup (E)	Kulkarni and Satyanarayana Shetty, 1983a
		Dog - pup (E)	Kulkarni and Satyanarayana Shetty, 1983b

Intermediate Host	Larva	Final Host	Reference
Buffalo (N)	*Coenurus cerebralis*		Gupta and Chowdhury, 1985
Sheep (N) Goats (N) Pigs (N)	*Coenurus cerebralis*		Varma and Malaviya, 1989b
Goat (N)	*Coenurus cerebralis*		Gogoi *et al.*, 1991
Goats	*Coenurus cerebralis*		Gogoi *et al.*, 1992
Goat (N)	*Coenurus cerebralis*		Juyal *et al.*, 1993
Buffalo (N)	*Coenurus cerebralis*		Varma and Malaviya, 1994
Goat (N)	*Coenurus cerebralis*		Varma *et al.*, 1994
Barbari goat (N)	*Coenurus cerebralis*		Sharma *et al.*, 1995
Himalayan thar (*Hemitragus jernlahicus*) (N) Markhor (*Capra falconeri herptneri*) (N)	*Coenurus cerebralis*		Charaborthy *et al.*, 1998
Goat (N)	*Coenurus cerebralis*		Sharma *et al.*, 2001.
Goat (N)	*Coenurus cerebralis*		Mandal *et al.*, 2004
Goat (N)	*Coenurus cerebralis*		Gahlot and Purohit, 2005
Goat (Black Bengal doe) (N)	*Coenurus cerebralis*		Debasis Jana and Mousumi Jana, 2006
Goat (N)	*Coenurus cerebralis*		Islam *et al.*, 2006
Goat (N)	*Coenurus cerebralis*, Protoscolices		Sarkar and Mitra., 2008

Intermediate Host	Larva	Final Host	Reference
Sheep Goats	*Coenurus cerebralis*		Gabhane *et al.*, 2009
Kid, Goats, Black Bengal (N)	*Coenurus cerebralis*		Jana and Jana, 2012
		Taenia hydatigena Palias, 1766	
Sheep (N) Goats (N)	*Cysticercus tenuicollis*		Thapar, 1956
Deer (N) Giraffe (N)	*Cysticercus tenuicollis*		Alwar and Lalitha, 1961
Goat (N)	*Cysticercus tenuicollis*		Tewari and Iyer, 1961
Sheep (N) Deer (N)	*Cysticercus tenuicollis*		Sharma Deorani, 1967a
Sheep (N) Deer (N)	*Cysticercus tenuicollis*		Sharma Deorani, 1967b
Goats (N)	*Cysticercus tenuicollis*		Deodar and Narasapur, 1968
Sheep (N)	*Cysticercus tenuicollis*		Mishra and Rupra, 1968
Pig (N)			Shrivastava and Shah, 1968
Sheep (N) Goats (N)	*Cysticercus tenuicollis*		Varma, 1970
Kids (N)	*Cysticercus tenuicollis*		Pathak and Gaur, 1971d
	Cysticercus tenuicollis		Pathak and Gaur, 1971e

Intermediate Host	Larva	Final Host	Reference
Sheep (N) Goats (N) Lambs (E)	Cysticercus tenuicollis Eggs		Varma et al., 1973
Sheep (N) Goats (N)	Cysticercus tenuicollis	Dogs (E)	Pythal, 1974
Rabbits (E) Sheep (N) Goats (N)	Cysticercus tenuicollis		Varma and Rao, 1974a
Sheep (N) Goats (N)	Cysticercus tenuicollis	Pup (E)	Verma et al., 1974b
Goats (N)	Cysticercus tenuicollis		Ray et al., 1977
Pigs (N)	Cysticercus tenuicollis		Upadhyay et al., 1977
Sheep (N) Goats (N)	Cysticercus tenuicollis	Pup (E)	Verma et al., 1985
Sheep (N) Goats (N)	Cysticercus tenuicollis		Gaur, et al., 1980
Goats, (Kids) (E)	Cysticercus tenuicollis	Pups (E)	Pathak and Gaur, 1981a
Goats, (Kids) (E)	Cysticercus tenuicollis	Pups (E)	Pathak and Gaur, 1981b
Sheep (N) Goats (N) Pigs (N)	Cysticercus tenuicollis	Dog (N)	Pathak and Gaur, 1982a
Goats (N)	Cysticercus tenuicollis		Pathak and Gaur, 1982b

Intermediate Host	Larva	Final Host	Reference
Goats (E)	*Cysticercus tenuicollis*	Pup (E)	Pathak et al., 1982
Goats (N)	*Cysticercus tenuicollis*	Pup (E)	Pathak et al., 1983
Beisa orynx (N)	*Cysticercus tenuicollis*		Arora et al., 1984
Sheep (N) Goats (N)	*Cysticercus tenuicollis*	Dogs (E)	Ebenezer Raja and Ananta-raman, 1984
Black buck (*Antelope cervicapra*)	*Cysticercus tenuicollis*		Varma et al., 1984
Goats (N)	*Cysticercus tenuicollis*		Deka et al., 1985
Sheep (N) Goats (N)	*Cysticercus tenuicollis*	Pup (E)	Varma et al., 1985
Buffaloes (N) Sheep (N) Lambs (E) Goats (N) Kids (E) Pig (N)	*Cysticercus tenuicollis*	Pup (E)	Varma and Ahluwalia, 1986
Golden hamster (E) Sheep (N) Goats (N) Buffaloes (N)	*Cysticercus tenuicollis*	Pup (E)	Varma, et al., 1987
Cattle (N) Pig (N) Goats (N)	*Cysticercus tenuicollis*		Roy and Tandon, 1989
Goats (E) (N)	*Cysticercus tenuicollis*		Deka and Gaur, 1990

Intermediate Host	Larva	Final Host	Reference
Goats (N)	*Cysticercus tenuicollis*		Deka and Gaur, 1991
Biesa oryx (N)	*Cysticercus tenuicollis*		Arora, 1991
Lambs Kids (Immunization)	*Cysticercus tenuicollis*	Pup (E)	Varma *et al.*, 1991
Sheep (N)	*Cysticercus tenuicollis*		Prasanna *et al.*, 1992
Sheep (N) Goats (N)	*Cysticercus tenuicollis*		Deka and Gaur 1993a
Kids (N)	*Cysticercus tenuicollis*		Deka and Gaur, 1994a
Kids (E)	*Cysticercus tenuicollis*	Pups (E)	Deka and Gaur, 1994b
Animal (not specified)	*Cysticercus tenuicollis*	Pups (E)	Deka and Gaur, 1994c
Kids (N)	*Cysticercus tenuicollis*		Deka and Gaur, 1994d
Sheep (N)	*Cysticercus tenuicollis*		Swarnkar *et al.*, 1994
Sheep (N)	*Cysticercus tenuicollis*		Sridhar *et al.*, 1996
Spotted deer (*Cervus porcinus*) (N)	*Cysticercus tenuicollis*		Kolte *et al.*, 1998
Sheep (N)	*Cysticercus tenuicollis*		Harikrishan *et al.*, 1999
Sheep (N) Lambs (E)	*Cysticercus tenuicollis*		Panda *et al.*, 1999
Sheep (N) Goats (N)	*Cysticercus tenuicollis*		Raman and Rajavelu, 2000
Sheep (N)	*Cysticercus tenuicollis*		Chaudhri *et al.*, 2000

Intermediate Host	Larva	Final Host	Reference
Sheep (N) Goats (N)	Cysticercus tenuicollis		Dash, et al., 2000
Sheep (N) Goats (N)	Cysticercus tenuicollis		Jeyathilakan, et al., 2000
Goats (N)	Cysticercus tenuicollis		Kurkure et al., 2000.
Sheep (N) Goats (N)	Cysticercus tenuicollis		Raman and Rajavelu, 2000
Sheep (N) (E)	Cysticercus tenuicollis		Panda et al., 2000
Corriedale lamb (N)	Cysticercus tenuicollis		Darzi et al., 2002
Cattle (N) Goats (N) Pigs (N)?	Cysticercus tenuicollis		Dev Sharma et al., 2002
Sheep (N) Goats (N)	Cysticercus tenuicollis		Bhaskara Rao et al., 2003
Cattle (N)	Cysticercus tenuicollis		Mathur et al., 2003
Pig (N)	Cysticercus tenuicollis		Gaurat and Gatne, 2005
Goat, crossbred (N)	Cysticercus tenuicollis		Soundararajan et al., 2005
Sheep (N)	Cysticercus tenuicollis	—	Sharma et al., 2008
Goats (N)	Cysticercus tenuicollis	—	Deka et al., 2009
Goats (N)	Cysticercus tenuicollis		Nath et al., 2009a
Goats (N)	Cysticercus tenuicollis		Nath et al., 2009b

Intermediate Host	Larva	Final Host	Reference
Goats (N)	*Cysticercus tenuicollis*	—	Nath *et al.*, 2009c
Cattle (N) Buffaloes (N) Sheep (N) Pig (N)	*Cysticercus tenuicollis*	—	Gatne *et al.*, 2010
Sheep (N)	*Cysticercus tenuicollis*	—	Jeyathilakan *et al.*, 2010
Goats (N) (E)	*Cysticercus tenuicollis*	—	Nath *et al.*, 2010a
Goats (N) (E)	*Cysticercus tenuicollis*	—	Nath *et al.*, 2010b
Goats (N)	*Cysticercus tenuicollis*	—	Nath *et al.*, 2011
Sheep (N) Goats (N)	*Cysticercus tenuicollis*	—	Nimbalkar *et al.*, 2011
Buffaloes (N)	*Cysticercus tenuicollis*	—	Patil *et al.*, 2010
Sheep (N)	*Cysticercus tenuicollis*	—	Rashid *et al.*, 2011
Goats (N)	*Cysticercus tenuicollis*	—	Sanyal, 2011
Sheep (N)	*Cysticercus tenuicollis*	—	Shahnawaz *et al.*, 2011
Goats, Barbari (N)	*Cysticercus teunuicollis*	—	Londhe *et al.*, 2012
	***Taenia ovis* Cobbold, 1869**		
Sheep (N)	*Cysticercus ovis*		Dhabolkar *et al.*, 1968
Goat (N)	*Cysticercus ovis*		Chakraborty *et al.*, 1984

Intermediate Host	Larva	Final Host	Reference
	Taenia pisiformis Bloch, 1780		
Rabbit (N)	Cysticercus pisiformis		Singh *et al.*, 1992
Rabbit (N)	Cysticercus pisiformis		Katoch and Jithendran, 1999
	Taenia saginata Goeze, 1782		
Measly meat (N)	Cysticercus bovis		Rahimuddin, 1926
Cattle (N)	Cysticercus bovis		Mudaliyar and Alwar, 1947
Cattle (N)	Cysticercus bovis		Ahmed, 1948
Cattle (N)	Cysticercus bovis		Endrejat, 1964
Cattle (N)	Cysticercus bovis		Krishnan and Ranganathan, 1972
Cattle (N)	Cysticercus bovis		Nayak, *et al.*, 1973
Buffalo (N)	Cysticercus bovis		Pythal, 1974
Cattle (N) Buffalo (N)	Cysticercus bovis		Gaur, 1976
Buffalo (N)	Cysticercus bovis		Sreemannarayana and Christopher, 1977
Cattle (N)	Cysticercus bovis		Abraham, 1979
Cattle (N)	Cysticercus bovis	Dog (E) Cat (E)	Pythal, 1974
Cow (N)	Cysticercus bovis		Kolte *et al.*, 1981

Intermediate Host	Larva	Final Host	Reference
Cattle (N) Buffalo (N)	Cysticercus bovis		Pramanik and Bhattacharyya, 1984
Cattle (N) Buffalo (N)	Cysticercus bovis		Krishnan et al., 1984
Cattle (N) Buffalo (N)	Cysticercus bovis		Deka et al., 1985
Cattle (N) Goats (N)	Cysticercus bovis		Roy and Tandon, 1989
Cattle (N)	Cysticercus bovis		Deka., 1992
Cattle (N)	Cysticercus bovis		Deka et al., 1995
Cattle (N)	Cysticercus bovis		Moghaddar et al., 2001
Cattle (N)	Cysticercus bovis		Tamuli et al., 2005
Taenia saginata asiatica			
Pigs	Cysticercus bovis		Dhanalaksmi, 2003
Taenia solium (Linnaeus, 1758)			
Measly meat (N)	Cysticercus cellulosae		Rahimuddin, 1926
Dog (N)	Cysticercus cellulosae	Dog (N)	Ayyar, 1929
Pig (N)	Cysticercus cellulosae		Rao, 1935
Pig (N)	Cysticercus cellulosae		Bahaduri and Maplestone, 1937

Intermediate Host	Larva	Final Host	Reference
Pig (N) Dog (N)	Cysticercus cellulosae	Dog (N)	Mudaliyar and Alwar, 1947
Pig (N)	Cysticercus cellulosae		Ahamad, 1948
Dog (N)	Cysticercus cellulosae	Dog (N)	Krishna Murthy, 1949
Pig (N)	Cysticercus cellulosae		Alwar, 1958
Pig (N)	Cysticercus cellulosae		Ahluwalia, 1959
Pig (N)	Cysticercus cellulosae		Ahluwalia, 1960
Dog (N)	Cysticercus cellulosae	Dog (N)	Wilson, 1962
	Cysticercus cellulosae		Souri, 1963
	Cysticercus cellulosae		Ramamurthy, 1964
Pig (N)	Cysticercus cellulosae		Sinha, 1966
Pig (N)	Cysticercus cellulosae		Mishra and Larka, 1967
Pig (N)	Cysticercus cellulosae		Shrivastava and Shah, 1968
Pig (N)	Cysticercus cellulosae		Nageswar Rao, 1970
Pig (N)	Cysticercus cellulosae		Sinha, 1970
Dog (N)	Cysticercus cellulosae	Dog (N)	Joshi and Gupta, 1970
Pig (N)	Cysticercus cellulosae		Nageswar Rao, 1970 (M. V. Sc. Thesis)
Pig (N)	Cysticercus cellulosae		Gill, 1972

Intermediate Host	Larva	Final Host	Reference
Pig (N)	Cysticercus cellulosae		Ray et al., 1972
Pig (N)	Cysticercus		Sadana and Kalra, 1973
Rodent (N)	Cysticercus cellulosae		Ebenezer Raja, 1974
Pig (N)	Cysticercus cellulosae		Gogoi, 1974
	Cysticercus cellulosae		Ratnam, 1975
Pigs (N)	Cysticercus cellulosae		Upadhyay et al., 1977
Monkeys (*Macaca mulatta*) (N)	Cysticercus cellulosae		Saleque, 1978
Monkeys (*Macaca mulatta*) (N)	Cysticercus cellulosae		Saleque et al., 1978
Dog (N)	Cysticercus cellulosae	Dog (N)	Thandaveshwar et al., 1978
Pig (N)	Cysticercus cellulosae		Varma and Ahluwalia, 1981
Kid	Cysticercus cellulosae		Pathak et al., 1982
Pigs (N)	Cysticercus cellulosae		Varma, 1982
Pigs (N)	Cysticercus cellulosae		Pathak and Gaur, 1983a
Pigs (N)	Cysticercus cellulosae		Pathak and Gaur, 1983b
Pigs (N)	Cysticercus cellulosae		Gaur, 1984
Pigs (N)	Cysticercus cellulosae		Pathak et al., 1984a
Pigs (N)	Cysticercus cellulosae		Pathak et al., 1984b
Pigs (N)	Cysticercus cellulosae		Pathak et al., 1984c
Pigs (N)	Cysticercus cellulosae		Deka et al., 1985

Intermediate Host	Larva	Final Host	Reference
Pigs (N)	Cysticercus cellulosae		Pathak and Gaur, 1985a
Pigs (N)	Cysticercus cellulosae		Pathak and Gaur, 1985b
Pigs (N)	Cysticercus cellulosae		Pathak and Gaur, 1985c
Pigs (N)	Cysticercus cellulosae		Pramanik et al., 1985
Pigs (N)	Cysticercus cellulosae		Pathak and Gaur, 1986a
Pigs (N)	Cysticercus cellulosae		Pathak and Gaur, 1986b
Pigs (N)	Cysticercus cellulosae		Varma and Ahluwalia, 1986
Pigs (N)	Cysticercus cellulosae		Varma et al., 1986
Pigs (N)	Cysticercus cellulosae		Gaur, 1987
Pigs (N)	Cysticercus cellulosae		Kumar, 1987
Pig (N)	Cysticercus cellulosae		Kumar and Gaur, 1987a
Pig (N), (E)	Cysticercus cellulosae		Kumar and Gaur, 1987b
Pig (N)	Cysticercus cellulosae		Kumar et al., 1987
Rhesus monkeys (Macaca mulatta) (E)	Cysticercus cellulosae		Saleque et al., 1987
Pigs (N)	Cysticercus cellulosae		Bajpai, 1988
Piglets (E)	Eggs, Cysticercus cellulosae		Kumar and Gaur, 1988
Pigs (N)	Cysticercus cellulosae		Kumar et al., 1988
Pig (N)	Cysticercus cellulosae	Man (N)	Pathak and Gaur, 1988
Pig (N)	Cysticercus cellulosae		Pathak and Gaur, 1989a

Intermediate Host	Larva	Final Host	Reference
Pig (N)	*Cysticercus cellulosae*	Man (N)	Pathak and Gaur, 1989b
Pig (N)	*Cysticercus cellulosae*		Singh *et al.*, 1988
Pig (N)	*Cysticercus cellulosae*		Roy and Tandon, 1989
Pig (N)	*Cysticercus cellulosae*		Varma and Ahluwalia, 1989b
Pig (N)	*Cysticercus cellulosae*		Deka and Gaur, 1990b
Pig (N)	*Cysticercus cellulosae*		Deka *et al.*, 1990
Pig (N)	*Cysticercus cellulosae*	Man, gravid proglottids (N)	Pathak and Gaur, 1990
Pig (N)	*Cysticercus cellulosae*	Pups (E)	Varma and Ahluwalia, 1990a
Pig (N)	*Cysticercus cellulosae*		Varma and Ahluwalia, 1990b
Pig (N)	*Cysticercus cellulosae*		Bishar, 1991
Pig (N)	*Cysticercus cellulosae*		Bhatia, 1991
Pig (N)	*Cysticercus cellulosae*		Kumar *et al.*, 1991
Pig (N)	*Cysticercus cellulosae*		Plain, 1991
Pig (N)	*Cysticercus cellulosae*		Shinde *et al.*, 1991
Pig (N)	*Cysticercus cellulosae*		Deka Dilip, 1992
Golden hamsters (E)	*Cysticercus cellulosae*		Varma and Ahluwalia, 1992
Pig (N)	*Cysticercus cellulosae*		Plain *et al.*, 1992
Pig (N)	Metacestode	—	Shinde, 1992

Intermediate Host	Larva	Final Host	Reference
Pig (N)	Cysticercus cellulosae		Deka and Gaur, 1993a
Pig (N)	Cysticercus cellulosae		Deka and Gaur, 1993b
Pig (N)	Cysticercus cellulosae		Deka and Gaur, 1993c
Pig (N)	Cysticercus cellulosae		Deka and Gaur, 1993d
Pig (N)	Cysticercus cellulosae		Shinde et al., 1993a
Pig (N)	Cysticercus cellulosae		Shinde et al., 1993b
Pig (N)	Cysticercus cellulosae		Kumar and Gaur, 1994
Pig (N)	Cysticercus cellulosae		Pathak, 1994
Pig (N)	Cysticercus cellulosae		Banerjee et al., 1994
Pig (N)	Cysticercus cellulosae		Pathak et al., 1994
Pig (N)	Cysticercus cellulosae		Malla et al., 1995
Pig (N)	Cysticercus cellulosae		Deka, et al., 1995
Pig (N)	Cysticercus cellulosae		D'Souza and Hafeez Md, 1995
Pig (N)	Cysticercus cellulosae		Malla et al., 1996
Pig (N)	Cysticercus cellulosae		Pathak, 1996
Pig (N)	Cysticercus cellulosae		D'Souza, 1998
Pig (N)	Cysticercus cellulosae		D'Souza and Hafeez, 1998a
Pig (N)	Cysticercus cellulosae		D'Souza and Hafeez, 1998b
Pig (N)	Cysticercus cellulosae		D'Souza and Hafeez, 1998c

Intermediate Host	Larva	Final Host	Reference
Pig (N)	Cysticercus cellulosae		D'Souza and Hafeez, 1998d
Pig (N)	Cysticercus cellulosae		Sreenivasa Murthy et al., 1999.
Pig (N)	Cysticercus cellulosae		Dev Sharma et al, 2000
Pig (N)	Cysticercus cellulosae		Jagannath and D' Souza, 2000
Pig (N)	Cysticercus cellulosae		Anand Prakash, 2001
Pig (N)	Cysticercus cellulosae		Rajkhowa and Bandopaydhyay, 2001
Pig (N)	Cysticercus cellulosae		Sreenivasa Murthy et al., 2001
Pig (N)	Cysticercus cellulosae		Prasanna et al., 2001a
Pig (N)	Cysticercus cellulosae		Prasanna et al., 2001b
Pig (N)	Cysticercus cellulosae		Prasanna et al., 2001c
Pig (N)	Cysticercus cellulosae		Sharma, 2001
Pig (N)	Cysticercus cellulosae		Paneer Selvam et al., 2002
Pig (N)	Cysticercus cellulosae	Man (N)	Prasad et al., 2002
Pig (N)	Cysticercus cellulosae		Avapal et al., 2003

Intermediate Host	Larva	Final Host	Reference
Pig (N)	*Cysticercus cellulosae*		Bhaskara Rao et al., 2003
Pig (N)	*Cysticercus cellulosae*		Dhanalaksmi, 2003
Pig (N)	*Cysticercus cellulosae*		Hafeez, et al., 2004a
Pig (N)	*Cysticercus cellulosae*		Hafeez, et al., 2004b
Pig (N)	*Cysticercus cellulosae*		Paneer Selvam et al., 2004
Pig (N)	*Cysticercus cellulosae*		Rupinderjit Singh Avapal, 2003
Pig (N)	*Cysticercus cellulosae*		Sharma et al., 2004a
Pig (N)	*Cysticercus cellulosae*		Sharma et al., 2004b
Pig (N)	*Cysticercus cellulosae*		Gaurat and Gatne, 2005
Pig (N)	*Cysticercus cellulosae*		Dhanalaksmi et al., 2005a
Pig (N)	*Cysticercus cellulosae*		Sharma et al., 2005a
Pig (N)	*Cysticercus cellulosae*		Sharma et al., 2005b
Pig (N)	*Cysticercus cellulosae*		Prakash et al., 2007
Pig (N)	*Cysticercus cellulosae*	—	Prasad, 2006
Pig (N)	*Cysticercus cellulosae*		Juyal et al., 2008
Pig (N)	*Cysticercus cellulosae*		Shukla et al., 2008
Pig (N)	*Cysticercus cellulosae*		Mandalikar et al., 2009
Pig (N)	*Cysticercus cellulosae*		Deka et al., 2009
Pig (N)	*Cysticercus cellulosae*		Gatne et al., 2010

Intermediate Host	Larva	Final Host	Reference
Pig (N)	*Cysticercus cellulosae*		Borkataki *et al.*, 2010
Pig (N)	*Cysticercus cellulosae*		Bhadrige *et al.*, 2011
Pig (N)	*Cysticercus cellulosae*		Sree Devi *et al.*, 2011a
Pig (N)	*Cysticercus cellulosae*		Sree Devi *et al.*, 2011b
Pig (N)	*Cysticercus cellulosae*		Sree Devi *et al.*, 2011c
Pig (N)	Metacestode	—	Kalai *et al.*, 2012
Pig (N)	*Cysticercus cellulosae*		Rout and Saikumar, 2012
Taenia taeniaeformis (Batsch, 1786)			
Nesocia bandicota Rattus rattus	*Cysticercus fasciolaris*		Mudaliyar and Alwar, 1947
	Cysticercus fasciolaris		Pujathi, 1950 (Vider Rao and Anantaraman, 1966)
Rats (N)	*Cysticercus fasciolaris*		Niphadkar and Rao, 1966
Rodents (N) Bandicoots (N)	*Cysticercus fasciolaris*		Rao and Anantaraman, 1966
Rats	*Cysticercus fasciolaris*	Cat (E)	Singh and Rao, 1966
Albino rats (E)	*Cysticercus fasciolaris*		Singh and Rao, 1967a
Albino rats (E)	*Cysticercus fasciolaris*	Cat (E)	Singh and Rao, 1967b
Rats Guinea pigs Hamster	*Cysticercus fasciolaris*	Cat (E)	Singh and Rao, 1968

Intermediate Host	Larva	Final Host	Reference
Albino rats (E)	Cysticercus fasciolaris	Cat (E)	Banerjee and Singh, 1969a
Albino rats (E)	Cysticercus fasciolaris	Cat (E)	Banerjee and Singh, 1969b
Albino rats (E)	Cysticercus fasciolaris	Cat (E)	Banerjee and Singh, 1969c
Albino rats (E)	Cysticercus fasciolaris	Cat (E)	Banerjee and Singh, 1969d
Laboratory animals (E)	Cysticercus fasciolaris		Rao, 1971
Pig (N)	Cysticercus fasciolaris		Singh, 1970
Rodents (N)	Cysticercus fasciolaris		Ebenezer Raja, 1974
Golden hamster (*Mesocricetus auratus*)	Cysticercus fasciolaris		Tripathi and Ray, 1976
Funambulus pennant	Cysticercus fasciolaris		Anjana Parihar and Nama, 1978
Rat (E)	Cysticercus fasciolaris		Lal, et al., 1983
Rat (E)	Cysticercus fasciolaris		Jain et al., 1988
Rat (E)	Cysticercus fasciolaris		Jain et al., 1989a
Rat (E)	Cysticercus fasciolaris		Jain et al., 1989b
Rat, Rattus rattus (N)	Cysticercus fasciolaris		Joshi and Sharma, 1988
Rat (E)	Cysticercus fasciolaris		Gupta et al., 1992
Wild rat (N)	Cysticercus fasciolaris		Somvanshi, et al., 1994
Rats (N)	Cysticercus fasciolaris		Bhattacharya et al., 1998
Rat (E) Mice (E)	Cysticercus fasciolaris		Jitendran and Somvanshi, 1999

Intermediate Host	Larva	Final Host	Reference
Rat (E), *Rattus norvegicus*	*Cysticercus fasciolaris*		Preet and Prakash, 2000a
Rat (E), Sprague Daweley	*Cysticercus fasciolaris*		Preet and Prakash, 2000b
Wild rat (*Bandicota bengalensis*) (N)	*Cysticercus fasciolaris*		Singla *et al.*, 2003
Rattus rattus (N) *Rattus nitidus* (N) *Rattus norvegicus* (N) *Bandicota bengalensis* (N) *Berylmys mackenziei* (N) *Berylmys bowersi* (N) *Niviventer fulvescens* (N) *Mus musculus* (N)	*Cysticercus fasciolaris*		Malsawmtluange and Tandon, 2008
Rattus rattus (N) *Rattus nitidus* (N) *Rattus norvegicus* (N) *Bandicota bengalensis* (N) *Berylmys mackenziei* (N) *Berylmys bowersi* (N) *Niviventer fulvescens* (N) *Mus musculus* (N)	*Cysticercus fasciolaris*		Malsawmtluange and Tandon, 2009b
House rat (*Rattus rattus*) (N) Lesser bandicoot rat (*Bandicota bengalensis*) (N) Indian gerbil (*Tatera indica*) (N) (all are reservoir hosts)	*Cysticercus fasciolaris*		Singla *et al.*, 2008
Bandicota bengalensis (N)	*Cysticercus fasciolaris*		Neena Singla *et al.*, 2008

Intermediate Host	Larva	Final Host	Reference
Rat (E)	*Cysticercus fasciolaris*	Cat (N)	Shukla et al., 2008a
Rat (E)	*Cysticercus fasciolaris*	Cat (N)	Shukla et al., 2008b
Rat (E)	*Cysticercus fasciolaris*		Sreekrishnan et al., 2008
Wild rat (*Bandicota bengalensis*) (N)	*Cysticercus fasciolaris*		Singla et al., 2009
Rat, laboratory (N) Wild rats (N)	*Cysticercus fasciolaris*		Goswami et al., 2010
Rat, *Rattus rattus* (N)	*Cysticercus fasciolaris*	Cat (N)	Neena Singla et al., 2010
Laboratory rat (*Rattus norvegicus*) (E)	*Cysticercus fasciolaris*		Kolte et al., 2011
Bandicota bengalensis (N)	*Cysticercus fasciolaris*		Neena Singla et al., 2012
Bandicota bengalensis (N)	*Cysticercus fasciolaris*		Neena Singla et al., 2013

2.9 FAMILY: DIPHYLLOBOTHRIIDAE

Diphyllobothrium sp.

Intermediate Host	Larva	Final Host	Reference
Freshwater fish: 2nd IH *Heteropneustus fossilis*	*Pleurocercoid*		Uppal, 1974
Spirometra sp.			
Fish	*Sparagnum* (*Pleurocercoid*)	Cats (N)	Saleque et al, 1990
Tropidonotus piscator	*Sparagnum* (*Pleurocercoid*)		Raju, 1974
Mesocyclops leukharti- 1st IH Tadpoles of *Rana* sp.-2nd IH	*Pleurocercoid*	Cats (N)	Rajavelu and Ebenzer, 1995

I.B PHYLUM: NEMATHELMINTHES

3. CLASS: NEMATODA

Final hosts include both primary and secondary hosts; Abbreviations used: E- Experimental; N-Natural

Intermediate Host	Larva	Final Host	Reference
3.1 FAMILY: HETERAKIDAE			
Ascaridia galli (Schrank, 1788) Freeborn, 1923			
Grass-hoppers*:		Chicks (E)	David Jacob et al., 1970
Oedaleus abruptus (N) (E)			
Spasthosternum prasiniferum (N)			
* Usual mode is direct transmission with no intermediate hosts.			
3.2 FAMILY: SUBULURIDAE			
Subulura brumpti (Lopez Neyra, 1922) Cram, 1926			
Beetle:	Encysted larva	Chicks (E)	Arora and Rai, 1972a
Gonocephalum depressum			
Gonocephalum depressum	Encysted larva	Chicks (E)	Arora and Rai, 1972b
Gonocephalum depressum	Encysted larva	Chicks (E)	Arora and Rai, 1972c
Gonocephalum depressum	Encysted larva	Chicks (E)	Arora and Rai, 1972d
Alphitobius diaperinus (N) (E)	Encysted larva	Chicks (E)	Karunamoorthy et al., 1994

Intermediate Host	Larva	Final Host	Reference
		***Subulura minetti* Bhalerao, 1941**	
Beetle: *Gonocephalum depressum* (N)	Encysted larva	Chicks (E)	Srivastava and Pande, 1967
Gonocephalum depressum (N)	Encysted larva	Chicks (E)	Mathur and Pande, 1967
	3.3 FAMILY: TRICHINELLIDAE		
	***Trichinella spiralis* (Owen, 1835) Railliet, 1895**		
	Larvae	Cat (N)	Maplestone and Bhaduri, 1942
Pig (N) Squirrel (N)	Cyst	—	Bhalerao, 1947
	Cyst	Cat (N)	Kalapesi and Rao, 1954
Rats (N)			Niphadkar *et al.*, 1968
Rat (N)	Larva	Wild civet cat (*Viverricula indica bengalensis*) (N)	Scad and Chowdhury, 1967
	Larva	Civet cat (*Viverricula indica bengalensis*) (N) Albino rats (E)	Scad *et al.*, 1967
Pigs (N)			Deodhar *et al.*, 1968

Intermediate Host	Larva	Final Host	Reference
Pigs (N)			Niphadkar et al., 1968
Pigs (N)			Parmeter et al., 1968
Pigs (N)			Ramamurthi and Ranganathan, 1968
Bandicoots (*Bandicota bengalensis*) (N)	Larvae Adult female	Bandicoots (*Bandicota bengalensis*) (N)	Niphadkar, 1973
Xenopsylla cheopsis (N) *Bandicota indica* (N) *Rattus rattus* (N)	Encapsulated larva		Ranade and Balchandra, 1976
House schrews (N)			Niphadkar, 1977
Rats (E)			Pardhan, 1978
Albino rats (E) *Bandicota bengalensis* (N)		Larva	Pradhan and Niphadkar, 1978
Domestic pigs (N)			Niphadkar et al., 1979
Animals (E)			Pethe, 1992
Mice (E)			Temjenmongla and Yadav, 2005
Swiss Albino mice (E)			Yadav and Temjenmongla, 2008
Boar, wild (N)		Man (N)	Sharma, 2011

Intermediate Host	Larva	Final Host	Reference
	3.4 FAMILY: STEPHANURIDAE		
	Stephanurus dentatus (Diesing, 1839)		
Earthworm: *Eutypheus waltoni* *Pheretima* sp.- Transport host		—	Sinha, 1967
	3.5 FAMILY: ANCYLOSTOMATIDAE		
	Ancylostoma caninum (Ercolani, 1859) Hall, 1913		
Cyclops (E)-Paratenic host	Larva	—	Banerjee and Omprakash, 1972
	3.6 FAMILY: PROTOSTRONGYLIDAE		
	Varestrongylus pnuemonicus (Bhalerao, 1932)		
Terrestrial snail: *Macrochlamys cassida*	II stage larva	—	Bhalerao and Kapoor, 1944
Macrochlamys cassida	II stage larva	Kid (E)	Bhalerao and Kapoor, 1945
Macrochlamys cassida	IIstage larva	Guinea pig (E)	Sharma Deorani, 1965
	3.7 FAMILY: METASTRONGYLIDAE		
	Metastrongylus apri (Godoelst, 1923)		
Earthworm: *Megascolex mauritii* (N)	Larvae	Pig (E)	Subramaniam *et al.*, 1971

Intermediate Host	Larva	Final Host	Reference
		Metastrongylus salmi (Godoelst, 1923)	
Earthworm: *Perionyx excavatus* (E)		Pig (E)	Bhattacharya *et al.*, 1971
Earthworm: *Dichogaster bolaui* (E) *Drawida pellucida var pallida* (E) *Lampito mauritii* (E)	Third stage larva	Pig (E)	Thomas and Peter, 1971
Earthworm: *Dichogaster bolaui* (E) *Drawida pellucida var pallida* (E) *Lampito mauritii* (E)	Third stage larva	Pig (E)	Thomas and Peter, 1975

3.8 FAMILY: SPIROCERCIDAE

Cyanthospirura chaubaudi Gupta and Pande, 1981

Intermediate Host	Larva	Final Host	Reference
Wall lizard: *Hemidactylus flaviviridis* (N)	Encapsulated larvae	Pups (E)	Gupta and Pande, 1981

3.9 FAMILY: SPIRURIDAE

Habronema sp.

Intermediate Host	Larva	Final Host	Reference
Empid (Diptera, new vector)	*Habronema muscae* Carter, 1865		Muralidhar and Rao, 1991.
Musca domestica			Anon, 1861 cited by Chowdhury, 2001)

Intermediate Host	Larva	Final Host	Reference
House fly: *Musca domestica* (N)	Infective larvae		Carter, 1861
House fly: *Musca domestica* (N)	Infective larvae		Jayalaksmi Jagannathan, 1980
Draschia megastoma (Rudolphi, 1819) Chitwood and Wehr, 1934			
Musca domestica			Gupta, 1970
Spiruroids species			
	Encysted larve of three forms		Srivastava and Pande, 1967
Spirocerca lupi Rudolphi, 1809			
Paratenic hosts – Vertebrates:	Cyst	—	Pujatti, 1953
Amphibian:			
Bufo melanostictus			
Lizards:			
Calotes versicolor			
Hemidactylus gleadovii			
Chamelion calcaratus			
Snakes:			
Zamensis fasciolatus			
Zamensis gracilis			
Helicops schistosus			
Drophis micterzans			
Naja tripudians			
Vivipera russelli			
Echis carinata			

Intermediate Host	Larva	Final Host	Reference
Birds: Eudinamys scolopaceus Froncolinas pondicera Athene brama Tyto alba javanica Milvous migrans govinda Corvus splendens splendens Corvus coronodles culminatus Mammals: Rattus rattus Loris tardigradus Vesperugo noctula (insectivorous bat)	Juvenile stages (unconfirmed)	Donkey (N) Ponies (N) Hill goat (N)	Pande et al., 1961
Gymnopleurus koenigi (N), (E) Euoniticellus (Oniticellus) pallipes (possible) (N)	Encysted larvae	Dog (E)	Anantaraman and Jayalaksmi, 1963
Calotes versicolar (N) Grey schrew: Suncus murinus (N)	Encysted larvae	—	Anantaraman and Sen, 1966
Dung beetles: Oniticellus pallens (N) Oniticellus pallipes (N) Onthophagus deflexicollis (N) Onthophagus quadridentatus (N)	Encysted larvae	Rabbits (E)	Chowdhury and Pande, 1969

Intermediate Host	Larva	Final Host	Reference
Scarabaeid beetles (N) *Calotes versicolar* (N) Grey schrew: *Suncus murinus* (N)		Dogs (E)	Sen and Anantaraman, 1971
Beetles (E) Chicks (E)	Infective juveniles	Pups (E) Cats (E) Albino rats (E) Kid (E)	Chhabra and Singh, 1972a
		Kids (E) Lambs (E)	Chhabra and Singh, 1972b
Baby rhesus monkey (E) Cat (E) Rabbit (E) Guinea pig (E) Albino rat (E) Swiss albino mice (E) Wall lizard (*Hemidactylus* sp.) (E) Toads (*Bufo melanostictus*) (E) Squirrel (E) Chicks (E)	Infective juveniles and cysts	Pups (E)	Chhabra and Singh, 1972c
Beetles (E) Cat (E) Albino rat (E) Kid (E)	Juveniles	Pups (E)	Chhabra and Singh, 1972d
Beetles (E)	Juveniles	Pups (E)	Chhabra and Singh, 1972e

Intermediate Host	Larva	Final Host	Reference
Vectors (E)	Juveniles	Pups (E)	Chhabra and Singh, 1972f
Coprophagus beetle:			Chhabra and Singh, 1973
Onitis philemon (E)			
Catharsius pitchecius (E)			
Onthophagus bonasus (E)			
Onthophagus gazelle (E)			
Onthophagus dama (E)			
Onthophagus mopsus (E)			
Euoniticellus pallipes (E)			
Hybosorus orientalis (E)			
Hister (Peranus) maindronii (E)			
Hister (Pachylister) lutarius (E)			
Scarites indus (E)			
Onitis philemon (E)		Pups (E)	Chhabra and Singh, 1977
Catharsius pitchecius (E)			
Onthophagus bonasus (E)			
Duck (N)			Bhatia *et al.*, 1979
Probably paratenic host			
Indian barn owl:			Joseph, 1979
Tyto alba javanica- Partenic host			
Garden lizards	Cyst containing Larvae		Joju Johns *et al.*, 2010
(*Calotes versicolor*) (N)			

3.10 FAMILY: ASCAROPIDAE

Ascarops strongylina (Rudolphi, 1819)

Intermediate Host	Larva	Final Host	Reference
Pond snake: *Natrix piscator* (N) Groung squirrel: *Funambulus palmarum* (N) (Paratenic hosts)	Encysted larvae	Rabbits (E)	Gupta, 1969
Dung beetle: *Aphodius moestus* (N) *Aphodius* sp. (N) *Gymnopleurus parvus* (N) *Oniticellus pallens* (N) *Oniticellus (Euoniticellus) pallipes* (N) *Onthophagus catta* (N) *Onthophagus cervus* (N) *Onthophagus deflexicollis* (N) *Onthophagus quadridentatus* (N) *Onthophagus* sp. (N)	Encysted larvae	Rabbits (E) Guinea pig (E)	Chowdhury and Pande, 1969
White mice (E) paratenic host			Varma *et al.*, 1976
Dung beetles: *Hister (S. S) corax* (N), (E) *Hister (Peranus) maindronni* (N), (E) *Hybosonus orientalis* (E) *Oniticellus pallipes* (N) *Onitis philemon* (E) *Onthophagus catta* (N), (E)	II stage larva	Guinea pig (E)	Varma *et al.*, 1977

Intermediate Host	Larva	Final Host	Reference
Onthophagus falsus (N) *Onthophagus gazelle* (N), (E) *Onthophagus mopsus* (N) *Onthophagus quadridentatus* (N) (E) *Onthophagus ramosellus* (N) *Trox granulatus* (E)			

Gongylonema pulchrum (Molin, 1857)

Intermediate Host	Larva	Final Host	Reference
Coprophagus beetles: *Onthophagus deflexicollis* (N) *Onthophagus catta* (N) *Onthophagus quadridentatus* (N) *Oniticellus pallens* (N)	Cyst and infective larvae	Rabbits (E)	Chowdhury and Pande, 1968b
Onthophagus deflexicollis (N) *Onthophagus catta* (N) *Onthophagus quadridentatus* (N) *Onthophagus* sp. (N) *Oniticellus pallens* (N) *Aphodius moestus* (N) *Catharasius* sp. (N) Other 3 species of unidenfied beetles (N)	Infective larvae		Gupta, 1970b

Gongylonema ingluvicola Ransom, 1904

Intermediate Host	Larva	Final Host	Reference
Cockroach: *Pycnoscelus surinamensis* (E)	Encysted larvae in muscles	Fowl (E)	Sunderam, 1971

Intermediate Host	*Larva*	*Final Host*	*Reference*
***Physocephalus sexalatus* (Molin, 1860)**			
Dung beetles: *Catharasius* sp. (N) *Gymnopleurus parvus* (N) *Oniticellus pallens* (N) *Oniticellus (Euonticellus) pallipes* (N) *Onthophagus catta* (N) *Onthophagus centricornis* (N) *Onthophagus cervus* (N) *Onthophagus deflexicollis* (N) *Onthophagus quadridentatus* (N) *Onthophagus* sp. (N)	2nd stage larva	Rabbits (E)	Chowdhury and Pande, 1968a
Onthophagus catta (N) *Onthophagus duporti* (N) *Onthophagus cervus* (N) *Onthophagus* sp. (N)	3rd stage larvae		Thomas and Peter, 1971
Onthophagus catta (N) *Onthophagus duporti* (N) *Onthophagus cervus* (N) *Onthophagus* sp. (N)	3rd stage larvae		Thomas and Peter, 1975
***Simondsia paradoxa* (Cobbold, 1864)**			
Beetle: Coprophagus beetles (N) *Onthophagus* sp. (N)	Encysted larvae	Piglets	Gupta and Pande, 1970

Intermediate Host	Larva	Final Host	Reference
Oniticellus pallens (N)			
Paratenic hosts:			
Lizards (N)			
Birds (N)			
Hemidactylus flaviviridis (N)			
Crocidura coerulea (N)			
Insectivorous mammals (N)			
Beetles:	Encapsulated larvae	Piglets (E)	Gupta and Pande, 1981
Arthrodeis bipartitus (N)			
Oniticellus pallens (N)			
Onthophagus sp. (N)			
Onthophagus boriasu (N)			
Onthophagus cata (N)			
Onthophagus deflexicollis (N)			
Onthophagus quadridentatus (N)			
Paratenic hosts:			
Vertebrates			
Varanus lizards			
Varanus monitor (N)			
Hemidactylus flaviviridis (N)			
Birds: *Upupa epope* (N)			
An unidentified species (N)			
Mammal:			
Crocidura coerulea (N)			

Intermediate Host	Larva	Final Host	Reference
3.11 FAMILY: ACUARIIDAE			
Acuaria (Cheilospirura) hamulosa (Deisinge, 1851)			
Grass-hoppers: *Spasthosternum parasiniferum* (E) *Oxya nitidula* (E)	Encysted 3rd stage larvae	Chicks (E)	Sunderam, 1971
Spasthosternum parasiniferum (E) *Oxya nitidula* (E)			Sunderam, 1977
Acuaria (Dispharynx) spiralis Molin, 1858			
Isopod: *Porcellio laevis* (E) *Cubaris chittoni* (E)		Fowl (E)	Sunderam, 1971
Isopod: *Porcellio laevis* (E) *Cubaris chittoni* (E)			Sunderam, 1977
Isopod: *Porcellio laevis* (E)	Infective juveniles	Chicks (E)	Ramaswamy and Sunderam, 1979
Isopod: *Porcellio laevis* (E)	Infective juveniles	White leghorn, male (E)	Ramaswamy and Sunderam, 1985
Echinuria uncinata (Rudolphi, 1819)			
Daphnia pulex (E) *Daphnia magna* (E)		Duck (E)	Chandrasekharan, 1977

3.12 FAMILY: RICTULARIIDAE

Rictularia caherensis (Jagerskiold, 1904)

Intermediate Host	Larva	Final Host	Reference
Lizard (N) Wall lizard: *Hemidactylus flaviviridis* (N)	Encysted larvae	Kitten (N)	Srivastava, 1940
	Re-encysted larvae	Guinea pigs (N)	Gupta and Pande, 1970

3.13 FAMILY: TETRAMERIDAE

Tetrameres anatis Chandrasekharan and Peter, 1969

Intermediate Host	Larva	Final Host	Reference
4 species of grasshoppers		Duck (E) Chicks (E)	Chandrasekharan, 1967
Grasshoppers: *Oedaleus abruptus* *Spasthosternum prasiniferum*		Duck (E) Chicks (E)	Chandrasekharan and Peter, 1969
Spasthosternum prasiniferum		Duck (E)	Chandrasekharan, 1977

Tetrameres mohtedai Bhalerao and Rao, 1944

Intermediate Host	Larva	Final Host	Reference
Beetles: *Liatongus cinctus* *Sphaeridium quinquemaculatus*			Mukherji and Sinha, 1965
6 varieties of grasshoppers (E)		Chicks (E)	Sunderam *et al.*, 1963

Intermediate Host	Larva	Final Host	Reference
Grasshoppers: *Spasthosternum parasiniferum* (E) *Oxya nitidula* (E)	Developing larvae in haemocoel Encysted 3rd stage larvae in fat bodies	—	Sunderam, 1971
Spasthosternum parasiniferum (E) *Oxya nitidula* (E)			Sunderam, 1977
Isopoda: *Porcellio laevis*	First stage larvae Second stage larvae Third stage larvae (encysted)	Chicken (E)	Ramaswamy and Sunderam, 1979
Grasshoppers: *Attractomorphacrenulata* *Heirogeuphus banian* *Oxyvelax chinensis* *Oxya nitidula* *Oedaelus abraptus* *Spasthosternum parasiniferum*	Infective larvae	White leghorn chicken (E)	Radhakrishna Reddy and Rao, 1984
Grasshoppers: *Attractomorpha crenulata* *Heirogeuphus banian* *Oxyvelax chinensis* *Oxya nitidula* *Oedaelus abraptus* *Spasthosternum parasiniferum* *Acrida turreta* *Acrida zyyancea* *Crytacanthacris ranacea* *Conocephalus maculates*	Infective larvae	White leghorn chicken (E)	Radhakrishna Reddy and Rao, 1985

Intermediate Host	Larva	Final Host	Reference
Conocephalus indicus			
Critoganus oxyplerus			
Ducetica japonica			
Phyllocoreoreia ramakrishnae			
Catantops sp.			
Colemania sphenerioida			
Palanger retnacea			
Satrophyllia rugosa			
Beetle:			
Coccinella septempunctuta			

3.14 FAMILY: PHYSALOPTERIDAE

Physaloptera brevispiculum Von Linstow, 1906

Intermediate Host	Larva	Final Host	Reference
Paratenic host: Spiny tailed lizard	Encapsulated larvae	Pups (E), stages of developing juveniles Cats (E), 4th stage juveniles and adults Rabbits (E), encysted larvae Rats (E), encysted larvae	Gupta and Pande, 1970

3.15 FAMILY: THELAZIIDAE

Oxyspirura mansoni Cobbold, 1879

Intermediate Host	Larva	Final Host	Reference
Cockroach: *Pycnoscellus surinamensis* (E)	Infective larvae in cysts	Fowl (E)	Sunderam, 1971

Intermediate Host	Larva	Final Host	Reference
Thelazia sp.			
Musca vicina			Rao, 1970
Musca crassirostris	Larva		Narsi Reddy and Narayana Rao, 1982
Thelazia gulosa (Railliet and Henry, 1910)			
Musca domestica			Gupta, 1970
Thelazia rhodesii (Gurlt, 1831)			
Musca vicina			Rao, 1949

3.16 FAMILY GNATHOSTOMIDAE

Gnathostomum spinigerum Owen, 1836

Intermediate Host	Larva	Final Host	Reference
Cyclops: 1st IH: *Mesocyclops leuckarti*	2nd stage larvae	Dog (N)Cat (N)	Chellappa and Anantaraman, 1970
Fishes: 2nd IH: *Ophiocephalus striatus* (N) *Ophiocephalus argus* (N) *Clarius batrachus* (N)	Encysted infective larvae		Rai, 1976
Mesocyclops leuckarti (E)	2nd stage larvae	Dog (N)Cat (N)	Chellappa and Anantaraman, 1977
Mesocyclops leuckarti	First & second stage larvae		Baruah et al., 1988

Intermediate Host	Larva	Final Host	Reference
Mesocyclops leuckarti (E) *Ophiocephalus punctatus* (N) (E) *Anabas scandens* (E)	First & second stage larvae Third stage larvae		Gogoi and Baruah, 1988
Fishes: *Ophiocephalus punctatus* (N) *Ophiocephalus punctatus* (E) *scandens* (E) Genus not indicated Cyclops: *Mesocyclops leuckartii* (E)	Third stage larvae Second stage larvae		Nabaneel Baruah et al., 2010
1st IH: *Mesocyclops leuckarti* (E) 2nd IH: *Ophiocephalus striatus* (N) (E) *Anabas scandens* (E)	First & second stage larvae Third stage larva		Baruah et al., 2011

3.17 FAMILY: FILARIIDAE

Filarid (Undescribed)

Intermediate Host	Larva	Final Host	Reference
Ticks			Shastri and Gafoor, 1979
Hyalomma anatolicum anatolicum, nymph, adults (N) *Boophilus microplus*, Nymph Adult female (N)	Microfilarial forms Developing stages Third stage larvae Second stage larvae Microfilarial forms First, second and third stage larvae	Cattle (N) Buffalo (N) Cow (N)	Shastri and Gafoor, 1981*

Intermediate Host	Larva	Final Host	Reference
Hyalomma anatolicum anatolicum, engorged nymphs (N)	Third stage larvae First and second stage larvae Microfilariae	Adult cattle	Shastri, 1983
Dipetalonema (Acanthoceilonema) sp. (See below)			
Dirofilaria repens **Railliet and Henry, 1911**			
Armigeres sp.Culex sp. (N)	Second stage larvae Infective larvae		Lucy Sabu and Subramanian, 2006
Culex sp. (N)	Second & third stage larvae		Lucy Sabu and Subramanian, 2007
Dipetalonema (Acanthoceilonema) **sp.**			
Boophilus microplus (N) Hyalomma anatolicum anatolicum (N)	Third stage larvae	Cattle (N)	Shastri and Gafoor, 1981*
Hyalomma anatolicum anatolicum (N)	Third stage larvae		Shastri , 2001
*Filarid undescribed? (see above)			
Acanthoceilonema **sp. nov.**			
Hippobosca longipennis (N)	Infective larvae		Megat Abd Rani et al., 2011
Parafilaria bovicola **de Jesus, 1934**			
Musca vitripennis (N)	Larva	Bullock (N)	Sahai and Singh, 1971

Intermediate Host	Larva	Final Host	Reference
Artionema labiopapillosa (Aless, 1838)			
Culex sp. (N)	Microfilaria	Buffalo (N)	Malaviya, 1966
3.18 FAMILY: SETARIIDAE			
Setaria cervi (Rud, 1819)			
	Microfilaria	Rabbits (E)	Lakra and Singh, 1963
	Microfilaria	White rats (E)	Ansari, 1964
Aedes aegypti (E)	Developing larvae	White rats (E)	Wajihulla, 2001
Aedes aegypti (E)	Larvae	White rats (E)	Khan and Singh, 2003
Setaria digitata (Von Linstow, 1906)			
Aedes vittatus (E) *Armiageres obturbans* (E)	Microfilaria	Rabbits (E)	Lakra and Singh, 1963
Aedes vittatus (E) *Armiageres obturbans* (E)	Microfilaria	Rabbit (E) Bullock (N) Cow (N)	Varma *et al.*, 1971
Stephanofilaria sp.			
Stephanofilaria assamensis Pande, 1936			
Musca autumnalis (N)	Juveniles		Patnaik and Vinaykumar, 1972
Musca conduscens (N)	Larva	Bullock (N)	Srivasatav and Dutt, 1963
Musca conduscens (E)	Larva	Calves (E)	Patnaik, 1966

Intermediate Host	Larva	Final Host	Reference
Musca conduscens (E)	Larva	Calves (E)	Patnaik and Roy, 1966a
Musca conduscens (N)	Larva	Calves (E)	Patnaik and Roy, 1966b
Musca conduscens (N)	Larva		Sahai and Srivastava, 1966
Musca conduscens (E)	Larva	Calves (E)	Patnaik and Roy, 1967
Musca conduscens (E)	Preinfective larval stages		Srivastava, *et al.*, 1967
Musca conduscens (N)	Larva		Sharma Deorani and Tewari, 1968
Musca conduscens (N)	Larva		Dutta, 1972
Musca conduscens (N)	Larva	Cattle (N)Buffaloes (N)	Malaviya, 1972
Musca conduscens (E)	Larva	Calf (E)	Patnaik, 1973
Musca pattoni (N) *Musca conduscens* (N)	Larva		Chowdhury Phukan, 2003
Musca pattoni (N)	Larva		Chowdhury Phukan *et al.*, 2005
Stephanofilaria zaheeri Singh, 1958			
Musca planiceps (N)	Larva		Dutt, 1970
Musca autmnalis (N)	Juveniles		Patnaik and Kumar, 1972 I
3.19 FAMILY: DRACUNCULIDAE			
Dracunculus medinensis (Linnaeus, 1758)			
		Dogs (E)	Moorhy and Sweet, 1936a

Intermediate Host	Larva	Final Host	Reference
Cyclops (N)	Larvae		Moorthy and Sweet, 1936b
Cyclops, *Cyclops leuckarti* (N)	Larvae		Moorthy , 1938
		Dogs (E)	Moorthy and Sweet, 1938
Eight species of Cyclops *Mesocyclops leuckarti* (N)			Mehta and Gupta, 1982
Cyclops: *Paracyclops fimbriatus* (N) *Thermocyclops oithonoides* (N) *Mesocyclops leuckarti sensu lato* (N) *Microcyclops varicans* (N)	Infective larvae		Saroj Bapna, 1985

I. C PHYLUM: ACANTHOCEPHALA

Final hosts include both primary and secondary hosts; Abbreviations used: E- Experimental; N-Natural

4.1 FAMILY: POLYMORPHIDAE

Intermediate Host	Larva	Final Host	Reference
	Polymorphus magnus **Skrjabin, 1913**		
Gammarus pulex (N)		Ducklings (E)	Fotedar *et al.*, 1977

4.2 FAMILY: OLIGACANTHORHYNCHIDAE

Onicicola sp.			
Partridges:	Encysted juveniles	Dog or carnivores	Padmavathi, 1967
Francolinus pondicerianus (N)			
Quails:			
Coturnix coturnix (Paratenic hosts)			

Section II

ARTHROPOD PARASITES

II. PHYLUM: ARTHROPODA
5. CLASS: PENTASTOMIDA

Final hosts include both primary and secondary hosts; Abbreviations used: E- Experimental; N-Natural

5.1 FAMILY: POROCEPHALIDAE

Intermediate Host	Larva	Final Host	Reference
		Linguatula serrata Forlich,1789	
Camel (N)	Nymphs		Gaiger, 1909
Camel (N)	Nymphs		Gaiger, 1911
Camel (N)	Larvae		Leese, 1911
Pig (N)			Ramanujachari and Alwar, 1954
Pig (N)			Alwar, 1958
Sheep (N) Goats (N) Buffaloes (N) Rabbits (N)	Adult (N)Nymphs (N)	Dog (N)	Gill *et al.*, 1968
Goats (N)	Nymphs (N)		Ramakrishna Reddy *et al.*, 1971
Domestic animal (N)			Gill, 1972
Goats (N)	Larvae		Lal Krishna *et al.*, 1973
Goats (N)			Singh *et al.*, 1973
Rodents (E)			Ebenzer Raja, 1974

Intermediate Host	Larva	Final Host	Reference
Indian palm squirrel: *Funambulus palmarum* (N)			Chauhan and Bhatia, 1978
Cattle (N)	Nymphs (N)		Muraleedharan and Syed Zaki, 1975
Goats (N) Sheep (N)	Nymphal stages (N)		Banerjee *et al.*, 2005
Buffaloes (N)	Nymphal stages (N)		Sivakumar *et al.*, 2005
Goats (N) Cattle, Crossbred (N) Buffaloes (N)	Encapsulated nymphal stages (N)		Bindu Lakshman *et al.*, 2006
Goats (N)	Nymphs (N)		Sangwan *et al.*, 2007
Goats (N) Sheep (N) Buffaloes (N)	Nymphs (N)		Ravindran *et al.*, 2008
Small ruminats (N)			Stuti Vatsya, 2011

Section III

PROTOZOAN PARASITES

III A. PHYLUM: SARCOMASTIGOPHORA

Final hosts include both primary and secondary hosts; Abbreviations used: E- Experimental; N-Natural

6.1 FAMILY: TRYPANOSOMATIDAE

Intermediate Host	Larva (Mode of Infection)	Final Host	Reference
	Trypanosoma evansi **(Steel, 1885) Balbaini, 1888**		
Horse fly (E) *Tabanus orientis* (Probable)		Horse (E) Dog (E) Guinea pigs (E) Rabbit (E)	Rogers, 1901
Tabanus tropicus (E)			Lingard, 1906
Tabanus nemoralis (E)			Sergent and Sergent, 1906
Lyperosia minuta (N) *Tabanus* (E) *Haematopota* (E) *Stomoxys* (E)		Horse	Leese,1909
Tabanus (E) *Stomoxys* (E)		Horse	Leese, 1912
Tabanus orientis (E) *Tabanus tropicus* (E) *Tabanus subcallosus* (E) *Stomoxys calcitrans* (E)		Horse Guinea pigs (E)	Baldrey, 1911

Intermediate Host	Larva (Mode of Infection)	Final Host	Reference
Tabanus albimedius (E) *Tabanus rubidus* (E) *Tabanus striatus* (E) *Musca crassirostris* (E)			Fletcher, 1916
Tabanus rubidus (E) *Tabanus virgo* (E) *Tabanus persis* (E) *Tabanus* sp. (E)		Dog Rabbit Buffalo Camel	Cross and Patel, 1921a
Ornithodoros crossi (E) *Ornithodoros lahorensis* (E)		Camels Rabbit (E)	Cross and Patel, 1921b
Ornithodoros tholozani (E) *Tabanus nemocallosus* (E) *Tabanus striatus* (E) *Tabanus rubidus* (E)			Cross and Patel, 1921c
Tabanus albimedius (E) *Tabanus hilaris* (E) *Tabanus striatus* (E) *Tabanus rubidus* (E)			Cross and Patel, 1922a
Ornithodoros tholozani			Cross and Patel, 1922b
Tabanus tenes (E)			Cross and Patel, 1922 c
Tabanus rubidus (E)			Cross and Patel, 1922d
Ornithodoros crossi (E)		Dogs (N) Rabbits (E) Rats (E)	Cross, 1923

Intermediate Host	Larva (Mode of Infection)	Final Host	Reference
Tabanus albimedius (E) (N)			Cross and Patel, 1923
Ornithodoros tholozani (E)			Cross and Abdulla Khan, 1923
Ornithodoros crossi (E) *Ornithodoros lahorensis* (E)		Dogs (N) Rabbits (E) Rats (E)	Kahan Singh, 1925
Tabanus bicallosus (E) (N) *Tabanus virgo* (E) (N) *Tabanus ditaeniatus* (E) (N) *Tabanus macer* (E) *Ctenocephalides felis* (E) (N)			Kahan Singh, 1926
Ornithodoros tholozani (E)			Kahan Singh, 1929
Ornithodoros tholozani (E)		Guinea pigs	Sen, 1938
Ornithodoros tholozani (E)		Guinea pigs (incapable of transmission from infected guinea pigs to healthy rabbits through this tick)	Sen, 1946
Arthropods (N) Vectors other than *Tabanus*			Basu, 1947
Tabanus (N)			Balarama Menon, 1952

Intermediate Host	Larva (Mode of Infection)	Final Host	Reference
Tabanus ditaeniatus (N)			
Tabanus macer (N)			
Tabanus nemocallosus (N)			
Tabanus rubidus (N)			
Tabanus striatus (N)			
Tabanus tropicus (N)			
Tabanus virgo (N)			Basu et al., 1952
Stomoxys calcitrans (E)		Dog (E)	Chaudhury et al., 1966
		Guinea pigs (E)	
Hyalomma anatolicum anatolicum (E)		Rabbits (E)	Sridhar et al., 1989
Trypanosoma theileri, Laveran, 1902			
Hyalomma anatolicum anatolicum, nymphs, adults (N) (E)	Amastigotes Sphaeromastigotes Epimastigotes Trypomastigotes	Cross-bred bull (N) (E)	Shastri and Deshpande, 1981
Hyalomma anatolicum anatolicum (E)		Cattle (N)	Shastri et al., 1981
Leishmania donovani (Laveran and Mansil, 1903)			
Phlebotomus argentipes (E)	Developmental forms		Christophers et al., 1925
Phlebotomus species (N)	Larval form	Man (N)	Sinton, 1925
Phlebotomus argentipes (E)	Developmental forms	Mamalian host (E)	Shortt, 1928
Phlebotomus argentipes (E)	Developmental forms	Hamsters (E)	Shortt et al., 1931

Intermediate Host	Larva (Mode of Infection)	Final Host	Reference
Phlebotomus argentipes (E)	Developmental forms	Hamster (E)	Shortt *et al.*, 1931
Phlebotomus argentipes (E)	Developmental forms	Hamster (E)	Shortt *et al.*, 1932
Phlebotomus argentipes (E)	Developmental forms		Napier *et al.*, 1933a
Phlebotomus argentipes (E)	Developmental forms	Hamster (E)	Napier *et al.*, 1933b
Phlebotomus argentipes (E)	Developmental forms	Hamster (E)	Smith *et al.*, 1936
Phlebotomus argentipes (E)	Developmental forms	Hamster (E) Mice (E)	Smith *et al.*, 1940
Phlebotomus argentipes (E)	Developmental forms	Hamster (E) Mice (E)	Smith *et al.*, 1941
Phlebotomus argentipes (E)	Developmental forms	Man (E)	Swaminath *et al.*, 1942
Phlebotomus argentipes (E)	Infection detected by non radioactive probe		Dinesh *et al.*, 2000
Phlebotomus argentipes (E)	Infection detected by DOT immunoblot for antigen	Balb/c mice (E)	Kumar *et al.*, 2000
			Sukhbir Kaur *et al.*, 2001

Leishmania tropica (Wright, 1903) Luhe, 1906

Intermediate Host	Larva (Mode of Infection)	Final Host	Reference
Phlebotomus argentipes (E)		Man (E)	Napier *et al.*, 1933a
Phlebotomus papatasii (E)		Man (E)	Alder and Ber, 1941

III B. PHYLUM: APICOMPLEXA

7.1 FAMILY: SARCOCYSTIDAE

Intermediate Host	Larva (Mode of Infection)	Final Host	Reference
	Sarcocystis		
Cattle (N)	Sarcocysts		Sen, 1951
Cattle (N) Buffaloes (N) Sheep (N)	Sarcocysts, Sarcospores		Ramanujachari and Alwar, 1951
Cattle (N) Buffaloes (N)	Sarcocysts		Thirunavukkarasu, 1964
Pig (N)	Sarcocysts		Deodhar et al., 1968
Cattle (N) Sheep (N)	Sarcocysts		Prasad et al., 1969
Cattle (N) Buffaloes (N)	Sarcocysts		Singh and Singh, 1970
Buffaloes (N)	Sarcocysts		Purohit, 1970
Goats (N)	Sarcocysts		Iyer, 1971
Animals (N)	Sarcocysts		Sathia Singh, 1971 (Thesis)
Cattle (N) Buffalo calves (N)	Sarcocysts		Purohit and D'Souza, 1973
Cattle (N)	Sarcocysts		Shukla and Victor, 1974

Intermediate Host	Larva (Mode of Infection)	Final Host	Reference
Cattle (N) Buffaloes (N) Buffalo calves (E)	Sarcocysts antibodies (CFT) Sarcocystic spores		Shukla and Victor, 1976
Bullock (N)	Sarcocysts		Sreemannarayana and Christopher, 1977
Cattle (N) Pig (N)	Sarcocysts		Srivastava et al., 1977
Goats (N)	Sarcocysts		Chhabra and Mahajan, 1978
Cattle (N)	Sarcocysts		Das and Borkakoty, 1978
Sheep (N)	Sarcocysts		Gupta et al., 1979
Sheep (N)	Sarcocysts		Swarnkar et al., 1979
Goats (E)	Sporocysts		Pethkar and Shah, 1981
Goats (E)	Sporocysts		Pethkar and Shah, 1982
Cattle (N) Buffaloes (N) Pigs (N)	Sarcocysts		Sahai et al., 1981
Buffaloes (N) Goats (N)	Sarcocysts		Shah, 1981
Goats (N)	Sarcocysts		Gupta, 1982
Goats (N)	Sarcocysts		Pethkar and Shah, 1982a

Intermediate Host	Larva (Mode of Infection)	Final Host	Reference
Cattle (N) Buffaloes (N) Pigs (N)	Sarcocysts		Sahai et al., 1982
Cattle (N)	Sarcocysts		Deshpande et al., 1983
Buffaloes (N) Sheep (N) Goats (N) Pigs (N)	Sarcocysts	Dogs (E)	Gupta, 1983
Bovines (N) Ovines (N)			Gopalakrishna Rao and Rama Rao, 1983
Cattle (N)	Sarcocysts	Pup (E) Kitten (E)	Sahai et al., 1983
Goats (N)	Sarcocysts		Shah, 1983
Pigs (N)	Sarcocystis-like zoites, Cystozoites, Immature metrocysts Sporocysts	Dogs, young (E)	Gupta and Gautam, 1984
Pigs (N)	Sarcocysts		Gupta and Iyer, 1984
Cattle (E)	Sarcocysts		Jain and Shah, 1985
Goats	Sarcocysts		Pethkar, 1985
Goats (N)	Sarcocysts		Agrawal et al., 1986
Buffaloes (N)	Sarcocysts		George Verghese et al., 1986

Intermediate Host	Larva (Mode of Infection)	Final Host	Reference
Buffaloes (N)	Macrocysts	Dogs (E)	Jain et al., 1986
Goats (N)	Cysts		Agrawal et al., 1987
Cattle (N) Buffaloes (N)	Sarcocysts		Deshpande et al., 1987
	Sarcocysts	Leopard (Panthera pardus) (N)	Somvanshi et al., 1987
Cattle (N) Buffaloes (N) Sheep (N)	Sarcocysts (Rainey's corpuscles)		Gopalakrishna Rao and Rao, 1987
Goats (N)	Cysts		Juyal, 1987
Cattle (N) Buffaloes (N)	Sarcocysts		Juyal and Bhatia, 1987
Wild ruminants (N)	Sarcocysts		Acharjyo and Rao, 1988
Goats (N)	Sarcocysts	Man (E), self	Chaudhry and Shah, 1988
Goats (N)	Sarcocysts		Gupta and Singh, 1988
Goats (N)	Sporocysts	Dog (N)	Shastri, 1988
Goats (N)	Cysts Cryptozoites		Juyal et al., 1988b
Goats (N)	Microcysts Cystozoites	Dog (E)	Juyal et al., 1989
Pony (N)	Cyst Cystozoites Metrozoites		Achutan and Ebenezer Raja, 1990

Intermediate Host	Larva (Mode of Infection)	Final Host	Reference
Buffaloes (N)	Sarcocysts		Biswas et al., 1990
Pigs (N)	Sarcocysts		Saleque and Bhatia, 1991
Cattle (N)	Sarcocysts		Biswas et al., 1992
Sambar deer (*Cervus unicolor*) (N)	Sarcocysts		Gangadharan et al., 1992
Cattle (N)	Sarcocysts		Ahamad Pandit, 1993
Mare (N)	Sarcocysts		Juyal et al., 1994
Cattle (N) Buffaloes (N) Sheep (N) Goats (N)	Sarcocysts	Dogs (N)	Mohanty et al., 1994a
Cattle (N) Buffaloes (N) Sheep (N) Goats (N)	Sarcocysts		Mohanty et al., 1994b
Goats (N)	Sarcocysts Bradyzoites		Wadajkar et al., 1994
Buffaloes (N)			Sandhu et al., 1995
Goats (N), cooked meat	Sarcocysts		Aulakh et al., 1996
Cattle (N)	Sarcocysts antibodies		Pandit et al., 1996
Camel (N)	Sarcocysts with Cystozoites		Ranga Rao et al., 1997

Intermediate Host	Larva (Mode of Infection)	Final Host	Reference
Cattle (N)	Bradyzoites		Singh et al., 1997
Sheep (N)	Micro-sporocysts		Goldy Srivastava and Jain, 2000
Pigs (N)	Cysts		Avapal et al., 2001
Pigs (N)	Cysts, sporozoites		Srinivasa Rao and Hafeez, 2001
Pigs (N)	Sarcocysts		Srinivasa Rao and Hafeez, 2002a
Pigs (N)	Cysts, oocysts/sporocysts	Pups (E)	Srinivasa Rao and Hafeez, 2002b
Pigs (N)	Seroprevalence		Avapal et al., 2002
Pigs (N)	Cysts, zoites		Avapal et al., 2003a
Chital (*Axis axis*)	Sporocysts	Dhole (*Cuon alpinus*) suspected	Jog et al., 2003
Cattle (N)	Bradyzoites		Singh et al., 2004
Cow, Sahiwal (N)	Sporocysts		Neelu Gupta, 2005
Heifer (N)	Sarcocysts		Tamuli et al., 2005
Chicken (N)	Sarcocysts		Bineesh et al., 2006
Bovines (N)	Sarcocysts		Patra et al., 2006
Mithun (*Bos frontalis*) (N)	Sarcocysts		Swapna Susan Abraham et al., 2009
Buffaloes (N) Beef (N)	Macrocysts		Jayashree Shit et al., 2010

Intermediate Host	Larva (Mode of Infection)	Final Host	Reference
Cattle (N) Sheep (N) Goats (N)	Oocysts Sporocysts (in faeces) Sporozoites on sporulation Macrocysts (in tissues)	Cattle (N) Sheep (N) Goats (N)	Palanivel et al., 2011 (Findings of oocysts in faeces had been questioned by Sahahardar, R. A.2011, Indian Vet J. 88: 117).
Black buck (*Antelopes cervicapra*)	Sarcocysts		Powar et al., 2011
Cattle Buffaloes (N)	Macrocysts Bradyzoites		Banothu Dasmabai et al., 2012b
Buffaloes (N)	Macrocysts		Mithalesh Kumari et al., 2012
Sarcocystis arieticanis			
Sheep (N)	Sarcocysts, cytozoites		Saleque et al., 1992
Sheep (N)	Micro-sporocysts	Dog	Goldy Srivastava and Jain, 2000
Sheep (N)	Sarcocysts	Dogs (Pups)	Goldy Shrivastava and Jain, 2001
Sheep (N)	Sarcocysts		Shahardar and Pandit, 2010
Sarcocystis axicuonis			
Chital (*Axis axis*) Dhole (*Cuon alpines*)	Sporocysts	—	Jog et al., 2003
Spotted deer or chital (*Axis axis*) Indian wild dog or dhole (*Cuon alpines*)	Sporocysts	—	Jog and Watve, 2005
Chital (*Axis axis*) Dhole (*Cuon alpines*)	Sporocysts	—	Jog et al., 2005

Intermediate Host	Larva (Mode of Infection)	Final Host	Reference
	Sarcocystis bovicanis		
Cattle (N)	Sarcocysts		Dafedar, 2005 (M.V.Sc. Thesis)
Cattle (N)	Bradyzoites		Mamatha, 2006 (Ph.D. Thesis)
Cattle (N)	Sarcocysts		Mamatha and D'Souza, 2006a
Cattle (N)	Sarcocysts		Mamatha and D'Souza, 2006b
Cattle (N)	Sarcocysts		Mamatha and D'Souza, 2007
Cattle (N)	Sarcocysts		Dafedar and D'Souza, 2008
Cattle (N)	Macrosarcocysts		Mamatha *et al.*, 2008a
Cattle (N)	Sarcocysts		Mamatha *et al.*, 2008b
Cattle (N)	Sarcocysts		Mamatha *et al.*, 2008c
Cattle (N)	Sarcocysts (cystozoites, microcysts bradyzoites)		Mamatha *et al.*, 2009
Cattle (N)	Microcysts Macrocysts		Dafedar *et al.*, 2011
	Sarcocystis bovifelis		
Cattle (N)	Sarcocysts		Dafedar, 2005 (M.V.Sc. Thesis)
Cattle (N)	Bradyzoites		Mamatha, 2006
Cattle (N)	Sarcocysts		Mamatha and D'Souza, 2006a
Cattle (N)	Sarcocysts		Mamatha and D'Souza, 2006b

Intermediate Host	Larva (Mode of Infection)	Final Host	Reference
Cattle (N)*	Sarcocysts		Dafedar and D'Souza, 2008
Cattle (N)	Macrosarcocysts		Mamatha et al., 2008a
Cattle (N)	Sarcocysts		Mamatha et al., 2008b
Cattle (N)	Sarcocysts		Mamatha et al., 2008c
Cattle (N)	Sarcocysts (cystozoites, macrocysts bradyzoites)		Mamatha et al., 2009
Cattle (N)	Microcysts Macrocysts		Dafedar et al., 2011
		Sarcocystis bovihominis (vide *Sarcocystis hominis*)	
Cattle (N)	Microcysts Macrocysts		Dafedar et al., 2011
		Sarcocystis capracanis Fischer, 1979	
Goats (N)	Sarcocysts		Singh and Singh, 1970
Goats (N)	Sarcocysts		Chhabra and Mahajan, 1978
Goats (N)	Sarcocysts	Dogs (E)	Pethkar, 1978
Goats (N)	Sarcocysts	Dogs (E)	Gupta and Gautam, 1982
Goats (N)	Sarcocysts		Pethkar and Shah, 1980
Goats (N)	Sarcocysts		Pethkar and Shah, 1981
Goats (N)	Sarcocysts		Gupta and Gautam, 1982
Goats (N)	Sarcocysts		Pethkar and Shah, 1982a

Intermediate Host	Larva (Mode of Infection)	Final Host	Reference
Goats (N)	Sarcocysts	Dogs (E)	Pethkar and Shah, 1982b
Goats (N) Lambs (E)- negative	Sarcocysts Sporocysts	Dogs (E)	Pethkar and Shah, 1982c
Goats (N)	Sarcocysts		Pethkar and Shah, 1982d
Goats (N)	Sarcocysts	Pups (E)	Pethkar and Shaw, 1983
Goats (N)	Sarcocysts		Srivastava et al., 1985a
Goats (N)	Sarcocysts	Pups (E)	Shah, et al, 1986
Goats (N)	Sarcocysts		Agrawal et al., 1987
Goats (N)	Sarcocysts		Aulakh et al., 1987
Goats (N) (E)	Sarcocysts, Microcysts Bradyzoites	Dogs (E)	Juyal, 1987 (Ph. D. Thesis)
Goats (N)	Microcysts		Juyal and Bhatia, 1987
Goats (N)	Cystozoites		Juyal et al., 1988a
Goats (N)	Sarcocysts	Dogs (E)	Juyal et al., 1988b
Goats, Black Bengal (E)	Sarcocysts		Kumar et al., 1988
Goats (N)	Microcysts Cystozoites Sporocysts	Pups (E)	Juyal et al., 1989a
			Juyal et al., 1989b

Intermediate Host	Larva (Mode of Infection)	Final Host	Reference
Goats (N)	Microcysts Cryptozoites		Juyal et al., 1989c
Goats (N)	Sporocysts		Juyal et al., 1989e
Goats (N)	Sporocysts		Juyal et al., 1989f
Goats (N)	Cysts		Saleque et al., 1990a
Goats (N)	Sarcocysts		Shastri, 1990
Goats (N)	Sarcocysts		Singh and Shah, 1990 a
Goats (N)	Sarcocysts		Singh and Shah, 1990b
Goats (N)	Sarcocysts		Singh et al., 1990a
Goats (N)	Sarcocysts	Pups (E)	Singh et al., 1990b
Goats (N)	Sarcocysts		Daya Shankar, 1991
Goats (N)	Sarcocysts	Dogs (E)	Lal Singh, 1991
Goats (N)	Sarcocysts		Sharma et al., 1991
Goats (N)	Sarcocysts	Dogs (E)	Lal Singh et al., 1992a
Goats (N)	Sarcocysts	Dogs (E)	Lal Singh et al., 1992b
Goats (N)	Microsarcocysts	Dog (E)	Lal Singh et al., 1993
Goats (N)	Sarcocysts	Pups (E)	Sharma and Shah, 1992
Goats, Black Bengal (N)	Sporocysts, bradyzoites		Saha and Ghosh, 1992
			Singh, 1992

Intermediate Host	Larva (Mode of Infection)	Final Host	Reference
Goats (N)	Sporocysts		Singh et al., 1992
Goats (N)	Metrozoites Cystozoites		Sharma and Chaudhary, 1994
Goats (N)	Sarcocysts Bradyzoites		Wadajkar et al., 1994
Goats (N)	Sarcocysts		Mohanty et al., 1995a
Goats (N)	Sarcocysts		Mohanty et al., 1995b
Goats (N)	Sarcocysts		Mohanty et al., 1995c
Goats (N)	Sarcocysts		Mohanty et al., 1995d
Goats (N) Kids (E)	Sarcocyst Sporocysts	Pup (E)	Wadajkar et al., 1995a
Kids (E)	Sarcocyst	Pup (E)	Wadajkar et al., 1995b
Goats (N)	Sarcocysts	Dogs (E)	Chauhan and Agarwal, 1997
Goats (N)	Sarcocysts	Dogs (E)	Venu and Hafeez Md., 1999
Goats (N)	Sarcocysts		Jumde et al., 2000
Goats (N)	Sarcocysts		Venu and Hafeez Md., 2000
Goats (N)	Microsarcocysts	Dog (E)	Venu et al., 2000
Goats (N)	Microsporocysts		Singh et al., 2004
Goats (N)	Sarcocysts	Dog (E)	Mohanty et al., 2005
Goats (N)	Sarcocysts		Dafedar et al., 2008
Goats (N)	Sarcocysts		Dafedar and D'Souza, 2009

Intermediate Host	Larva (Mode of Infection)	Final Host	Reference
	Sarcocystis cruzi (Syn. *S. bovicanis*)		
Cattle (N)	Sarcocysts	Dogs (E)	Jain, 1982
Cattle (N)	Sarcocysts		Sahai et al., 1982
Cattle (N)	Sarcocysts		Juya et al., 1982
Cattle (N)	Sarcocysts		Srivastava et al., 1982
Cattle (N)	Sarcocysts	Dogs (E) Cat (E)	Sahai et al., 1983
Cattle (N)	Metrocytes and bradyzoites		Jain and Shah, 1985a
Cattle (N)	Sarcocysts		Saha et al., 1986
Cattle (N)	Sarcocysts Sporocysts	Dogs (E)	Jain and Shah, 1986c
Cattle (N) Buffaloes (N)	Metrocytes and bradyzoites	Dog (E)	Jain and Shah, 1987
Cattle (N)	Sarcocysts		Chaudhury and Shah, 1987
Cattle (N) Buffaloes (N) Sheep (N)	Sarcocysts Raineys corpuscles		Gopalakrishna Rao and Rama Rao, 1987
Cattle (N)	Sarcocysts		Jain and Shah, 1988
Monkeys (N)	Sarcocysts		Srivastava, 1988
Cattle (N)	Sarcocysts	Pups (E)	Pandit, 1993

Intermediate Host	Larva (Mode of Infection)	Final Host	Reference
Cattle (N)	Sarcocysts		Pandit et al., 1993
Cattle (N)	Sarcocysts		Pandit and Bhatia, 1994a
Cattle (N)	Sarcocysts	Pups (E)	Pandit and Bhatia, 1994b
Cattle (N)	Microcysts		Pandit et al., 1994a
Cattle (N)	Sarcocysts		Mohanty et al., 1995a
Cattle (N)	Sarcocysts		Mohanty et al., 1995b
Cattle (N)	Sarcocysts		Mohanty et al., 1995c
Deer, barasingha (N)	Sarcocysts		Shrivastava *et al.*, 1999
Cattle (N)	Sarcocysts	Dogs (E)	Venu and Hafeez Md., 1999
Cattle (N)	Sarcocysts		Venu and Hafeez Md., 2000
Cattle (N)	Sarcocysts	Dog (E)	Venu et al., 2000
Cattle (N)	Sarcocysts	Dog (E)	Mohanty et al., 2005
Cattle (N)	Sarcocysts		Shahardar and Pandit, 2010
Bovines (N)	Macrocysts Bradyzoites		Banothu Dasmabai et al., 2012a
Sarcocystis equicanis			
Horse (N)	Sarcocysts		Juyal et al., 1990
Horse (N)	Sarcocysts	Dog (E)	Juyal et al., 1993
Mare (N)	Sarcocysts		Juyal et al., 1994

Intermediate Host	Larva (Mode of Infection)	Final Host	Reference
		Sarcocystis ovicanis	
Sheep (N)	Sarcocysts		Hussain et al., 1986a
Sheep (N)	Sarcocysts		Hussain et al., 1986b
Sheep, lambs (N)	Sarcocysts		Hussain et al., 1986c
Sheep (N)	Sarcocysts	Pups (E)	Hussain et al., 1986d
		Sarcocystis fusiformis **(Railliet and Henry, 1897)**	
Buffaloes (N)	Sarcocysts		Chauhan et al., 1977
Buffaloes (N)	Sarcocysts		Chauhan et al., 1978
Buffaloes (N)	Sarcocysts Sporocysts	Dog (E)	Deshpande and Shastri, 1982
Buffaloes (N)	Sarcocysts		Saha et al., 1982
Buffalo (N) Buffalo calves (E)	Sarcocysts	Cats (E)	Achuthan , 1983
Buffaloes (N)	Sarcocysts	Dogs (E) Cat (E)	Saha et al., 1983
Buffaloes (N)	Sarcocysts		Chaudhry et al., 1985
Buffaloes (E)	Sarcocysts		Saha et al., 1985
Buffaloes (N)	Sarcocysts	Dogs (E) Cat (E)	Srivastava et al., 1985c

Intermediate Host	Larva (Mode of Infection)	Final Host	Reference
Buffaloes (N)	Sarcocysts		Chaudhry et al., 1986
Buffaloes (N)	Microcysts		Ghoshal et al., 1986
Buffaloes (N)	Sarcocysts	Cat (E)	Jain and Shah, 1986a
Buffaloes (N)	Sarcocysts	Cat (E)	Jain and Shah, 1986b
Buffaloes (N)	Sarcocysts		Saha et al., 1986
Buffaloes (N)	Sarocysts	Cats (E)	Ghoshal et al., 1987a
Buffaloes (N)	Sarocysts	Cats (E)	Ghoshal et al., 1987d
Cattle (N) Buffaloes (N)	Sporocyst/oocysts	Cats (E)	Jain and Shah, 1987
Buffaloes (N)	Sarcocysts		Ghosal et al., 1988b
Buffaloes (N)	Macrocysts Microcysts		Juyal et al., 1988a
Buffaloes (N)	Macrocysts Sporocysts/oocysts	Cats (E)	Ghoshal, 1989a
Buffalo calves (E)	Sporocysts (fed) Sarcocysts Metrocysts Cystozoites		Ghoshal, 1989b
Buffaloes (N)	Sarcocysts		Juyal and Bhatia, 1989b
Buffaloes (N)	Sarcocysts		Khulbe et al., 1989a
Buffaloes (N)	Sarcocysts		Khulbe et al., 1989b

Intermediate Host	Larva (Mode of Infection)	Final Host	Reference
Buffaloes (N)	Sarcocysts		Gupta, 1990
Buffaloes (N)	Sarcocysts		Juyal et al., 1990
Buffaloes (N)	Sarcocysts		Khulbe et al., 1990
Buffaloes (N)	Sarcocysts		Saleque et al., 1991
Buffaloes (N)	Sarcocysts, zoites	Dogs (E)	Gupta et al., 1992
Buffaloes (N)	Sarcocysts		Juyal and Bhatia, 1992
Buffaloes (N)	Sarcocysts		Saleque et al., 1992
Buffaloes (N)			Saleque et al., 1992
Buffaloes (N)	Sarcocysts, zoites		Gupta et al., 1993a
Buffaloes (N)	Sarcocysts, zoites		Gupta et al., 1993b
Buffaloes (N)	Sarcocysts, zoites		Gupta et al., 1994
Buffaloes (N)	Marcrocysts	Cats (E)	Deshpande and Pethkar, 1995
Buffaloes (N)	Sarcocysts		Gupta et al., 1995
Buffaloes (N)	Sarcocysts		Mohanty et al., 1995a
Buffaloes (N)	Sarcocysts		Mohanty et al., 1995b
Buffaloes (N)	Sarcocysts		Mohanty et al., 1995c
Buffaloes (N)	Sarcocysts		Mohanty et al., 1995d
Buffaloes (N)	Sarcocysts	Dog (E)	Mohanty et al., 2005
Buffaloes (N)	Sarcocysts		Shahardar and Pandit, 2010
Water buffaloes (N)	Sarcocysts		Rajat Garg, 2012

Intermediate Host	Larva (Mode of Infection)	Final Host	Reference
	Sarcocystis gigantea		
Sheep (N)	Sarcocysts		Shahardar and Pandit, 2010
	***Sarcocystis hircicanis*, Heydorn and Uterholzner, 1983**		
Goats (N)	Sarcocysts	Pups (E)	Shaw et al., 1986
Goats (N)	Sarcocysts		Srivastava et al., 1990
Goats (N)	Sarcocysts		Srivastava et al., 1991
Goats (N)	Sarcocysts	Pups (E)	Sharma and Shah, 1992
Goats (N)	Microsarcocysts	Pups (E)	Dayashankar and Bhatia, 1993a
Goats (N)	Sarcocysts Bradyzoites		Wadajkar et al., 1994
Goats (N)	Sarcocysts		Chauhan and Agrawal, 1997
Goats (N)	Micro-sporocysts		Singh et al., 2004
Goats (N)	Sarcocysts		Dafedar et al., 2008
Goats (N)	Sarcocysts		Shahardar and Pandit, 2010
	***Sarcocystis hirsuta* (Moule, 1887)**		
Cattle (N)	Sarcocysts	Dog (E)	Jain, 1982
Cattle (N)	Sarcocysts		Saha et al., 1986
Cattle (N)	Metrocytes and bradyzoites	Cat (E)	Jain and Shah, 1987

Intermediate Host	Larva (Mode of Infection)	Final Host	Reference
Cattle (N)	Sarcocysts		Jain and Shah, 1988
Cattle (N)	Sarcocysts	Kitten (E)	Pandit, 1993
Cattle (N)	Sarcocysts		Pandit et al., 1993
Cattle (N)	Sarcocysts		Pandit and Bhatia, 1994a
Cattle (N)	Sarcocysts	Kittens (E)	Pandit and Bhatia, 1994b
Cattle (N)	Microcysts Macrocysts		Pandit et al., 1994a
Cattle (N)	Sarcocysts		Pandit et al., 1994b
Cattle (N)	Sarcocysts		Pandit and Bhatia, 1996
Cattle (N)	Sarcocysts	Kittens (E)	Pandit and Bhatia, 2002
Bovines (N)	Macrocysts Bradyzoites of Sarcocystis hirsutia		Banothu Dasmabai et al., 2012a
Sarcocystis hominis (Vide S.bovihominis)			
Cattle (N)	Sarcocysts		Shahardar and Pandit, 2010
Cattle (N)	Sarcocysts		Dafedar, 2005 (M.V.Sc. Thesis)
Cattle (N)	Sarcocysts		Mamatha, 2006 (Ph.D. Thesis)
Cattle (N)	Sarcocysts		Jain, 1992
Cattle (N)	Sarcocysts		Jain and Shah, 1986b

Intermediate Host	Larva (Mode of Infection)	Final Host	Reference
Cattle (N)	Metrocytes and bradyzoites		Jain and Shah, 1987a
Cattle (N)	Sarcocysts		Jain and Shah, 1988
Cattle (N)	Sarcocysts		Pandit, 1993
Pigs (N)	Sarcocysts		Saleque and Bhatia, 1991
Cattle (N)	Sarcocysts		Pandit and Bhatia, 1994
Cattle (N)	Microcysts Macrocysts		Pandit et al., 1994b
Cattle (N)	Sarcocysts		Singh et al., 2003
Cattle (N)	Sarcocysts		Dafedar and D'Souza, 2008
Bovines (N)	Macrocysts Bradyzoites of *Sarcocystis hominis* (?)		Banothu Dasmabai et al., 2012a
Sarcocystis levinei Dissanaike and Kan, 1978			
Buffaloes (N)	Microcysts Sporocysts/Oocysts	Dogs (E)	Ghoshal, 1978a
Buffaloes (N)	Microcysts	Dogs (E)	Ghoshal, 1978b
Buffaloes (N)	Microcysts	Pups (E)	Ghoshal and Shaw, 1978
Buffaloes (N)	Sarcocysts		Ghoshal and Shaw, 1979
Buffaloes (N)		Pups (E)	Deshpande and Shastri, 1982
Buffaloes (N)			Deshpande et al., 1982

Intermediate Host	Larva (Mode of Infection)	Final Host	Reference
Buffaloes (N)	Sarcocysts		Juyal et al., 1982
Buffaloes (N)		Pups (E)	Deshpande et al., 1983
Buffaloes (N)	Sarcocysts		Jain, 1985 (Thesis)
Buffaloes (N)	Microcysts	Pups (E)	Jain et al., 1985
Buffaloes (N)	Microcysts		Ghoshal et al., 1986
Buffaloes (N)	Sporocysts	Dogs	Jain and Shah, 1986c
Buffalo calves, male (E)	Sporocysts	Pups	Jain et al., 1986a
Buffalo calves, male (E)	Sporocysts		Jain et al., 1986b
Buffaloes (N)	Sarcocysts		Saha et al., 1986
Buffaloes (N)	Sarcocysts	Dogs (E)	Srivastava et al., 1986
Buffaloes (N)	Sarcocysts	Dogs (E)	Srivastava et al., 1987
Buffaloes (N)	Sarcocysts	Dogs (E)	Ghoshal, et al., 1987b
Buffaloes (N)	Sarcocysts		Ghoshal et al., 1987c
Buffaloes (N)	Metrocytes and bradyzoites	Cat (E)	Jain and Shah, 1987b
Buffaloes (N)	Sarcocysts	Dogs (E)	Ghoshal, et al., 1988a
Buffaloes (N)	Sarcocysts		Jain, 1988
Buffaloes (N)	Macrocysts Microcysts		Juyal et al., 1988a
Buffaloes (N)	Microcysts Sporocysts	Dogs (E)	Ghoshal, 1989a

Intermediate Host	Larva (Mode of Infection)	Final Host	Reference
Buffalo calves (E)	Sporocysts (fed), Sarcocysts, Metrocysts, Cystozoites		Ghoshal, 1989b
Buffaloes (N)	Sarcocysts	Dogs (E)	Venu and Hafeez Md., 1999
Buffaloes (N)	Sarcocysts		Venu and Hafeez Md., 2000
Buffalo (N)	Sarcocysts	Dog (E)	Venu et al., 2000
Buffalo (N)	Microcysts		Rohini and Hafeez Md., 2005
Buffaloes (N)	Sarcocysts		Shahardar and Pandit, 2010
Sarcocystis miescheriana (Khun, 1865) Labbe, 1889			
Pigs (N)	Sarcocysts		Sahai et al., 1982
Pigs (N)	Sarcocysts-like zoites, Metrocytes (Immature), Cytozoites, Sporocyst	Dog (E)	Gupta and Gautam, 1984
Pigs (N)	Microcysts		Agnihotri et al., 1987
Pig (N)	Sarcocysts		Solanki et al., 1988
Pig (N)	Sarcocysts		Solanki, 1989
Pigs (N)	Sarcocysts	Pup (E)	Saleque et al., 1990b
Pigs (N)	Sarcocysts	Dog (E)	Solanki et al., 1990

Intermediate Host	Larva (Mode of Infection)	Final Host	Reference
Pigs (N)	Sarcocysts Microsarcocysts		Salecue and Bhatia, 1991
Pigs (N)	Sarcocysts Macrogamont Microgamont Metrocytes Cytozoites	Dog (E)	Solanki et al., 1991a
Pigs (E)	Endogenous stages Sporocysts	Dogs (E)	Solanki et al., 1992
Pigs (N) (E)	Sarcocysts	Dogs (E)	Khatsar et al., 1993
Piglets (E)	Sarcocysts Sporocysts	Pups (E)	Dey et al., 1995
Pigs (N)	Sarcocysts		Hemaprasanth, 1995a
Pigs (N)	Sarcocysts		Hemaprasanth, 1995b
Pigs (N) (E)	Sarcocysts	Pups (E)	Hemaprasanth and Bhatia, 1995a
Pigs (N) (E)	Sarcocysts	Pups (E)	Hemaprasanth and Bhatia, 1995b
Pigs (N)	Sarcocysts/sporulated oocysts	Pups (E)	Devi et al., 1998
Pigs (N)	Sarcocysts		Avapal et al., 2002
Pigs (N)	Sarcocysts		Avapal et al., 2003a
Pigs (N)	Sarcocysts		Avapal et al., 2003b

Intermediate Host	Larva (Mode of Infection)	Final Host	Reference
Sarcocystis muris			
Swiss albino mice (N)	Sarcocysts		Nikita Sharma et al., 2012
Sarcocystis ovicanis (Syn. S. tenella)			
Sheep	Sarcocysts		Hussain et al., 1986a
Sheep	Sarcocysts		Hussain et al., 1986b
Lambs (E)	Sarcocysts		Hussain et al., 1986c
Sheep (N)	Microcysts	Pups (E)	Hussain et al., 1987
Sheep (N)	Sarcocysts		Mohanty et al., 1995a
Sheep (N)	Sarcocysts		Mohanty et al., 1995b
Sheep (N)	Sarcocysts		Mohanty et al., 1995c
Sheep (N)	Sarcocysts		Mohanty et al., 1995d
Sheep (N)	Sarcocysts	Dogs (E)	Venu and Hafeez Md., 1999
Sheep (N)	Sarcocysts	Dog (E)	Mohanty et al., 2005
Sarcocystis porcifelis			
Pig (N)	Sarcocysts		Hemaprasanth and Bhatia, 1995
Sarcocystis sinensis			
Water buffaloes (N)	Sarcocysts		Rajat Garg, 2012

Sarcocystis suihominis (Tadros and Laarman, 1976; Heydorn, 1977)

Intermediate Host	Larva (Mode of Infection)	Final Host	Reference
Pigs (N), claiming as first record from India	Sarcocysts-like zoites Metrocytes (Immature) Cytozoites Sporocyst	Dog (E)	Gupta and Gautam, 1984
Pigs (N)	Sarcocysts Microsarcocysts		Saleque and Bhatia, 1991
Pigs (N) (E)	Sarcocysts Macrogamont Microgamont Metrocytes Cytozoites		Solanki et al., 1991
Pigs (N)	Sarcocysts		Hemaprasanth, 1995a
Pigs (N)	Sarcocysts		Hemaprasanth, 1995b
Pigs (N) (E)	Sarcocysts		Hemaprasanth and Bhatia, 1995a
Pigs (N) (E)	Sarcocysts	Pups (E)	Hemaprasanth and Bhatia, 1995b
Pigs (N)	Sarcocysts		Devi et al., 1998
Pigs (N)	Sarcocysts		Avapal , 2001
Pigs (N)	Sarcocysts		Avapal et al., 2003a
Pigs (N)	Sarcocysts		Avapal et al., 2003b

Intermediate Host	Larva (Mode of Infection)	Final Host	Reference
Pigs (N)	Sarcocysts		Avapal et al., 2003c
Pigs (N)	Sarcocysts		Singh et al., 2003
		Sarcocystis tenella (Railliet, 1886)	
Sheep (N)	Sarcocysts, cytozoites		Saleque et al., 1992
Lambs (E)	Sarcocysts		Chowdhury, 1999
Sheep (N)	Sarcocysts		Venu and Hafeez Md., 2000
Sheep (N)	Sarcocysts	Dog (E)	Venu et al., 2000
Sheep (N)	Sarcocysts	Dogs (Pups)	Goldy Shrivastava and Jain, 2001
		Toxoplasma cuniculi	
Rabbit (N)			Krishnan and Lal, 1933
	Toxoplasma gondii (Nicolle and Manceaux, 1908) Nicolle and Manceaux, 1909		
Domestic fowl, eggs (N)			Pande et al., 1961
Camels (N)			Gill and Prakash, 1969
Rhipicephalus sanguineus (E) Rabbit (E)		Rabbits (E)	Gill, 1970, cited by Gill et al., 1971
Goats (N)			Gill and Prakash, 1970a
Sheep (N)			Gill and Prakash, 1970b
Pigs (N)			Gill and Prakash, 1971a

Intermediate Host	Larva (Mode of Infection)	Final Host	Reference
Cattle (N)			Gill and Prakash, 1971b
Buffaloes (N)			Gill, 1972
Sheep (N) Goats (N)			Chhabra and Gautam, 1972
Sheep (N) Goats (N)			Sharma and Gautam, 1972
Goats (N)			Bharadwaj, 1974
Camels (N) Pigs (N)			Sharma and Gautam, 1974
Rhesus monkey (N)	Antibodies		Chhabra et al., 1976
Human patients (N)	Antibodies	Mice (E)	Mahajan et al., 1977
Earthworms (*Pheretima posthuma*) (N), Indirect evidence for transport hosts?	Pooled digestive tract suspension	Mice (E) Rise of post-inoculative titre	Chhabra et al., 1978
Buffaloes (N)			Chhabra and Mahajan, 1978a
Pig (N)	Serological		Chhabra and Mahajan, 1978b
Human patients (N)		Mice (E)	Chhabra et al., 1978
Pig (N)			Chhabra and Mahajan, 1979a
Zebu cattle (N)			Chhabra and Mahajan, 1979b
Dogs (N)	Antibodies		Chhabra and Mahajan, 1979c

Intermediate Host	Larva (Mode of Infection)	Final Host	Reference
Human patients (N)	'Por' (porcine) strain	Swiss albino mice (E)	Chhabra and Mahajan, 1979d
Pigs (N)	Tachyzoites from mice		
Pigs (N)			Chhabra *et al.*, 1979a
Human patients (N)		Mice (E)	Chhabra *et al.*, 1979b
Human patients (N)		Mice (E)	Chhabra *et al.*, 1979c
Sheep (N)			Gupta *et al.*, 1979
Horses (N)			Bhandari *et al.*, 1980
Mules (N)			
Horse (N)			Chhabra and Gautam, 1980
Donkeys (N)			
Mice, Swiss albino, virulent	Tachyzoites		Chhabra *et al.*, 1980
strain (E)			
Lambs			Gupta *et al.*, 1980a
Goats (N)			Chhabra *et al.*, 1981
Sheep (N)			Gupta *et al.*, 1981a
Mice (E)			Gupta *et al.*, 1981b
Sheep (N)			Chhabra and Mahajan, 1982
Goats (N)			
Animals (N)	Antibodies, ILA & IHA tests		Chhabra *et al.*, 1982a
Goats, pregnant (E)			Chhabra *et al.*, 1982b

Intermediate Host	Larva (Mode of Infection)	Final Host	Reference
Chicks (E)			Gautam, et al., 1982a
Buffalo calves, male (E)	Oocysts of a virulent strain, HA antibodies		Gautam, et al., 1982b
			Korwar and Bhoop Singh, 1982
Goats (N)	Serological		
Sheep (N) Goats (N)	Serological		Korwar et al., 1982
Wister rats	Oocysts	Kitten (N) (E)	Shastri and Mandakhalikar, 1982
Wild rat (Rattus rattus) (N) Swiss albino mice (E)	Serological Tachyzoites		Mir et al., 1982
Sheep (N) Goats (N)			Ali et al., 1983
Sheep (N) Goats (N)			Chhabra et al., 1983
Sheep (N)			Pillai and Khader, 1983
Sheep (N) Goats (N)			Srivastava et al., 1983a
Mice (E)	RH strain		Srivastava et al., 1983b
Sheep (N)			Srivastava et al., 1984
Mice (E)	Trophozoites		Wagle et al., 1984

Intermediate Host	Larva (Mode of Infection)	Final Host	Reference
Sheep (N) Goats (N) Cattle (N) Buffaloes (N) Horses (N) Pigs (N) Dogs (N) Cat (N) Bandicoot rats (N)	Serological	Dogs (N) Cats (N)	Chhabra *et al.*, 1985
	Serological	Human beings (livestock and pet owners Lab. animal handlers)	Gupta *et al.*, 1985
Sheep (N)	Serological		Bhoop Singh and Msolla, 1986
Cattle (N) Sheep (N) Goats (N) Pigs (N)			Nene *et al.*, 1986
Sheep (E)	Tachyzoites		Verma *et al.*, 1988a
Sheep (E)	Tachyzoites		Verma *et al.*, 1988b
Ewes, aborted (N)	Antibodies		Verma *et al.*, 1988c
Mice (E)			Verma *et al.*, 1988d
Mice (E)	Tachyzoites		Goyal *et al.*, 1988
Mice (E)	Tachyzoites		Goyal *et al.*, 1989

Intermediate Host	Larva (Mode of Infection)	Final Host	Reference
Ewe (N)			Verma *et al.*, 1989
Rabbit (E) Guinea pig (E)	Oocysts	Kitten (*Felis catus*) (N)	Shastri, 1990
		Dog (E)	Mandhakhalikar, 1992
Wistar rats (E)	Sarcocysts Oocysts from kitten	Kitten (N)	Shastri and Ratnaparkhi, 1992
Wister rats (E) Albino mice (E)	Sarcocysts Oocysts from kitten	Kitten (N)	Shastri and Mandhakhalikar, 1992
Stray cats (N)	Sporulated oocysts Oral inoculation of sporulated oocysts in experimental kittens failed, but their tissues passed oocysts on feeding to two other kittens	Kitten (E)	Shastri and Ratnaparkhi, 1992
Goat (N) Sheep (N)			Dubey *et al.*, 1993
Poultry (N)			Gaikwad, 1993
Sheep (N) Goats (N)			Shastri *et al.*, 1993
Albino mice (E)	Oocysts Tachyzoites Bradyzoites	Pups (E)	Mandhakhalikar *et al.*, 1994a
Rabbit (E)	Oocysts		Mandhakhalikar *et al.*, 1994b

Intermediate Host	Larva (Mode of Infection)	Final Host	Reference
Albino mice (E), Rats (E)	Oocysts	Kitten (N)	Mandhakhalikar et al., 1994c
Sheep (N), Goats (N)	Antibodies		Jithendran and Vaid, 1996
Cow (N), Goat (N), Sheep (N)	Antibody		Mirdha and Samantray, 1996
Domestic rats (N), Wild Rats (N), Pigeon (N), Mice (E), Chicken (E)	Oocysts	Cat (N)	Devada, 1996, 1998
Domestic rats (N), Wild rats (N), Pigeons (N)	Tissue cysts		Devada and Anandan, 1998
Chicken	Seropositive		Devada et al., 1998
Sheep (N), Goats (N)	Seropositive		Mirdha et al., 1999
Laboratory animals (N)			Somvanshi and Singh, 1999
Goats (N)	Antibodies		Syamala and Devada, 1999
Sheep (N)			Vijaykumar, 1999
Chicken (N), Chicken (E)	Tachyzoites		Devada and Anandan, 2000a

Intermediate Host	Larva (Mode of Infection)	Final Host	Reference
Chicken (E) Chicken (N)	Tachyzoites Oocysts		Devada and Anandan, 2000b
Chicken, country (N) Swiss mice (E)	Oocysts Positive to IFAT Tachyzoites	Cat, stray (N) (E)	Sreekumar et al., 2001a
Chicken, free range (N) Swiss mice (E)	Seropositive by IFAT, fed to cats Oocysts, Tachyzoites, Bradyzoites	Cat (E)	Sreekumar et al., 2001b
Chicken, free range (N)	Antibodies by MAT	Cats (E)	Sreekumar et al., 2003
Goats (N)	Antibodies		Vijaya Bharathi et al., 2003
Mice (E) Sheep (N) Pig (N)	Tachyzoites		Thimma Reddy, 2003
Goat, Barbari (N)	By PCR		Sreekumar et al., 2004
Mice (E) Sheep (N) Pig (N)	Tachyzoites		Thimma Reddy and Hafeez Md, 2004
Goats (E)	Sporulated oocysts	Cat (N)	Udayakumar, 2004
Human (N) Goats (N) Cattle (N) Buffaloes (N) Pig (N)	Serological (Modified Agg. Test)		Malik et al., 2005

Intermediate Host	Larva (Mode of Infection)	Final Host	Reference
Goats (N)	Antibodies (Modified Agg. Test)	Man (N)	Syamala et al., 2005
Zoo animals Feline group: Lion (N) Tiger (N) Leopard (N) Reptile group: Snake (N) Python (N) Tortoise (N)	Serological	Persons working with zoo (N)	Jani et al., 2006
Chicks	Sporulated oocysts		Raote et al., 2007
She-buffaloes (N)			Selvaraj et al., 2007
Goats (N)	Antibodies (Carbon immune assay)		Syamala et al., 2007
Sheep (N) Cattle (N) Buffaloes (N)	Antibodies		Sharma et al., 2008
Goats (N)	Antibodies (Latex Aggl. Test)		Syamala et al., 2008
Goats (N) Rabbits (N) Human (N)	Antibodies		Velumurugan et al., 2008
Swiss Albino mice (E)	Tachyzoites of RH strain		Hira Ram, et al., 2009

Intermediate Host	Larva (Mode of Infection)	Final Host	Reference
Swiss albino mice (E)			Kumar et al., 2009
Goats (N)	Tachyzoites	—	Udaya Kumar et al., 2009
	Sporulated oocysts		
		Cat (N)	
Mice (E)			Udaya Kumar et al., 2010a
			Udaya Kumar et al., 2010b
Swiss albino mice (E)	Tachyzoites		Dhananjay Kumar, et al., 2010
Swiss albino mice (E)			Kumar et al., 2010
Cattle (N)	Tachyzoites based	—	Tewari et al., 2010a
Cattle (N)	Tachyzoites based	—	Tewari et al., 2010b
Swiss albino mice (E)	Tachyzoites (cryopreserved)		Kumar et al., 2011
Swiss albino mice (E)	Tachyzoites (cryopreseved)		Pravin et al., 2011
Domestic animals (N)	Tachyzoites (cryopreseved)		Singh et al., 2011
Mice (E)	Tachyzoites (RH strain)		Sudan et al., 2011a
Mice (E)	Tachyzoites (RH strain)		Sudan et al., 2011b
Goats (E) (N)	Antibodies		Udayakumar et al., 2011
Swiss albino Mice (E)	Tachyzoites		Pravin et al., 2012
Mice (E)	Tachyzoites		Sucilathangam et al., 2012
Mice (E)	Tachyzoites		Sudan et al., 2012
Mice (E)	Tachyzoites		Tiwari et al., 2012
Domestic animals (N)	Tachyzoites (cryopreseved)		Singh et al., 2013

7.2 FAMILY: HAEMOGREGARINIDAE/HEPATOZOOIDAE

Hepatozoon canis (James, 1906)
(Syn: *Leucocytozoon canis; Haemogregarina canis*)

Intermediate Host	Larva (Mode of Infection)	Final Host	Reference
Rhipicephalus sanguineus (E)			Christophers, 1906
Rhipicephalus sanguineus (E)			Christophers, 1907
Rhipicephalus sanguineus (E)			Christophers, 1912
Rhipicephalus sanguineus (E)	Round to oval bodies, irregularly round forms (gamonts) Oval forms that tapered at one end and became bigger (sporocysts)	Dog (E)	Harikrishnan et al., 2001
Rhipicephalus sanguineus (N)	Ground up tick supernates (GUTS) Schizonts Developmental stages	*In vitro* cell line (E)	Harikrishnan and Ponnudurai, 2006
Rhipicephalus sanguineus (N)	Mature oocysts Sporulated oocysts Sporocysts Sporozoites	Dog (E)	Harikrishnan et al., 2008a
Rhipicephalus sanguineus (N)	Gamonts Developing forms Meronts	Dog, crossbred (E), oral feeding of macerated ticks	Harikrishnan et al., 2008b

7.3 FAMILY: PLASMODIIDAE

Plasmodium gallinaceum Brumpt, 1935

Intermediate Host	Larva (Mode of Infection)	Final Host	Reference
Aedes aegypti (E)		-	Shortt et al., 1940
Aedes aegypti (E)		Chicken (E) Guinea pig (E) Rabbit (E)	Ray et al., 1956
Aedes aegypti (E)		Chicken (E) Guinea pig (E)	Ghosh and Ray, 1957
Haemoproteus sp.			
New vector			Ponnudurai et al., 2010

Haemoproteus columbae Celli and Sanfelice, 1831

Intermediate Host	Larva (Mode of Infection)	Final Host	Reference
Lynchia sp. (N) *Lynchia* sp. (E)	Ookinetes Zygotes Oocysts Sporozoites Zygotes sporozoites	Pigeon (N) Pigeon (E)	Adie, 1915

7.4 FAMILY: BABESIIDAE

Babesia bigemina (Smith and Kilborne, 1893)

Intermediate Host	Larva (Mode of Infection)	Final Host	Reference
Boophilus microplus (E) *Hyalomma anatolicum anatolicum* (E)		Calf (E)	Chaudhuri et al., 1975

Intermediate Host	Larva (Mode of Infection)	Final Host	Reference
Boophilus microplus (N) (E) Boophilus annulatus (E) Haemaphysalis bispinosa (N) (E)	Sporozoites		Achutan et al., 1980
Boophilus annulatus (N)		Calf (E)	Rajamohanan, 1982
Boophilus microplus (E) larvae engorged to adults		Cattle (N) Calves (E)	Ravindran et al., 2006
Boophilus microplus (E) larvae engorged to adults		Cattle (N) Calves (E)	Ravindran et al., 2008
Babesia canis (Paina and Galli-Vallerio, 1895)			
Rhipicephalus sanguineus			Christophers, 1907
Rhipicephalus sanguineus (E) Rhipicephalus turanicus (E)	Sporozoites	Dogs (E)	Achutan et al., 1980
Babesia equi (Laveran, 1901) (Transferred this genus to Theileria equi by Mehlhorn and Schein, 1998)			
Hyalomma sp. (E)		Donkey (E)	Malhotra et al., 1979
Hylomma anatolicum anatolicum (E)		Donkey (E)	Sanjeev Kumar et al., 2007
Hyalomma anatolicum anatolicum		Donkey (E)	Sanjeev Phogat, 1999
Babesia gibsoni (Pattoni, 1910) **(Piroplasma gibsoni)**			
Rhipicephalus sanguineus (E)		Dog (E)	Sen, 1933
Haemaphysalis bispinosa (N) (E)		Dog (E)	Rao, 1926
Haemaphysalis bispinosa (E)		Dog (E)	Swaminathan and Shortt , 1937

7.5 FAMILY: THEILERIIDAE

Theileria sp. in buffaloes

Intermediate Host	Larva (Mode of Infection)	Final Host	Reference
Haemaphysalis bispinosa (E)		Calves, buffalo (E)	Shastri *et al.*, 1985

Theileria annulata (Dschunkowsky and Luhs, 1904) Wenyon, 1926

Intermediate Host	Larva (Mode of Infection)	Final Host	Reference
Ixodid ticks: *Hyalomma aegyptium* (*H. savignyii*) (E)		Cattle (E)	Ray, 1950
Hyalomma anatolicum anatolicum (E)		Calves (E), Crossbred of Haryana x Holstein, Jersey, Red Dane and Brown Swiss	Anon, 1970-73
Hyalomma anatolicum anatolicum (E)		Calves (E)	Bhattacharyulu *et al.*, 1972
Hyalomma detritum (E)		Cattle (E)	Gill *et al.*, 1974a
Hyalomma dromedarii (E)			Gill *et al.*, 1974b
Hyalomma anatolicum anatolicum (E)	Intra-erythrocytic stages, pre-infective stages, infective stages	Calves (E)	Bhattacharyulu *et al.*, 1975a
Hyalomma anatolicum anatolicum (E) *Hyalomma dromedarii* (E)		Calves (E)	Bhattacharyulu *et al.*, 1975b
Ticks		Calves, buffalo (E)	Das and Sharma, 1975

Intermediate Host	Larva (Mode of Infection)	Final Host	Reference
Hyalomma anatolicum anatolicum (E), *Hyalomma dromedarii* (E), *Hyalomma marginatum isaaci* (E), *Hyalomma detritum* (N)			Gill and Bhattacharyulu, 1976
Hyalomma anatolicum anatolicum (E), *Hyalomma dromedarii* (E)	Tick induced infection	Calves (E)	Gill *et al.*, 1976a
Hyalomma anatolicum anatolicum (E), *Hyalomma dromedarii* (E)		Calves (E)	Gill *et al.*, 1976b
Hyalomma anatolicum anatolicum (E)		Calves (E)	Srivastava and Sharma, 1976a
Hyalomma anatolicum anatolicum (E)		Calves (E)	Srivastava and Sharma, 1976b
Hyalomma anatolicum anatolicum (E)		Calves (E)	Srivastava and Sharma, 1976c
Hyalomma anatolicum anatolicum (E)		Calves (E)	Srivastava and Sharma, 1976d
Hyalomma anatolicum anatolicum (E)		Calves (E)	Srivastava and Sharma, 1976e
Hyalomma anatolicum anatolicum (E), adult ticks		Calves (E)	Dhar and Gautam, 1977a
Hyalomma anatolicum anatolicum (E), adult ticks		Calves (E)	Dhar and Gautam, 1977b
Hyalomma anatolicum anatolicum (E)	Tick bites	Calves (E)	Gill *et al.*, 1977a
Hyalomma dromedarii (E), adult ticks	Tick tissue stabilates	Calves (E), HF x Sahiwal or Tharparkar, male cow calves	Gill *et al.*, 1977b
Hyalomma anatolicum anatolicum (E)	Tick bites	Calves (E)	Gill *et al.*, 1977c

Intermediate Host	Larva (Mode of Infection)	Final Host	Reference
Hyalomma anatolicum anatolicum (E)		Calves (E)	Sharma et al., 1977a
Hyalomma anatolicum anatolicum (E)		Calves (E)	Sharma et al., 1977b
Hyalomma anatolicum anatolicum (E)		Calves (E)	Sharma et al., 1977c
Hyalomma anatolicum anatolicum (E)	Infective stages		Singh, 1977
Hyalomma anatolicum anatolicum (E)		Calves (E)	Sharma, 1978
Hyalomma anatolicum anatolicum (E)		Calves (E)	Dhar and Gautam, 1978
Hyalomma anatolicum anatolicum (E)		Calves, crossbred (HF x Zebu) (E)	Gill et al., 1978
Hyalomma anatolicum (E)			Srivastava and Sharma, 1978
Hyalomma anatolicum anatolicum (E), adults	Through infected ticks	Calves (E)	Dhar and Gautam, 1979a
Hyalomma anatolicum anatolicum (E)		Calves (E)	Dhar and Gautam, 1979b
Hyalomma anatolicum anatolicum (E)	Infective stages	Calves (E)	Jagadish, 1977
Hyalomma anatolicum anatolicum (E)	Ground up tick supernate (GUTS)	Calves, CB (E)	Jagadish, 1979
Hyalomma anatolicum anatolicum (E)	Ground up tick supernate (GUTS)	Calves (E)	Singh et al., 1979a
Hyalomma anatolicum anatolicum (E), nymphs, adults	Ground up tick supernate (GUTS)	Calves (E)	Singh et al., 1979b
Hyalomma anatolicum anatolicum (E)	Ground up tick supernate (GUTS)	Calves (E)	Singh et al., 1979c

Intermediate Host	Larva (Mode of Infection)	Final Host	Reference
Ticks	Stabilates Ludhiana and Hissar strains Uruli-Kanchan, Bangalore and Jaipur strains	Calves, immunized/untreated	Gill et al., 1980
Hyalomma anatolicum anatolicum (E)	Ground up tick supernate	Calves (E)	Jagadish et al., 1980a
Hyalomma anatolicum anatolicum (E)	Developing stages	Calves, male, (Haryana x Jersey) (E)	Jagadish et al., 1980b
Hyalomma anatolicum anatolicum (E)	Ground up tick supernate (GUTS)	Calves (E)	Khanna et al., 1980
Hyalomma anatolicum anatolicum (E)	Infective stages	Calves (E)	Naithani and Subramanian, 1980
Hyalomma anatolicum anatolicum (E)	Ground up tick supernate	Crossbred male (Bos taurus x Bos indicus)	Sastry et al., 1980
Hyalomma anatolicum anatolicum (E)	Tick stabilates, Sporozoites Irradiated sporozoites	Calves (E)	Samantary et al., 1980
Hyalomma anatolicum anatolicum (E)	Ground up tick supernate (GUTS)	Calves (E)	Singh et al., 1980
	Ground up tick supernate (GUTS)	Calves, crossbred (E)	Gill et al., 1981
Hyalomma anatolicum anatolicum (E)	—	—	Dhar, et al., 1982
Hyalomma anatolicum anatolicum (E)	Tick supernate	Calves (E)	Gautam et al., 1982

Intermediate Host	Larva (Mode of Infection)	Final Host	Reference
Hyalomma anatolicum anatolicum (E), infective nymphs and adults (from HAU, Hissar)		Calves (E), crossbred (stray organisms & schizonts in lymph glands, failed to induce clinical infection)	Ebenezer Raja et al., 1983
Hyalomma anatolicum anatolicum (E)	Ground up tick supernate (GUTS)	Calves, mixed breed (E)	Khanna et al., 1983a
Hyalomma anatolicum anatolicum (E)	Ground up tick supernate (GUTS)	Calves, mixed breed (E)	Khanna et al., 1983b
Hyalomma anatolicum anatolicum (E)	Attachment	Calves cross bred Kankrej x Jersey	Lal and Soni, 1983
Hyalomma anatolicum anatolicum (E)	Ground up tick supernate (GUTS)	Calves (E)	Manickam et al., 1983
Hyalomma anatolicum anatolicum (E)			Patel and Avsathi, 1983
	GUTS	Calves (E)	Bansal and Sharma, 1986
Hyalomma anatolicum anatolicum (E)	GUTS	Calves (E), crossbred	Dhar et al., 1986
Hyalomma anatolicum anatolicum (E)	Acini positive of Th. masses-suspension of salivary gland	Cattle (N) Buffaloes (N) Calves, cross-bred (E)	Sangwan et al., 1986
Ticks	Four isolates	Calves, Native cross-bred (E)	Subramanian et al., 1986
Ticks		Calves (E)	Yadav and Sharma, 1986

Intermediate Host	Larva (Mode of Infection)	Final Host	Reference
Hyalomma anatolicum anatolicum (E)	Sporoblast stage	Calves (E)	Dhar et al., 1987
Hyalomma anatolicum anatolicum (E)	Ground up tick supernate (GUTS)	*Bos taurus x Bos indicus,* CB (E)	Mallick et al., 1987
Hyalomma anatolicum anatolicum (E)	Sporozoites	Calves, cross-bred (E)	Ray et al., 1987
Hyalomma anatolicum anatolicum (E)	Sporozoites	Calves, cross-bred (E)	Sharma et al., 1987
Hyalomma anatolicum anatolicum (E)		Haryana cows & H. F.bulls (N)	Subramanian et al., 1987
Hyalomma anatolicum anatolicum (E)	Fed on blood containing Madras Hisar strains	Calves, Cross-bred (E)	Anandan et al., 1988
Ticks	Ground-up ticks	Calves, cross-bred (E)	Bhattacharyulu and Singh, 1976
Hyalomma anatolicum anatolicum (N) (E)		Calves (E)	Datta et al., 1988
Hyalomma anatolicum anatolicum (E)	Groundup tick supernate (GUTS)	Calves, cross-bred (E)	Dhar et al., 1988a
Hyalomma anatolicum anatolicum (E)	Groundup tick supernate (GUTS), Hisar isolate	Calves, cross-bred, *Bos taurus x Bos indicus* (E)	Dhar et al., 1988b
Hyalomma spp.			Khurana et al., 1988
		Calves, cross-bred, male (E)	Mehta et al., 1988
Ticks , nymph & adults	Groundup tick tissue	Calves, cross-bred (E)	Sandhu et al., 1988
Hyalomma anatolicum anatolicum (E)	Sporozoites	—	Sangwan et al., 1988

Intermediate Host	Larva (Mode of Infection)	Final Host	Reference
Hyalomma anatolicum anatolicum (E)	GUTS Cell culture infection	Cross-bred calves, male (E)	Shukla and Sharma, 1988
	Groundup tick tissue tissue suspension (GUTTS)	Calves, cross-bred, male (E)	Bansal and Sharma, 1989
Ticks		Calves, cross-bred (E)	Rao *et al*, 1989
Hyalomma anatolicum anatolicum (E) *Hyalomma dromedarii* (E)		Calves, cross-bred (E)	Sangwan *et al*, 1989
Hyalomma anatolicum anatolicum (E) (N)		Cross-bred calves (E)	Bhattacharyulu *et al*, 1990
Report	Sporozoites	Calves, cross-bred	Dhar *et al*., 1990
Hyalomma anatolicum anatolicum (E)	GUTS	Cross-bred calves (E)	Sharma and Mishra, 1990
Hyalomma anatolicum anatolicum (N) *Hyalomma anatolicum anatolicum* (N) *Hyalomma dromedarii* (N) *Hyalomma marginatum isaaci* (N) *Hyalomma detritum* (N)		Cattle (N) Buffaloes (N)	Singh, 1990
Hyalomma anatolicum anatolicum (E)		Calves, cross-bred, male	Ashok Kumar *et al*., 1991
	GUTS (cryopreseved)	Cross bred calves, *Bos taurus x Bos indicus* (N)	Chaudhri and Subramanian, 1991a
	GUTS (cryopreseved)	Calves cross bred, *Bos taurus x Bos indicus* (N)	Chaudhri and Subramanian, 1991b

Intermediate Host	Larva (Mode of Infection)	Final Host	Reference
Hyalomma anatolicum anatolicum (N)		Cattle, local (N)	Das and Sharma, 1991a
Hyalomma anatolicum anatolicum (N)		Cattle (N)	Das and Sharma, 1991b
Hyalomma anatolicum anatolicum (E)		Calves, cross-bred, male	Momin *et al.*, 1991
Hyalomma anatolicum anatolicum (E), nymph, adults	GUTS, Cell culture	Calves, cross-bred, male	Shukla and Sharma, 1991
Hyalomma anatolicum anatolicum (E)			Singh *et al.*, 1991
Hyalomma anatolicum anatolicum (E)	Macroschizonts Sporozoites	Calves, cross-bred, *Bos taurus* x *Bos indicus* (N)	Chaudhri and Subramanian, 1992a
Hyalomma anatolicum anatolicum (E)	Macroschizonts infected lymphoblasts Sporozoites	Calves cross bred, *Bos taurus* x *Bos indicus* (N)	Chaudhri and Subramanian, 1992b
Hyalomma anatolicum anatolicum (E)	Sporozoites Cytomeres		Das and Sharma, 1992
Hyalomma anatolicum anatolicum (E)	GUTS	Calves, cross-bred (E)	Dhar *et al.*, 1993
Hyalomma anatolicum anatolicum (E)	Sporozoites		Das and Sharma, 1993a
Hyalomma anatolicum anatolicum (E)	Sporozoites	Calves (E)	Das and Sharma, 1993b
Hyalomma anatolicum anatolicum (E)	Macroschizonts	Calves, cross-bred (E) (N)	Ray and Bansal, 1993
Hyalomma anatolicum anatolicum (E)	GUTS, ODE Anand Strain	Calves (E)	Singh *et al.*, 1993
Hyalomma anatolicum anatolicum (E) *Hyalomma marginatum isaaci* (E)		Calves, Jersey cross-bred, bull (E)	Sundar *et al.*, 1993

Intermediate Host	Larva (Mode of Infection)	Final Host	Reference
Hyalomma anatolicum anatolicum (E)	Sporozoites	Calves, cross-bred (E)	Bansal and Ray, 1994
Hyalomma anatolicum anatolicum (E)	Schizonts		Sahoo and Mishra, 1994
Hyalomma anatolicum anatolicum (E)	Ground-up tick supernate	Calves, cross-bred (E)	Das and Sharma, 1994a
Hyalomma anatolicum anatolicum (E)	Sporozoites	Calves, crossbred, male (E)	Das and Sharma, 1994b
	GUTS	Calf (E)	Ray *et al.*, 1994
Hyalomma anatolicum anatolicum (N)	Schizonts		Sahoo and Misra, 1994
Hyalomma anatolicum anatolicum (N)	Schizonts	Cattle (N)	Sahoo and Misra, 1994
Hyalomma anatolicum anatolicum (N)	Sporoblasts		Sangwan *et al.*, 1994
Hyalomma	Ground-up-tick tissue sporozoites	Calves, Cross-bred (E)	Khan, 1997
Hyalomma anatolicum anatolicum (N)	Ground-up-tick tissue sporozoites	Calves, cross-bred (E)	Bansal and Ray, 1997
Hyalomma anatolicum (E)	Ground-up-tick tissue sporozoites	Calves, cross-bred (E)	Bansal and Ray, 1998
Hyalomma anatolicum anatolicum (E)	Ground-up-tick tissue stabilates (GUTS)	Calves, crossbred (HFX Sahiwal) (E)	Sandhu *et al.*, 1998
Hyalomma anatolicum anatolicum (E)	GUTS	Cross-bred (E)	Kumar and Malik, 1999a
Hyalomma anatolicum anatolicum (E)	GUTS	Cross-bred Jersey x Tharparkar (E)	Kumar and Malik, 1999b

Intermediate Host	Larva (Mode of Infection)	Final Host	Reference
Hyalomma anatolicum anatolicum (E)	—	Calves, cross-bred (E), male	Saluja et al., 1999
Hyalomma anatolicum anatolicum (N)		Cross-bred calves (E)	Sangwan and Sangwan, 1999
Hyalomma anatolicum anatolicum (E), larvae, nymphs, adults	—	Cross-bred calves (E)	Harikrishnan et al., 2000
Hyalomma anatolicum anatolicum (E)	Ground-up tick supernate	Calves, cross-bred (E)	Dinesh Patel et al., 2001
Hyalomma marginatum issaci (E)	—	Calves (E), cross-bred (Jersey cross, Brown and White)	Harikrishnan et al., 2001
Hyalomma anatolicum anatolicum (N)		Cattle (N)	Sangwan and Malhotra, 2001
Hyalomma anatolicum anatolicum (E)	Ground-up-tick tissue stabilate	Calves, cross-bred (E)	Singh et al., 2001
Ticks	Theileria annulata positive	Cattle (N)	Rup Ram et al., 2001
Hyalomma anatolicum anatolicum (E)		Bovine calves (E)	Das, 2003 (Ph. D. Thesis)
Hyalomma anatolicum anatolicum (N)	Different stages in salivary gland acini		Das and Ray, 2003a
Hyalomma anatolicum anatolicum (N) (engorged nymphs and unfed adults)	Salivary gland acini positive for sporoblasts	Cattle (N)	Das and Ray, 2003b
Hyalomma anatolicum anatolicum (E)	GUTS	Calves (E)	Saravanan et al., 2003
Hyalomma anatolicum anatolicum (E) (N)	GUTS/Salivary gland homogenates	Cattle (N) Calves HF (E)	Rup Ram et al., 2004

Intermediate Host	Larva (Mode of Infection)	Final Host	Reference
Hyalomma anatolicum anatolicum (N)	"Immature and mature parasites"		Das et al., 2005
Rhipicephalus haemaphysaloides (N)	Harbouring *T. annulata*	Cattle (N)	Raghorte et al., 2006a
Rhipicephalus (N)	Harbouring *T. annulata*	Cattle (N)	Raghorte et al., 2006b
Ticks	Sporozoites	Calves, bovine, cross-bred (E)	Ray et al., 2006
Hyalomma marginatum isaaci (E)	Sporozoites		Das et al., 2007
Hyalomma anatolicum anatolicum (N) *Hyalomma* sp. (N)	Salivary gland acini positive for *T. annulata*		Sangwan, 2007
Ticks	*T. annulata* positive	Calves, cross-bred (N)	Sangwan and Sangwan, 1999
Ticks	*T. annulata* positive	Calves, cross-bred (N) male	Sangwan and Sangwan, 2007
Mithuns (*Bos frontalis*)	Antibodies		Rajkhowa et al., 2008
Hyalomma anatolicum anatolicum (E)	GUTS		Saravanan et al., 2010
Hyalomma anatolicum anatolicum (E)	Sporoblasts Sporozoite	Cattle (N) Buffaloes (N)	Haque et al., 2010
Hyalomma anatolicum anatolicum (N) adults, males and females	Sporoblasts	Cattle, cross-bred (N)	Haque et al., 2011
Hyalomma anatolicum anatolicum (N) adults		Cattle, cross-bred (E)	Jeybal et al., 2012

Intermediate Host	Larva (Mode of Infection)	Final Host	Reference
	***Theileria equi* Mehlhorn and Schein, 1998** vide *Babesia equi* (Laveran, 1901) for IH recorded from India		
	Theileria hirci* Dschunkowsky and Urodschevich, 1924		
Rhipicephalus spp. (E) *Hyalomma anatolicum anatolicum*, nymphs and adults (E)		Sheep (E)	Sisodia and Gautam, 1980
	Theileria orientalis* Yakimoff & Soundatschenenkoff, 1931		
Haemaphysalis bispinosa (E)		Calves cross-bred (E)	Shastri *et al.*, 1988
	***Theileria ovis* Rodhain, 1916**		
Rhipicephalus haemaphysaloides (E)			Gill *et al.*, 1980
ORDER: RICKETTSIALE			
	Anaplasma marginale Theiler, 1910		
Ornithodoros savignyi (N)		Calves (E)	Ebenezer *et al.*, 1983
Hyalomma (E)		Calves, Gir x Red Dane (E)	Misraulia *et al.*, 1987
Boophilus annulatus (N)		Calf (E)	Jagannath, 1988
	Ehrlichia		
		Dogs (E)	Manohar and Ramakrishnan, 1984

PART II
BIBLIOGRAPHY

SECTION I–HELMINTH PARASITES

1. CLASS: TREMATODA

1.1 FAMILY: DICROCOELIIDAE

Bhalerao, G. D., 1946. Proc. 33rd Indian Sci. Cong. (Bangalore), 3, p.120.

Gomathinayagam, S., Rajavelu, G., Selvaraj, J. and Sakthivel, S. M., 2004. A note on the occurrence of *Eurytrema pancreaticum* in a Kangayam bull. J. Vet. Parasitol., 18: 183-184.

Srivatsava, H. D., 1944. A study of the life history of *Dicrocoelium dendriticum*, the small liver fluke of Indian ruminants. Proc. 31st Indian Sci. Cong. (Delhi), Part III, Sect. III, p.113.

1.2 FAMILY: OPISTHORCHIDAE

Mohandas, A., 1971. Contributions to the cercarial fauna of Kerala. Ph. D. Thesis submitted to University of Kerala. (Not indexed).

Mohandas, A., 1974. Studies on the freshwater cecariae of Kerala 1. Incidence of infection and seasonal variation. Folia Parasitol.(Praha), 21: 311-317.

Mohandas, A., 1976. Studies on the freshwater cecariae of Kerala V. Paramphistomatoid and opisthorchoid cercariae. Vest. Cs. Spol. Zool., 40: 196-205.

Pande, B. P. and Shukla, R. P., 1974. Experimentally induced *Opisthorchis caninus* infections in hamsters and rhesus monkey with a note on the role of larval digeneans in fisheries production. Indian J. Exp. Biol., 12: 184-191.

Pande, V. and Premvati, G., 1974. Experimental infection and development of metacercarial cysts of *Opisthorchis caninus* (Trematoda: Opisthorchidae) in albino rats and mice. Indian J. Anim. Sci., 44: 572-580.

Pramod Kumar, Pande, B. P. and Rai, P., 1983. Studies on opisthorchiasis in experimental pups with special reference to diagnosis and its zoonotic importance. Seventh Natl. Cong. Parasitol., Ravi Shankar University, Raipur, 26-28, Dec., 1986, Abstr. B-26, p.82.

Rai, P., 1969. Notes on the histopathology of opisthorchiid, plagiorchiid and isoparaorchiid metacercoidal invasion of some Indian freshwater fishes. Indian J. Anim. Sci., 39: 177-183.(not indexed).

Rai, P., 1966. Preliminary note on the life-cycle on the common liver-fluke of Indian domestic carnivores. Indian Vet. J., 43: 688.

Rai, P. and Pande, B. P., 1965. On the metacercaria of the common liver- fluke of Indian domestic carnivores, *Opisthorchis caninus*, and its experimental infection in guinea-pig. Z. f. Parasitenkd., 26: 18-23.

Rekha Devi, K., Narain, K. and Mahanta, J., 2005. Food-borne parasitic zoonosis: Status of metacercarial infection in fishes of Assam. 17th Natl. Cong. Parasitol., Regional Medical Research Centre, N. E. Region (ICMR), Dibrugarh-786001, Assam, India. Abstr. No. 087-P., p. 40.

Sahai, B. N., 1967. Sudies on the biology and control of common trematode of the dog. Ph. D. Thesis submitted to Agra University, Agra, India. (Not indexed).

Sahai, B. N., 1969-70. A survey of trematode infection in aquatic snails. J. Vetcol., Assam Agri. Univ., 10: 9-12.

Sahai, B. N. and Srivasata, H. D., 1972. On the morphology and life cycle of *Opisthorchis noverca*, Braun, 1902 liver fluke of carnivores mammals. Proc 41ˢᵗ Natl. Acad. Sci., India (Biol. Sect.), p. 39.

Sahai, B. N. and Srivasata, H. D., 1978. Morphology and life-history of *Opisthorchis noverca*, Braun, 1902, a trematode parasite of dogs and pigs in India. Indian J. Anim. Sci., 48: 113-122.

Vijay Sachdeva, Chauhan, P. P. S. and Agrawal, R. D., 1994. Development of opisthorchid metacercariae in experimental animals. Sixth Natl. Cong. Vet. Parasitol., 22-24 Oct., 1994, Dept. Parasitol., Coll. Vet. Sci., JNKVV, Jabalpur-482 001, M. P., India, Abstr. S-2: 11, pp. 13-14.

1.3 FAMILY: FASCIOLIDAE

Ani Bency Jacob, Mooyottu, S., Singh, P., Pathania, S., Anees, C., Raina. O. K. and Verma, A. K., 2011. *Fasciola gigantica* egg induced granuloma in the liver of experimentally infected cross bred calf. Indian J. Vet. Pathol., 35(1): 75-76.

Ani Bency Jacob, Singh, P., Raina, O. K. and Verma, A. K., 2012. Effect of deoiled mahua seed cake extract on juvenile flukes of *Fasciola gigantica*. Indian Vet. J., 89(5): 15-16.

Banerjee, D. P. and Singh, 1991. Serodiagnosis of experimental *Fasciola gigantica* infection in lambs. Indian J. Anim. Sci., 61: 268-269.

Buckley, J. J. C., 1939. Observations on *Gastrodiscoides hominis* and *Fasciolopsis buski* in Assam. J. Helminthol., 17: 1-12.

Bhalerao, G. D., 1933. A preliminary note on the life history of the common liver fluke in India, *Fasciola gigantica*. Indian J. Vet. Sci., 3: 120- 121.

Bhatnagar, P. K. and Gupta, D. K., 1995. Epizootiological studies on *Fasciola gigantica* in *Lymnaea auricularia* during the annual cycle. 7th Natl. Cong. Vet. Parasitol., 19-21 August., 1995. Dept. Parasitol., Madras Vet. Coll., Madras -7, Abstr. Sl 33, p.90.

Bhatnagar, P. K. and Prasad, A., 2001. Studies on the cercarial emergence of *Fasciola gigantica* from *Lymnaea auricularia*. 12th Natl. Cong. Vet. Parasitol., Aug. 26-27, 2001, Dept. Parasitol., Coll. Vet. Sci., Acharya Ranga Agri. Univ., Tirupati-517 502., India, **Abstr.** S-2: 10, p.49.

Bhatia, B. B., Upadhyaya, D. S. and Juyal, P. D. 1989. Epidemiology of *Fasciola gigantica* in buffaloes, goats and sheep in Tarai region of Uttar Pradesh. J. Vet. Parasitol., 3: 25-29.

Bhatia, A. K. and Roy Choudhury, G. K., 1985. Comparative efficacy of Ancylol [R] and Distodin [R] against fascioliasis in experimentally infected cow calves. Indian Vet. J., 62: 405-406.

Chandra, R., 1972. Studies on the biology, immunity and control of *Fasciola gigantica* infection. Ph. D, Thesis, Agra Univ., Agra, India.

Das, M., 1976. Studies on *Lymnaea auricularis* var *rufescens*, the intermediate host of *Fasciola gigantica* (Cobbold, 1885). M. V. Sc. Thesis submitted to Agra Univ., Agra, India.

Das, M. and Subramanian, G., 1981. Comparative susceptibility of *Lymnaea auricularis* var *rufescens* from different localities to infection with *Fasciola gigantica*, Cobbold, 1885. J. Res. Assam Agri. Univ., 2: 52.

Deo, P. G., Tandon, K. C., Kumar, V. and Srivastava, H. D., 1967. Studies on the effects of feeds deficient in trace elements, vitamin-A, protein, calcium and phosphorous on the natural resistance of sheep to common liver fluke, *Fasciola gigantica*, Cobbold, 1885. Indian J. Vet. Sci., 37: 351- 359.

Dixit, A. K., Pooja Dixit, Sharma, R. L., 2008. Immunodiagnostic / protective role of cathepsin L cysteine proteinases secreted by *Fasciola* species. Vet. Parasitol., 154: 177-184. (Review; not indexed).

Dixit, A. K., Yadav, S. C. and Sharma, R. L., 2002. 28 kDa *Fasciola gigantica* cysteine proteinase in the diagnosis of prepatent ovine fascioliasis. Vet. Parasitol., 109: 233-247.

Dixit, A. K., Yadav, S. C. and Sharma, R. L., 2004. Experimental bubaline fasciolosis: Kinetics of antibody response using 28 kDa *Fasciola gigantica* cysteine proteinase as antigen. Trop. Anim. Hlth. Prod., 36: 49-54.

Dixit, A. K., Yadav, S. C., Mohini Saini and Sharma, R. L., 2003. Purification and characterization of 28 kDa cysteine proteinase for immunodiagnosisof tropical fascioliasis. J. Vet. Parasitol., 17: 5-9.

Dutt, S. C. and Bali, H. S., 1980. Snails of Punjab and their trematode infections. J. Res. Punjab Agri. Univ., 17: 222-228.

Edith, R., 2004. Comparative immunoprotection profile of infection specific and excretory and secretary protein of *Fasciola gigantica* against tropical fascioliasis in buffaloes. M. V. Sc. Thesis submitted to the Deemed Univ., Indian Vet. Res. Inst., Izatnagar, U.P., India, pp.1-96. (not indexed).

Edith, R., Godara, R., Sharma, R. L. and Thilagar, M. B., 2010a. Serum enzyme and haematological profile of *Fasciola gigantica* immunized and experimentally infected riverine buffaloes. Parasitol. Res., 106: 947-956. Vide DOI 10.1007/s 00436-010-1741-I.

Edith, R., Godara, R., Sharma, R. L. and Thilagar, M. B., 2012. *Fasciola gigantica* induced adrenal dysfunction and its pathological significance in riverine buffaloes (*Bubalus bubalis*). Buff. Bull., 31: 51-62.

Edith, R., Sharma, R. L., Godara, R. and Thilagar, M. B., 2010b. Experimental studies on anaemia in riverine buffaloes (*Bubalus bubalis*) infected with *Fasciola gigantica*. Comp. Clin. Pathol., Vide DOI 10.1007/s 00580-010-1109.

Edith, R., Thilagar, M. B., Godara, R. and Sharma, R. L., 2010c. Tropical liver fluke-induced stress in experimentally infected and immunised buffaloes. Vet. Rec., 167: 571-575.

Edith, R., Thilagar, M. B., Godara, R. and Sharma, R. L., 2011. Alterations in serum enzyme profile of profile of riverine buffaloes (*Bubalus bubalis*) experimentally infected with *Fasciola gigantica*. J. Parasitol., 25: 50-55.

Ganga, G., 2001. Experimental studies on bubaline fascioliasis. Ph. D. Thesis (Vet. Med.) submitted to the Deemed Univ., Indian Vet. Res. Inst., Izatnagar-243 122, Barelly, India. (Not indexed).

Ganga, G., Varsney, J. P.and Sharma, R. L., 2004a Oxidative stress in *Fasciola gigantica* infected buffaloes. J.Vet. Parasitol., 18: 71-72.

Ganga, G., Varsney, J. P. and Sharma, R. L., 2004b. Effect of *Fasciola gigantica* excretory-secretary antigen on rat haematlogical indices. J. Vet. Sci., 7: 123-125.

Ganga, G., Varsney, J. P., Sharma, R. L., Varsney, V. P. and Kalicharan, 2003. Hypothyroidism associated with bubaline fasciolosis. Indian J. Anim. Sci., 73: 640-651.

Ganga, G., Varsney, J. P., Sharma, R. L., Varsney, V. P. and Kalicharan, 2007. Effect of *Fasciola gigantica* infection on adrenaline and thyroid glands of reverine buffaloes. Res. Vet. Sci., 82: 61-67.

Ghosh, S., Neha Saxena, Niranjan Kumar and Gupta, S.C., 2009. Kinetics of antibody response in experimentally infected buffaloes with *Fasciola gigantica*. Indian J. Anim. Sci., 79: 537-540.

Ghosh, S., Preeti Rawat, Gupta, S. C. and Singh, B. P., 2005a. Comparative diagnostic potentiality of ELISA and dot ELISA in prepatent diagnosis of experimental *Fasciola gigantica* infection in cattle. Indian J. Exp. Biol., 43: 536-541.

Ghosh, S., Preeti Rawat, Veluswamy, R., Joseph David, Gupta, S. C. and Singh, B. P., 2005b. 27 kDa *Fasciola gigantica* glycoprotein for the diagnosis of prepatent fascioliasis in cattle. Vet. Res. Commun., 29: 123-135.

Ghosh, S., Gupta, S. C., Preeti Rawat, and Singh, B. P., 2006. Evaluation of antibody responses by ELISA in reverine buffaloes experimentally infected with *Fasciola gigantica*. J. Vet. Parasitol., 20: 13-16.

Ghosh, S., Saxena, N., Kumar, N. and Gupta, S. C., 2009. Kinetics of antibody response in experimentally infected buffaloes with *Fasciola gigantica*. Indian J. Anim. Sci., 79: 537-540.

Gupta, A., Dixit, A. K., Pooja Dixit and Mahajan, C., 2012. Prevalence of gastrointestinal parasites in cattle and buffaloes in and around Jablpur, Madhya Pradesh. J. Vet. Parasitol., 26: 186-188.

Gupta, R. P. and Paul, J. C., 1987. An unusual outbreak of fascioliasis (*Fasciola gigantica*) on a farm. Indian J. Anim. Res., 21: 41-42.

Gupta, R. P., Yadav, C. L. and Ruprah, N. S., 1986. The epidemiology of bovine fascioliasis (*F. gigantica*) in Haryana State. Indian Vet. J., 63: 187-190.

Gupta, S. C. and Chandra, R., 1987. Susceptibility of some laboratory animals to infection with *Fasciola gigantica*. J. Vet. Parasitol., 1: 19-21.

Gupta, S. C. and Singh, B. P., 2002. Fascioliasis in cattle and buffaloes in India. J. Vet. Parasitol., 16: 139-145. (Review, not listed in index table)

Gupta, S. C. and Yadav, S. C., 1992. Sexual maturity of *Fasciola gigantica* in experimentally infected rabbit, goat and buffaloes. Indian J. Parasitol., 16: 133-134.

Gupta, S. C. and Yadav, S. C., 1993. *Fasciola gigantica*: Somatic, and excretory and secretory antigens in the diagnosis of experimental fasciolosis. Indian J. Anim. Sci., 63: 025-1027.

Gupta, S. C. and Yadav, S. C., 1994a. Emergence of *Fasciola gigantica* cercariae from naturally infected *Lymnaea auricularia*. Indian J. Parasitol., 18: 53-56.

Gupta, S. C. and Yadav, S. C., 1994b. Antibody response of rabbit to different doses of *Fasciola gigantica* experimental infection. 6th Natl. Cong. Vet. Parasitol, 22-24, October, 1994, Dept. Parasitol., Coll. Vet. Sci., JNKVV, Jabalpur- 482 001, M. P., India, Abstr. S-3: 19, p.38.

Gupta, S. C. and Yadav, S. C., 1994c. Antibody response of rabbit to enhanced doses of *Fasciola gigantica* experimental infection. J. Vet. Parasitol., 9: 73-77.

Gupta, S. C., Chandra, R. and Yadav, S. C., 1989. Efficacy of triclabendazole against experimental *Fasciola gigantica* infection in sheep, goat, buffalo and rabbit: a comparative study. Indian Vet. J., 66: 680-682.

Gupta, S. C., Ghosh, S., David Joseph and Singh, B. P., 2003. Diagnosis of experimental *Fasciola gigantica* infection in cattle by affinity purified antigen. Indian J. Anim. Sci., 73: 63-966.

Gupta, S. C., Vinay Verma, Singh, P., Verma A. K. and Mehra, U. R., 2006. Clinical course of experimental *Fasciola gigantica* infection in buffalo calves maintained on different dietary protein levels. Proc. XVII Natl. Cong. Vet. Parasitol., November, 15-17, 2006, Dept. Parasitol., Rajiv Gandhi Coll. Vet. Sci. & Anim. Sci, Kurumbapet, Puducherry- 605 009, Abstr., S.II.38, p.72.

Hafeez, Md., 2003. Helminth parasites of public health importance - trematodes. J. Parasit. Dis., 69-75. (Not indexed).

Ingale, S. L., Singh, P, Raina O. K., Mehra, U. R., Verma, A. K., Gupta, S. C. and Mulik, S. V., 2008. Interferon-gamma and interleukin-4 expression during *Fasciola gigantica* primary infection in crossbred bovine calves as determined by real-time PCR. Vet. Parasitol., 152: 158-161.

Jayraw, A. K., 2005. Immunological and Nucleic Acid Based Diagnosis of *Fasciola gigantica*. Ph. D. Thesis submitted to the Deemed Univ., Indian Vet.Res. Inst., Izatnagar-243 122, India.

Jayraw, A. K, Singh B. P. and Raina O. K., 2005a. EITB based prepatent diagnosis of experimental bovine tropical fasciolosis. Proc. XVI Natl. Cong. Vet. Parasitol., December 6-8, 2005, Dept. Parasitol., Coll. Vet. Sci. & Anim. Husb., Indira Gandhi Agri. Univ., Anjora, Durg-491 001, Chhattisgarh. Ab-21, pp.62-63.

Jayraw, A. K, Singh B. P. and Raina O. K., 2005b. Dynamics of immunoglobulin isotype response in experimental bovine tropical fasciolosis. Proc. XVI Natl. Cong. Vet. Parasitol., December 6-8, 2005, Dept. Parasitol., Coll. Vet. Sci. & Anim. Husb., Indira Gandhi Agri. Univ., Anjora, Durg-491 001, Chhattisgarh. Ab-22, pp.63-64.

Jayraw, A. K, Singh B. P. and Raina O. K., 2005c. Detectionof circulating 52kDa antigen in serum of *Fasciola gigantica* infected bovine calves. Proc. XVI Natl. Cong. Vet. Parasitol., December 6-8, 2005, Dept. Parasitol., Coll. Vet. Sci. & Anim. Husb., Indira Gandhi Agri. Univ., Anjora, Durg-491 001, Chhattisgarh. Ab-23, p.64.

Jayraw, A. K, Singh B. P., Raina O. K., Udaya Kumar, M., 2009. Kinetics of serum immunoglobulin isotype response in experimental bovine tropical fasciolosis. Vet. Parasitol, 165: 155- 160.

Jedeppa, A., Raina O. K, Samanta, S., Nagar, G, Niranjan Kumar, Anju Varghese, Gupta S.C, and Banerjee P.S., 2009. Molecular cloning and characterization of a glutathione S-transferase in the tropical liver fluke, *Fasciola gigantica*. J. Helminthol., 85: 55-60.

Khajuria, J. K. and Bali, H. S., 1987. Effect of X-irradiation on the development of *Fasciola gigantica* (Cobbold, 1885). Indian J. Parasitol., 11: 207-210.

Khajuria, J. K. and Bali, H. S., 1987. Experimental *Fasciola gigantica* (Cobbold, 1885) infection in various animals species. Cheiron., 16: 81-85.

Kumar, V., 1980a. The digenetic trematodes, *Fasciolopsis buski*, *Gastrodiscoides hominis* and *Artyfechinostomum malayanum* as zoonotic infection in South Asian countries. Ann. Soc. Belge. Med. Trop., 60: 331-339. (Not indexed).

Kumar, V., 1980b. Zoonotic trematodiasis in south-east and far-east asian countries. Curr. Top. Vet. Med. Anim. Sci., 43: 106-118. (Not indexed).

Malaviya, H. C., 1985.The susceptibility of mammals to *Fasciolopsis buski*. J. Helminthol., 59: 19-22.

Malaviya, H. C. and Varma, T. K., 1989.On the life history of *Fasciolopsis buski*. Ninth Natl. Cong. Parasitol., Ujjain, India, A22.

Mandal, S., 1997. Evaluation of *Fasciola gigantica* antigens immunodiagnosis of fasciolosis in ruminants. M. V. Sc. Thesis submitted to the Deemed Univ., Indian Vet. Res. Inst., Izatnagar, U.P., India. (Not indexed).

Mandal, S., Yadav, S. C. and Sharma, R. L., 1998. Evaluation of *Fasciola gigantica* antigenic preparations in serodiagnosis of fasciolosis in sheep. J. Parasit. Dis., 22: 25-29.

Mehra, U. R., Verma, A. K., Dass, R. S., Sharma, R. L. and Yadav, S. C., 1999. Effects of *Fasciola gigantica* infection on growth and nutrient utilisation of buffalo calves. Vet. Rec., 145: 699-702.

Murthy, G. S. S. and D' Souza, P. E., 2008. Fasciolosis and amphistomosis in livestock. In: XI. Natl. Trng. Prog. on Trends and perspectives in the biology, ecology and control of parasitic diseases. 10th to 30th March, 2008, Cent. Adv. Stud., Dept. Parasitol., Vet. Coll. Univ., Agri. Sci., Hebbal, Bangalore-560 024. pp.10-18. (not included in Index list).

Nagar G., Raina O.K., Varghese Anju, Niranjan Kumar, Samanta, S., Prasad, A., Gupta, S. C., Banerjee, P. S., Singh, B. P., Rao, J. R., Tewari, A. K., Souvik Paul, Jayraw, A. K., Dinesh Chandra and Rajat Garg, 2010. *In vitro* excystment of *Fasciola gigantica* metacercariae. J. Vet. Parasitol., 24: 169-171.

Nambi Azhahia, P., Yadav, S. C., Raina, O. K. Sriveni, D.and Saini Mohini, 2005. Vaccination of buffaloes with *Fasciola gigantica* recombinant fatty acid binding protein. Parasitol. Res., 97: 129-135.

Neha Singh, Pradeep Kumar and Singh, D. K., 2012. Variant abiotic factor and the infection of *Fasciola gigantica* larval stages in vector snail, *Indoplanorbis exustus*. J. Biol. Earth Sci., 2012(2): B110-B117.

Niranjan Kumar, Ghosh, S. and Gupta, S. C., 2008a. Early detection of *Fasciola gigantica* infection in buffaloes by enzyme linked immunosorbant assay and dot enzyme-linked immunosorbant assay. Parasitol. Res., 103: 141-150.

Niranjan Kumar, Ghosh, S. and Gupta, S. C., 2008b. Detection of *Fasciola gigantica* infection in buffaloes by enzyme linked immunosorbant assay. Parasitol. Res., 104: 155-161.

Niranjan Kumar, Anju Varghese, Gaurav Nagar, Dinesh Chandra, Samanta, S., Gupta, S. C., Adeppa, J. and Raina, O. K., 2011. Vaccination of buffaloes with *Fasciola gigantica* recombinant glutathione S- transferase and fatty acids binding protein. Parasitol Res., On line DOI 10.1007/s00436-011-2507-0.

Pal, S. and Das Gupta, C. K., 2007. Cross reactive antigens of *Fasciola gigantica*, *Paramphistomum epiclitum* and *Gigantocotyle explanatum*. J. Vet. Parasitol., 21: 29-31.

Pal, S., Das Gupta, C. K., Sanyal, P. K. and Mandal., S. C., 2005. Detection of 44 kDa *Fasciola gigantica* antigen in an experimentally infected buffalo calf by Sandwich-ELISA., Proc. XIV Natl. Cong. Vet. Parasitol., December 6-8, 2005, Dept. Parasitol., Coll. Vet. Sci. & Anim. Husb., Indira Gandhi Agri. Univ., Anjora, Durg-491 001.,Chhattisgarh. Ab-29, p.67.

Patnaik, M. M., 1968. Notes on glycogen deposits of *Lymnaea auricularia* var *rufescens* and parasitic larval stages of *Fasciola gigantica* and *Echinostoma revolutum* – A histochemical study. Ann. Parasitol., 43: 449-456.

Patnaik, M. M. and Ray, S. K., 1968. Studies on geographical distribution and ecology of *Lymnaea auricularia* var *rufescens* the intermediate host of *Fasciola gigantica* in Orissa. Indian J. Vet. Sci., 38: 484-508.

Patnaik, M. M., 1971. Studies on geographical distribution and ecology of *Fasciola gigantica* with special reference to its intermediate host, *Lymnaea auricularia* var. *rufescens* in Orissa. Indian J. Helminth., 23: 115-134.

Pemola Devi, N. and Jauhri, R. K., 2008. Diversity and cercarial shedding of malco fauna collecting from water bodies of Ratnagiri District, Maharashtra. Acta Trop., 105: 249-252.

Pokhriyal, B. P., Jauhari, R. K. and Sudarshana, R., 1998. Trematode cercarial infection in the snail *Thiara (M.) tuberculata* (Mueller, 1774) in different localities of Deon Valley. J. Exp. Zool., India. 1(2): 107-110.

Pokhriyal, B. P., Mahesh, R. K. and Jauhari, R. K., 1996. Prevalence of trematode cercariae infection in the snail *Lymnaea (Pseudosuccinea) acuminata* Lamarck, 1822 in different localities of Dehradun-Valley. Global Meet Parasit. Dis., New Delhi, India 18-22 March, 1996. Abstr., J. Parasit. Dis., 20: 111.

Prasad, A., 1989. Experimental infection of *Lymnaea auricularia* race *rufescens* for mass harvesting of metacercariae of *Fasciola gigantica*. J. Vet. Parasitol., 3: 31-34.

Prasad, A. Ghosh, S. and Singh, R., 1999. Studies on the viability of metacercariae of *Fasciola gigantica*. J. Helminthol., 73: 163-166.

Preet, S. and Prakash, S., 2001a. Seasonal dynamics of cercarial emergence in *Fasciolopsis buski*. Proc. 15th Natl. Cong. Parasitol., Jodhpur, 1-3 October., 2001, Abstr. No 125, p.108.

Preet, S. and Prakash, S., 2001b. Cercarial emergence in *Fasciolopsis buski* (Lankester). J. Parasit. Dis., 25: 108-110.

Rai, R. B., Senai, S., Ahlawat, S. P. S., and Vijay Kumar, B., 1996. Studies on the control of fasciolosis in Andaman and Nicobar Islands. Indian Vet. J., 73: 822-825.

Raina, O. K., Sriveny, D. and Yadav, S. C., 2004. Humoral immune response against *Fasciola gigantica* fatty acid binding protein. Vet. Parasitol., 124: 65-72.

Raina, O. K., Yadav, S. C., Sriveny, D. and Gupta, S. C., 2006. Immuno-diagnosis of bubaline fasciolosis *Fasciola gigantica* cathepsin-L and recombinant cathepsin-L1-D proteinases. Acta Trop., 98: 145-151.

Raina, O. K., Gaurav Nagar, Anju Varghese, Prajitha, G., Asha Alex, Maharana, B. R. and Joshi, P., 2011. Lack of protective efficacy in buffaloes vaccinated with *Fasciola gigantica* leucine aminopeptidase and peroxiredoxin recombinant proteins. Acta Trop., 118: 217-222.

Raina, O. K., Tripathy, A., Sriveny, D., Samanta, S., Gupta, S. C., Singh, R., Tewari, A. K., Banerjee, P. S., Kumar, S. and Yadav, S. C., 2009. Immune responses to polythylenimine delivered plasmid DNA encoding of *Fasciola gigantica* fatty acid binding protein in mice and rabbits. J. Helminthol., 83: 275-283.

Raina, O. K., Yadav, S. C. Gupta, S. C., Dinesh Chandra, Samanta, S., Ghosh, S., Singh, B. P. and Tiwari, A. K., 2005. Serodiagnosis of *Fasciola gigantica* in ruminants using cathepsin-L cysteine proteinase. Proc. XIV Natl. Cong. Vet. Parasitol., December, 6-8, 2005, Dept. Parasitol., Coll. Vet. Sci.& Anim. Husb., Indira Gandhi Agri. Univ., Anjora, Durg-491 001, Chhattisgarh. Ab-27, p.66.

Rajat Garg, Yadav, C. L., Banrejee, P. S., Kumar, R. R. Banerjee, P. S., Stuti Vatsya and Rajesh Godara, 2006. Epidemiology of *Fasciola gigantica* in ruminants in different geo-climatic conditions of north in India. Trop. Anim. Health. Prod.,41: 1695-1700.

Rao, J. R., Sikdar, A., Deorani, V. P. S. and Jha, S. K., 1985. A report on the outbreak of fasciolosis in Andaman and Nicobar Islands. Cheiron, 14: 162-163.

Santra, P. K., Prasad, A. and Ghosh, S., 1999a. Chemotherapeutic response of triclabendazole by enzyme linked immunosorbant assay in experimental infection in sheep with *Fasciola gigantica*. Indian J. Anim. Sci., 69: 472-474.

Santra, P. K., Prasad, A. and Ghosh, S., 1999b. Efficacy of triclabendazole against experimental fasciolosis in lambs. J. Vet. Parasitol., 13: 111-114.

Sanyal, P. K., 1996. Kinetic disposition and clinical efficacy of triclabendazole against experimental bovine and bubaline fasciolosis. J. Vet. Parasitol., 10: 147-152.

Sanyal, P. K., 1998. Pharmacokinetics and efficacy of triclabendazole incorporated urea molasses block against experimental immature fasciolosis in cattle and buffaloes. J. Vet. Parasitol., 12: 25-29.

Sanyal, P. K. and Gupta, S. C., 1996a. Efficacy and pharmacokinetics of triclabendazole in buffalo with induced fascioliasis. Vet. Parasitol., 63: 75-82.

Sanyal, P. K. and Gupta, S. C., 1996b. The efficacy and pharmacokinetics of long term low level intraruminal administration of triclabendazole in buffalo with induced fascioliasis. Vet. Res. Comm., 63: 75-82.

Sanyal, P. K. and Gupta, S. C., 1998. Pharmacokinetics and efficacy of long term low level administration of triclabendazole in urea mollases blocks against induced bovine and bubaline fascioliasis. Vet. Parasitol., 76: 57-64.

Sahai, B. N., 1969-70. A survey of trematode infection in aquatic snails. J. Vetcol., 10: 9-12.

Sardey, M. R., 1982. A study on the prevalence of trematode cercaria in snails of Nagpur region. In: Proc. Sym. Vectors and Vector-borne Diseases. Trivandrum, Kerala, India. February, 26-28, 1982, pp.137-141.

Sharma, R. L., Dhar, D. N. and Raina, O. K., 1989. Studies on the prevalence and laboratory transmission of fascioliasis in animals in the Kashmir valley. British Vet. J., 145: 57-61.

Sharma, R. L., Godara, R. and Thilagar, M. B., 2011. Epizootiology, pathogenesis and immunoprophylactic trends to control tropical bubaline fascioliasis: an overview. J. Parasit. Dis., 35: 1-9. (Review, not listed in index table).

Shikari, R. N., Gupta, S. C. and Subodh Kishore, 1999. Studies on the protective efficacy of *Fasciola gigantica* antigens in rabbits. J. Vet. Parasitol., 13: 135-137.

Singh, A., Srivastava, S., Chandra Sehkar and Jaswant Singh, 2009. Prevalenc of trematodes in bovines and snails. Indian Vet. J., 86: 206-207.

Singh, B. P., Jayrao, A. K., Gupta, S. C., Raina, O. K. and Ghosh, S., 2005. Prevalenc of *Fasciola gigantica* infection amonst *Lymnaea auricularia* population to surrounding regions of Bareilly, India. Proc. XVI Natl. Cong. Vet. Parasitol., December, 6-8,

2005, Dept. Parasitol., Coll. Vet. Sci. & Anim. Husb., Indira Gandhi Agri. Univ., Anjora, Durg-491 001, Chhattisgarh. Ab-6, pp.19-20.

Singh, K. S., 1971a. Some observations on the pathology of experimental *Fasciola gigantica* infection in rabbit. Indian J. Anim. Sci., 41: 1218-1222.

Singh, K. S., 1971b. Early migration of *Fasciola gigantica*, Cobbold, 1885 in guinea pigs. Indian J. Anim. Sci., 41: 1223-1231.

Singh, K. S. and Malaki, A., 1963. Parasitological survey of Kumaun region. Part VIII. One known and two unknown cercariae from freshwater snails. Indian J. Helminth., 15: 54-69.

Singh, P., Verma, A. K., Ani Bency Jacob, Gupta, S. C. and Mehra, U. R., 2011. Haematological and biochemical changes in *Fasciola gigantica* infected buffaloes fed in diet containing deoiled mahua (*Bassia latifolia*) seed cake. J. Appl. Anim. Res. 39: 185-188.

Singh, P., Verma, A. K., Ani Bency Jacob, Gupta, S. C. and Mehra, U. R., 2012. Effect of supplementation of deoiled mahua (*Bassia latifolia*) seed cake on nutrient utilization in buffaloes during *Fasciola gigantica* infection. Indian Vet. J., 89(5): 31-33.

Smitha, S., Raina, O. K, Singh, B. P., Samanta, S., Velusamy, R., Dangoudoubiyam, S., Tripathi, A., Gupta, P. K., Sharma, B. and Meeta Saxena, 2010. Immune responses to polyethylenimine-mannose-delivered plasmid DNA encoding a *Fasciola gigantica* fatty acid binding protein in mice. J. Helminthol., 84: 149-155.

Srivastava, H. D., 1944a. The intermediate host of *Fasciola hepatica* in India. Proc. 31st Indian Sci. Cong. (Delhi), Part III, Sect. VIII, pp.113-114.

Srivastava, H. D., 1944b. The intermediate host of *Fasciola gigantica* of Indian ruminants. Proc. 31st Indian Sci. Cong. (Delhi), Part III, Sect. VIII, p.142.

Srivastava, P. S. and Singh, K. S., 1972a. Early migration of *Fasciola gigantica*, Cobbold, 1885 in guinea pig. Indian J. Anim. Sci., 42: 63-71.

Srivastava, P. S. and Singh, K. S., 1972b. Histopathological study of immature fascioliasis in guinea pig. Indian J. Anim. Sci., 42: 120-125.

Srivastava, P. S. and Singh, K. S., 1974. On the susceptibility of three species of laboratory animals to experimental infection with *Fasciola gigantica*, Cobbold, 1885. Indian J. Anim. Res., 8: 15- 20.

Sriveny, D., Raina, O. K., Yadav, S. C., Chandra, D., Jayaram, A. K., Singh, M., Velusamy, R. and Singh, B. P., 2006. Cathepsin L cysteine proteinase in the diagnosis of bovine *Fasciola gigantica* infection. Vet. Parasitol., 135: 25-31.

Sunita, K. and Singh, D. K., 2011. Fascioliasis control: *In vivo* and *in vitro* physiotherapy of snail vector to kill *Fasciola* larvae. J. Parasitol. Res., 2011: 1-7.

Sunita, K., Kumar, P. and Singh, D. K., 2012. Abiotic environment factors and infection of *Fasciola gigantica* in vector snail, *Lymnaea acuminata*. Researcher, 4(8): 49-53.

Sunita, K., Kumar, P. and Singh, D. K., 2013a.Phytotherapy of intermediate host snail by ferulic acid to kill the *Fasciola gigantica* larvae in different months of the year 2011-2012. Sci. J. Biol Sci., Home 2 (7). (Internet Version).

Sunita, K., Kumar, P. and Singh, D. K., 2013b. Seasonal variation in toxicity of umbelliferone against *Fasciola* larvae. J. Biol. Earth Sci., 3(1): B93-99.

Sunita, K., Kumar, P. and Singh, D. K., 2013c. Fasciolosis Control: Phytotherapy of host snail *Lymnaea acuminata* by allicin to kill *Fasciola gigantica* larvae. Ann. Rev. Res. Biol., 3 (4): 694-704.

Sunita, K., Kumar, P. and Singh, D. K., 2013d. *In vivo* phytotherapy of snail by plant derived active components in control of fascioliasis. Scient. J. Vet. Adv., 2: (5) 61-67.

Tamloorkar, S. L. Narladkar, B. W. and Despande, P. D., 2001. Prevalence of snail species in Marathwada region and their cercariae carrier status. Twelfth Natl. Cong. Vet. Parasitol., Aug 26-27, 2001, Dept. Parasitol., Coll. Vet. Sci., Acharya Ranga Agri. Univ., Tirupati-517 502, India. Abstr. S-2: 68, p.83.

Tandon, R. S., 1970. Observations on the germinal development in the life cycle of the Indian liver fluke, *Fasciola indica* Verma, 1953. H. D. Srivastava Commemoration Volume, pp.537-544.

Thapar, G. S. and Tandon, R. S., 1952. On the life history of liver fluke, *Fasciola gigantica*, Cobbold, 1885 in India. Indian J. Helminth., 4: 1-36.

Tripathi, A., Gupta P. K., Sharma, B. and Saxena, M., 2009. Immune responses to polyethylenimine-mannose-delivered plasmid DNA encoding a *Fasciola gigantica* fatty acid binding protein in mice. J. Helminthol., 84: 149-155.

Tripathi, J. C., Srivatava, H. D. and Dutt, S. C., 1973. A note on experimental infection of *Helicorbis coenosus* and pig with *Fasciolopsis buski*. Indian J. Anim. Sci., 43: 647-649.

Upadhyay, A. K. and Kumar, M., 2004. Immunodiagnostics in experimental fasciolosis in animals. Indian Vet. J., 81: 228- 229.

Varma, A. K., 1954. Studies on the nature, incidence, distribution and control of nasal schistosomiasis and fascioloiasis in Bihar. Indian J.Vet. Sci., 24: 11-34.

Veena Tandon, Bishnupada Roy and Prasad, P. K., 2012. Fasciolopsis. In: Molecular Detection of Human Parasitic Pathogens. Chapter 32. Dongyou Liub (Ed), Boca Raton, CRC Press, USA.

Velusamy, R., 2002. Biological and immunological studies of *Fasciola gigantica* infection. Ph. D. (Vet. Parasitol.) Thesis submitted to the Deemed Univ., Indian Vet. Res. Inst., Izatnagar-243 122, India.(Not indexed).

Velusamy, R., Dwivedi, P., Sharma, A. K., Singh, B. P. and Chandra, D. 2002. Pathomorphological changes in liver of calves experimentally infected with *Fasciola gigantica*. Indian J. Vet. Pathol., 26 (1&2): 35-37.

Velusamy, R., Singh, B. P. and Raina, O.K., 2004a. Detection of *Fasciola gigantica* infections in snail by polymerase chain reaction. Vet. Parasitol., 120: 85-90.

Velusamy, R., Singh, B. P., Gupta, S. C. and Chandra, D., 2004b. Prevalence of *Fasciola gigantica* infection in buffaloes and its snail intermediate host at Bareilly. J. Vet. Parasitol., 18: 171-173.

Velusamy, R., Singh, B. P., Rao, J. R. and Chandra, D., 2004c. Assessment of cell-mediated immune response in bovine calves experimentally infected with *Fasciola gigantica*. J. Vet. Parasitol., 18: 23-26.

Velusamy, R., Singh, B. P., Sharma, P. C. and Chandra, D., 2004d. Detection of circulating 54 kDa antigen in sera of bovine calves experimentally infected with *F. gigantica*. Vet. Parasitol., 119: 187-195. (Note: words, *F. gigantica*-retained as in the original title).

Velusamy, R., Singh, B. P., Ghosh, S. and Gupta, S. C., 2009. Affinity purified antigen in the diagnosis of experimental fasciolosis in bovines. Indian J. Anim. Sci., 79: 565-567.

Velusamy, R., Singh, B. P., Ghosh, S., Chandra, D., Raina, O. K., Gupta, S. C. and Jayraw, A. K., 2006. Prepatent detection of *Fasciola gigantica* infection in bovine calves using metacercarial antigen. Indian J. Exp. Biol., 44: 749-753.

Verma, A. K., 1981. Epidemiology, host-parasite relationship and control of fasciolisis in sheep, goats, cattle and buffaloes. Proc. Summer Institute, ICAR., p.39, Ranchi, Patna. (not indexed).

Yadav, B. B. and Karyakarte, P. P., 1979. Histological effects of larval digenea on nonspecific esterase activity in the digestive gland of *Lymnaea acuminata*. Indian J. Parasitol, 3: 193.

Yadav, C. L., Rajat Garg, Banrejee, P. S., Rajeev Ranjan Kumar, and Sanjay Kumar., 2006. Epidemiology of *Fasciola gigantica* in infection in cattle and buffaloes of western Uttar Pradesh, India. Proc. XVII Natl. Cong. Vet. Parasitol., November 15-17, 2006, Dept. Parasitol., Rajiv Gandhi Coll.Vet. & Anim. Sci, Kurumbapet, Puducherry- 605009, Abstr., S-II.33, p.69.

Yadav, C. L., Rajat Garg, Banrejee, P. S. Rajeev Ranjan Kumar, and Sanjay Kumar., 2009. Epizootology of *Fasciola gigantica* infection in cattle and buffaloes in Western Uttar Pradesh. J. Vet. Parasitol., 23: 135-138.

Yadav, C. L., Rajat Garg, Kumar, R. R., Banerjee, P. S. and Rajesh, G., 2007. Seasonal dynamics of *Fasciola gigantica* infection in cattle and buffaloes in Uttaranchal, India. Indian J. Animal Sci.,77: 133-135.

Yadav, P., Kumar, V., Hira Ram, Rajat Garg, Yadav, C. L. and Banrejee, P. S., 2001. Cercarial burden of *Lymnaea auricularia* collected from an endemic area of fasciolasis. 12th Natl. Cong. Vet. Parasitol., August, 26-27, 2001, Dept. Parasitol., Coll. Vet. Sci., Acharya Ranga Agri. Univ., Tirupati-517 502, India, Abstr. S-2: 36, p.65.

Yadav, S. C. and Gupta, S. C., 1988a. On the viability of *Fasciola gigantica* metacercariae ingested by *Lymnaea auricularia*. J. Helminthol., 62: 303-304.

Yadav, S. C. and Gupta, S. C., 1988b. Immune response of rabbits against *Fasciola gigantica* through homologous irradiated metacercariae. Proc. VIII Natl. Cong. Parasitol., Calcutta, 10-12 February, 1988, Abstr, p.89.

Yadav, S. C. and Gupta, S. C., 1992. Immunodiagnosis of *Fasciola gigantica* infection by Dot-ELISA in experimentally-infected kids. Proc. III Asian Cong. Parasitol., Lucknow, PS3.6, p.12.

Yadav, S. C. and Gupta, S. C., 1993. dot-ELISA and DID in the diagnosis of *Fasciola gigantica* infection in rabbits. J. Vet. Parasitol., 7: 51-54.

Yadav, S. C. and Gupta, S. C., 1994. Immunoblot analysis of diagnostic fraction of *Fasciola gigantica* antigen in experimentally infected goats. Sixth Natl. Cong. Vet. Parasitol., 22-24 Oct., 1994, Dept. Parasitol., Coll. Vet. Sci., JNKVV, Jabalpur-482 001, M. P., India, Abstr. S-3: 18, pp.37-38.

Yadav, S. C. and Gupta, S. C., 1995a. Immunodiagnosis experimental fasciolasis in animals. Abstr. 12th Natl. Cong. Parasit. Panaji, Goa, 23-25 January, 1995, J. Parasit. Dis., 19: 96.

Yadav, S. C. and Gupta, S. C., 1995b. Immunodiagnosistic moieties in somatic and excretory/secertory antigens of *Fasciola gigantica*. Indian J. Exp. Biol., 33: 824-828.

Yadav, S. C. and Gupta, S. C., 1996a. Identification of immunodiagnostic antigens of *Fasciola gigantica*. J. Parasit. Dis., 20: 201-202.

Yadav, S. C. and Gupta, S. C., 1996b. Immunodiagnostic moieties in somatic and excretory antigens of *Fasciola gigantica*. Global Meet Parasit. Dis., New Delhi, India 18-22 March, 1996. Abstr., J. Parasit. Dis., 20: 86.

Yadav, S. C., Kumar, S., Sharma, R. L. and Gupta, S. C., 1997a. Observation on escaping site of cercariae from redia of *Fasciola gigantica* in *Lymnaea auricularia*. J. Parasit. Dis. 21: 189-193.

Yadav, S. C., Saini, M., Raina, O. K., Nambi, P. A. Jadav, K. and Sriveny, D., 2005. *Fasciola gigantica* cathepsin L-cysteine proteinase in the detection of early experimental fasciolosis in ruminants. Parasitol. Res., 97: 527-534.

Yadav, S. C., Sharma, R. L., Kalicharan, A., Mehra, U. R., Dass, R. S. and Verma, A. K., 1999. Primary experimental infection of reverine buffaloes with *Fasciola gigantica*. Vet. Parasitol., 82: 285-296.

Yokananth, S., Ghosh, S, Gupta, S. C., Suresh, M. G. and Saravanan, B. C., 2005. Characterization of specific and cross-reacting antigens of *Fasciola gigantica* by immunoblotting. Parasitol. Res., 97: 41-48.

1.4 FAMILY: ECHINOSTOMATIDAE

Agrawal, R. D., 1971. Infective stages of two helminth parasites from intermediate-paratenic hosts with notes on their partial life cycles in experimental mammals and on two of the amphistome cercariae from two of the common snails, their encystment and development in experimental mammals. (Abstr. Thesis), Agra Univ. J. Res., 20: 137-138.

Agrawal, R. D. and Pande, B. P., 1972. Partial life cycle of echinostomal fluke of pigs: An experimental study. Indian J. Anim. Sci., 42: 194-198.

Arya, A. K., 1977. Biology of the *Echinostoma malayanum* (Leipur, 1911) an intestinal echinostome of man and pig. M. V. Sc., Thesis, Bihar Vet. Coll., Patna, Bihar, India.(not indexed).

Bandyopadhyay, B. K., Maji, A. K., Manna, B., Bera, D. K., Addy, M. and Nandy, A., 1995. Pathogenecity of *Artyfechinostomum oraoni* in naturally infected pig. Trop. Med. Parasitol., 46: 138-139.

Choubisa, S. L., 2002. Focus on seasonal occurrence of larval trematode (cercarial) parasites and their host specificity. J. Parasit. Dis., 26: 72-74.

Choubisa, S. L., 2008. Focus on pathogenic trematode cercaria infecting fresh water snails (Mollusca: Gastropoda) of tribal region of southern Rajasthan. India. J. Parasit. Dis., 32: 47-55. (vide Table I).

Chaudhuri, S. S., Gupta, R. P. and Yadav, C. L., 1982. Note on cercarial fauna of aquatic snails of Haryana State. Indian J. Anim. Sci., 52: 1273-1275.

Chaudhuri, S. S., Gupta, R. P., Kumar, S., Singh, J. and Sangwan, A. K., 1993. Epidemiology and control of infectionof cattle and buffaloes in eastern Haryana, India. Indian J. Anim. Sci., 63: 600-605.

Dhanumkumari, C., Hanumanth Rao, K. and Shyamasundari, K., 1991. Life cycle of *Echinochasmus baugulai* (Trematoda, Echinostomatidae). Int. J. Parasitol., 21: 259-263.

Dharmendra Nath, D., 1969. *Rana cyanophlyctis* as the second intermediary of *Artyfechinostomum sufrartyfex* (Echinostomatidae: Trematoda). Curr. Sci., 38: 342-343.

Dharmendra Nath, 1971. Studies on certain aspects in the life history, histopathology and taxonomy of some of the flukes parasitic in domestic poultry. Ph.D. Thesis submitted to Agra Univ., Agra, U. P., India. (Not indexed).

Dharmendra Nath, D., 1971. Occurrence of the metacercarial cyst of *Stephanoprora*, Odhner, 1909 and its experimental development in Indian domestic fowl. Indian J. Anim. Res., 5: 81-82.

Dharmendra Nath, D., 1972a. A note on the occurrence of metacercarial cysts in the Indian pond frog - *Rana cyanophlyctus*. Indian Vet. J., 49: 434-435.

Dharmendra Nath, D., 1972b. Observations on metacercaria of *Artyfechinostomum sufrartyfex* Lane, 1915 and its experimental development in a reptile. Indian Vet. J., 49: 767-771.

Dharmendra Nath, D., 1973a. A note on the echinostomatid metacercarial fauna encountered in Indian fresh water snails. Indian Vet. J., 50: 292-293.

Dharmendra Nath, D., 1973b. A note on an echinostome metacercaria occurring in the Indian pond-frog and fresh-water fish. Indian J. Anim. Sci., 43: 446-449.

Dharmendra Nath, D., 1973c. A note on the occurrence of an echinostome metacercaria in the pond-frog and fresh-water snail and fish. Indian J. Anim. Sci., 43: 669-671.

Dharmendra Nath, D., 1973d. Experimental development of *Echinoparyphyium flexum* in domestic poultry and a mammal. Indian J. Anim. Sci., 43: 1068-1074.

Dharmendra Nath, D., 1974a. Observations on the experimental development of *Echinochasmus corvus* Bhalerao, 1926, in domestic poultry with remarks on the validity of some other species. Indian J. Anim. Sci., 44: 198-203.

Dharmendra Nath, D., 1974b. A note on the metacercarial fauna encountered in Indian fresh water fishes. Indian Vet. J., 51: 481-483.

Dharmendra Nath, D., 1975. A note on pathology of *Echinoparyphyium flexum* (Linton, 1892) Dietz, 1960 infection in experimental white leghorn chicks. Indian J. Anim. Sci., 45: 505-507.

Dharmendra Nath, D. and Pande, B. P., 1970. Metacercarial cyst of *Echinochasmus corvus* Bhalerao, 1926 and its development in experimental birds. Z. Parasitenk., 34: 343-350.

Dutt, S. C. and Bali, H. S., 1980. Snails of Punjab State and their trematode infections. J. Res. Punjab agric. Univ. 12: 222-228.

Johnpaul, A., Raghunathan, M. B. and Selvanayagam, M., 2010. Population dynamicas of freshwater molluscan in the lentic eco-systems in and around Chennai. Rec. Res. Sci. & Tech., 2: 80-86.

Kumar, V., 1980a. The digenetic trematodes, *Fasciolopsis buski, Gastrodiscoides hominis* and *Artyfechinostomum malayanum* as zoonotic infection in South Asian countries. Ann. Soc. Belge. Med. Trop., 60: 331-339. (not indexed).

Madhavi, R., 1980. Comparison of parasitic fauna of *Aplocheilus panchax* and *A. melastigma*. J. Fish. Biol., 17: 349-358.

Madhavi, R., Rao, N. N. and Rukmini, C., 1989. The life history of *Echinochasmus bagulai* Verma 1935 (Trematoda, Echinostomatidae). Acta Parasitol. Pol., 34: 259-265.

Maji, A. K., Manna, B., Bandyopadhyay, B. K., Bera, D. K., Addy, M. and Nandy, A., 1995. On the life cycle of *Artyfechinostomum oraoni* Bandyopadhyay, Manna and Nandy, 1989: Embryogenesis and development in the intermediate host. Indian J. Med. Res., 102: 124-128.

Matta, S. C., 1965. Studies on larval trematodes with special reference to metacercariae and gaint eimerian schizonts in abomasums and small intestine of local sheep – a histological study. M. V. Sc. Thesis, Agra Univ., Agra, U. P., India, (Not indexed).

Matta, S. C. and Pande, B. P., 1966. Studies on some metacercariae in local snails. Indian J. Helminth., 18: 128-141.

Mohandas, A., 1971a. Contributions to the cercarial fauna of Kerala. Ph. D. Thesis submitted to the Univ. of Kerala, Trivandrum, Kerala. (Not Indexed).

Mohandas, A., 1971b. *Artyfechinostomum sufrartyfex* Lane, 1915, a synonym of *Echinostoma malayanum* Leiper, 1911 (Trematoda, Echinostomatidae). Acta Parasitol. Pol., 19: 361-368.

Mohandas, A., 1973. Studies on life history of *Echinostoma ivaniosi* n. sp. J. Helminthol., 47: 421-438.

Mohandas, A., 1974a. The pathological effect of larval trematodes on the digestive glands of four species of gastropods. Folia Parasitol. (Praha), 21: 219-224.

Mohandas, A., 1974b. Studies on freshwater cecariae of Kerala 1. Incidence of infection and seasonal variation. Folia Parasitol. (Praha), 21: 311-317.

Mohandas, A., 1974c. The present status of the genera *Artyfechinostomum, Neo-artyfechinostomum* and *Psuedoartyfechinostomum* and the validity of the species included in the three genera (Trematoda, Echinostomatidae). Riv. Parassitol., 35: 205-212.

Mohandas, A. and Nadakal, A. M., 1978. *In vivo* development of *Echinostomum malayanum* Leiper, 1911 with notes on population density, chemical composition and pathogenecity and in vitro encystment of the metacercariae (Trematoda, Echinostomatidae). Z. Parasitenkd., 55: 139-151.

Mohandas, A., 1981. Studies on fresh water cercariae of Kerala.VII. Echinostomatid cercariae. Proc. Indian Acad. Sci., 90: 433-444.

Mudaliar, S. V. and Alwar, V. S, 1947. A check-list of parasites (Class-Trematoda and Cestoda) in the Department of Parasitology, Madras Veterinary College laboratory. Indian Vet. J. 23: 423-434.

Mukherjee, R. P., 1966. Seasonal variations of cercarial infections in snails. J. Zool. Soc., India. 18: 39-49.

Mukherjee, R. P., 1986. Fauna of India and Adjacent Countries: Larval Trematodes, Part II – Parapleurolophocerca and Echinostome Cercariae. Zoological Survey of India, Calcutta. p.1-98.(Reference book, IH not indexed).

Muraleedharan, K., 1967. Studies on experimental infections with some of the intestinal flukes in laboratory-raised chicks and laying hens, on normal and deficient rations. M. V. Sc. Thesis submitted to Agra Univ., Agra, U. P., India. (Not indexed).

Muraleedharan, K. and Pande, B. P., 1967. Notes on an experimental infection of laboratory raised chicks maintained on normal and deficient feeds with *Echinoparyphium flexum* and a coccidium. Indian J. Anim. Hlth., 6: 197-213.

Muraleedharan, K., Prasanna Kumar, S., Hegde, K. S. and Alwar, V. S., 1977. Larval trematode infections in aquatic snails of Karnataka State – A preliminary study. Mysore J. agric. Sci., 11: 101-104.

Patnaik, M. M., 1968. Notes on glycogen deposits of *Lymnaea auricularia var rufescens* and parasitic larval stages of *Fasciola gigantica* and *Echinostoma revolutum* – A histochemical study. Annal. Parasitol., 43: 449-456.

Patnaik, M. M. and Ray, S. K., 1966. On the life history and distribution of *Echinostoma revolutum* (Froleich, 1802) in Orissa. Indian Vet. J., 43: 591-600.

Patnaik, M. M. and Ray, S. K., 1967. A histologic study of *Lymnaea auricularia* var *rufescens* infected with larval stages of *Echinostoma revolutum*. Jap. J. Med. Sci. Biol., 19: 253- 258.

Patnaik, M. M. and Ray, S. K., 1968. Studies on geographical distribution and ecology of *Lymnaea auricularia* var. *rufescens* the intermediate host of *Fasciola gigantica* in Orissa. Indian J. Vet. Sci., 38: 484- 508.

Pemola Devi, N. and Jauhri, R. K., 2008. Diversity and cercarial shedding of malco fauna collecting from water bodies of Ratnagiri District, Maharashtra. Acta Trop., 105: 249-252.

Peter, C. T., 1954. Studies on larval trematodes from fresh water snails in Madras. Proc. 41st Indian Sci.Cong. (Hyderabad), Part III, Abstr., pp.221-222.

Peter, C. T., 1955. Echinostome cercaria from Bareilly with development of a new species. Proc. 42nd Indian Sci. Cong. (Baroda), Part III, Sec. 7, Zool. and Entmol., Abstr. No. 41, p.285.

Peter, C. T., 1955. Studies on the cercarial fauna in Madras II. A new species of echinostome cercaria. Indian J. Vet. Sci., 25: 219-224.

Peter, C. T., 1957. Observations on precercarial development of *Echinostoma revolutum* (Froleich). Proc. 44th Indian Sci. Cong. (Calcutta), Part III, Sec. 9, pp.369-370.

Premvati, G. and Pande, V., 1974. On *Artyfechinostomum malayanum* (Leiper, 1911) Mendheim, 1943 (Trematoda: Echinostomatidae) with synonymy of allied species and genera. Proc. Helm. Soc. Wash., 41: 151-161.

Rai, D. N., 1966. Observations on the metacercariae encountered in snails and studies on the helminth parasites of fish-eating birds, snake birds and little cormorant. M. V. Sc. Thesis submitted to Agra Univ., Agra, U. P., India. (Not indexed).

Rai, D. N. and Pande, B. P., 1967. On the metacercariae in *Vivipara bengalensis* (Lamarck) race *mandiensis* Kobelt and observations on an experimental infection with echinostome form. Z. Parasitenkd., 28: 264-276.

Ramalingam, K., 1960. The morphology and life history of *Echinochasmus baugulai*, Verma 1935 (Trematoda, Echinostomatidae) with ecological observations on its larval forms. J. Mar. Biol. Ass. India, 2: 35-50.

Rao, M. A. N., 1933. A preliminary report on the adult trematode obtained from cercariae XXIII Sewell, 1922. Indian J. Vet. Sci., 3: 317- 320.

Sahai, B. N., 1969-70. A survey of trematode infection in aquatic snails. J. Vetcol., 10: 9-12.

Sewell, R. B. S., 1922. *Cercariae indicae*. Indian J. Med. Res., 10: 131-135.

Singh, U., 1977. Studies on echinostomes in India. 7. Morphology and life history of *Echinostoma luteola* n. sp. trematodea (Echinostomatidae). In: Abstr. 1st Natl. Cong. Parasitol.(Baroda), 24-26 February, 1977.

Singh, V., 1978. Morphology and life cycle of *Echinoparyphium* sp. (Trematoda: Echinostomatidae). Asian Cong. Parasitol., 23-26, February, 1978, Bombay, Abstr. IV-77, p.224.

Sreekumaran, P., 1966. Studies on the common trematodes of ducks. Thesis (M. Sc). Faculty of Vet. Sci., Univ. Kerala.

Sreekumaran, P. and Peter, C. T., 1973. Studies on the common trematodes of ducks in Kerala. IV. Life history of echinostomes. Indian Vet. J., 50: 1152.

1.5 FAMILY: HETEROPHYIDAE

Chakrabarti, K. K., 1974., Studies on some metacerceariae of the Indian fresh water fishes, *Channa punctatus* (Bloch) and *C. striatus* (Bloch). Rev. Iber. Parasitol., 34: 57-81.

Dhanumkumari, C, Hanumantha Rao, K. and Shyamasundari, K., 1991.Two species of heterophyid larval digeneans from a thiarid gastropod, *Thiara tuberculata* (Muller) from India. Bol. Chil. Parasitol., 46: 14-18.

Dhanumkumari, C., Hanumantha Rao, K. and Shyamasundari, K., 1993. Life history of *Centrocestus formosanus* (Nishigori, 1924) Trematoda, Heterophyidae) from India. Indian J. Parasitol., 17: 59-65.

Dharmendra Nath, 1971. Studies on certain aspects in life-history, histopathology and taxonomy of some of the flukes parasitic in poutry. Ph. D. Thesis submitted to Agra Univ. (Not indexed).

Dharmendra Nath, 1972a. Experimental development of *Centrocestus formosanus* (Nishigori, 1924) in Indian domestic poultry and notes on natural infection. Indian J. Anim. Sci., 42: 862-868.

Dharmendra Nath, 1972bc. Pathology of *Centrocestus formosanus* (Nishigori, 1924) infection in experimental pigeon. Indian J. Anim. Sci., 42: 952-954.

Dharmendra Nath, 1972c. Observations on the metacercarial cyst of *Haplorchis thaichui* (Nishigori, 1924) and its development in experimental hosts. Indian J. Anim. Sci., 42: 55-60.

Dharmendra Nath, 1973a. Observations on the metacercarial cyst of *Haplorchis taichui* and its development in experimental hosts. Indian J. Anim. Sci., 43: 55-60.

Dharmendra Nath, 1973b. Observations on the metacercaria of *Haplorchis yokogawai* (Katsuta, 1932) and its development in experimental hosts with remarks on some other Indian species. Indian J. Anim. Sci., 43: 649-655.

Dharmendra Nath, 1974. A note on metacercarial fauna encountered in Indian fresh-water fishes. Indian Vet. J., 51: 481-483.

Dharmendra Nath and Pande, B. P., 1970. Identity of three heterophyid metacercariae infesting some of the fresh-water fishes. Curr. Sci., 39: 325-326.

Madhavi, R., 1980. Comparison of parasitic fauna of *Aplocheilus panchax* and *A. melastigma*. J. Fish Biol., 17: 349-358.

Madhavi, R., 1986. Distribution of metacercaria of *Centrocestus formosanus* (Nishigori, 1924) (Trematoda, Heterophyidae) on the gills of *Aplocheilus panchax*. J. Fish Biol., 29: 685-690.

Madhavi, R., 2009. Fish-borne zoonotic haplorchine trematodes: the question of differentiation of species. In: Currents Trends in Parasitolgy, Veena Tandon, Arun K.Yadav and Bishnupada Roy. (Eds), Proc. 20th Natl. Cong. Parasitol.,

Shillong, India (November 3-5, 2008). Panima Publishing Corporation, New Delhi. pp.21-32.

Madhavi, R. and Rukmini, C., 1988. Population biology of the metacercariae of *Centocestus formosanus* (Nishigori, 1924) (Trematoda: Heterophyidae) in *Aplocheilius sanus* (Hamilton) from a fresh water stream. Proc. VIII Natl. Cong. Parasitol., Calcutta, 10-12 February, 1988, Abstr., p.11.

Madhavi, R. and Rukmini, C., 1991. Population biology of the metacercaria of *Centrocestus formosanus* (Trematoda, Heterophyidae) on the gills of *Aplocheilus panchax*. J. Zool., 223: 509-520.

Madhavi, R., Uma Devi, K. and Swarnakumari, V. G. M., 1997. Community structure of larval trematode fauna of the snail, *Thiara tuberculata* from a freshwater stream at Visakhapatnam, Andhra Pradesh. Curr. Sci., 72: 582-585.

Pande, B. P. and Shukla, R. P., 1972. Metacercarial cyst of *Haplorchis pumilio* and its development in experimental mammals and two other heterophyid infections of fresh-water fishes and their zoonotic significance. Indian J. Anim. Sci., 42: 971-978.

Pande, B. P. and Shukla, R. P., 1973. Experimental development of metacercarial cyst of several heterophyid fluke in hamster, histopathology of lesions and their role in human intestinal heterophyidiasis. Indian J. Anim. Sci., 43: 766- 774.

Pandey, K. C., 1966. Studies on metacercaria of the fresh-water fishes of India. 1. On the morphology of metacercaria of *Haplorchis yokogawai* (Katsuta, 1932) Chen, 1936, Proc. Natl. Acad. Sci., India (Section B), 36: 437-440.

Pande, V. (Vide Vibha Pande, below)

Premvati, G. and Vibha Pande, 1973. Chick as an experimental host of *Haplorchis pumilio* (Looss) Looss 1899. Proc. 60th Indian Sci. Cong., Part III, Sect 7, Zool and Entomol., Abstr., p.535.

Premvati, G. and Pande, V., 1974.On *Centrocestus formosanus* (Nishigori, 1924) Price, 1932 and its experimental infection in white leghorn chicks. Japanese J. Parasit., 23: 79-84.

Rai, P., 1976. Fish borne zoonoses with special reference to helminth diseases. Uttar Pradesh Vet. J., 4: 52-54.

Swarnakumari, V. G. M., 2001. Comparison of cercarial infection of the snail, *Thiara tuberculata* at different habitats. J. Parasitol. Appl. Anim. Biol., 10: 49-54.

Uma Devi, K. and Madhavi, R., 1997a.The effect of light and temperature on the cemergence of *Haplorchis pumilio* cercariae from the snail host, *Thiara tuberculata*. Acta Parasitol., 42: 12-17.

Uma Devi, K. and Madhavi, R., 1997b. Arginophillic papillae of *Haplorchis pumilio* cercariae (Trematoda: Heterophyidae). J. Parasitol. Appl. Anim. Biol., 10: 49-54.

Uma Devi, K. and Madhavi, R., 2000. Observations on the morphology and life cycle of *Procerovum varium* (Onji and Nishio, 1916) (Trematoda: Heterophyidae). System. Parasitol., 46: 215-225.

Uma Devi, K. and Madhavi, R., 2006. The life cycle of *Haplorchis pumilio* (Trematoda: Heterophyidae) from Indian region. J. Helminthol., 80: 327-332.

Vibha Pande and Premvati, G., 1977a. Development of metacercariae of *Haplorchis* spp. in chicks. Indian J. Parasitol., 1: 165-167.

Vibha Pande and Premvati, G., 1977b. On *Centrocestus formosanus* (Nishigori, 1924) Price, 1932 and its experimental infection in white leghorn chicks. Japanese J. Parasitol., 25: 79-84.

1.6 FAMILY: PLAGIORCHIDAE

Agarwal, R. D., 1970. Infective stages of two helminth parasites from intermediate/paratenic hosts with notes on their partial life-cycle in experimental mammals. On two amphistome cercariae from common snails, their encystment and development in experimental mammals. M. V. Sc. Thesis, Thesis submitted to Agra Univ., Agra, U. P., India. (Abstract Thesis, 1971). Agra Univ. J. Res. 20: 137-138.

Chaudhry, R. K., 1972. Studies on certain aspects of *Prosthogonimus ovatus* (Rudolphi, 1803) Luhe, 1899 of poultry and some of the stomach worms of pig. M. V. Sc. Thesis submitted to Agra Univ., Agra, U. P., India.(Not indexed, Vide Thesis Abstact, 1977, Agra Univ. J. Res., 23: 99.).

Chaudhry, R. K. and Ahluwalia, S. S. 1973. Note on alkaline and acid phoshatase activity in bursa Fabricii of white leghorn pullets during infection with *Prosthogonimus ovatus* (Rudolphi, 1803) Luhe 1899. Indian Vet J., 50: 549-550.

Chaudhry, R. K, Bhargava, D. N. and Ahluwalia, S. S., 1973. The first intermediate host of *Prosthogonimus ovatus* (Trematoda: Plagiorchidae). Indian Vet. J., 50: 492-493.

Dharmendra Nath, 1971. Studies on certain aspects in the life history, histopathology and taxonomy of some of the flukes parasitic in domestic poultry. Ph.D. Thesis submitted to Agra Univ., Agra, U. P., India. (Not indexed).

Dharmendra Nath and Pande, B. P., 1973. Experimental development of *Prosthogonimus ovatus* (Rud, 1803) Luhe 1899 in common quails, grey patridges and guinea fowls. Indian Vet J., 50: 465-475.

Haleem, M. A., 1969. Experimental studies on some of the common flukes parasites in domestic animals. M. V. Sc. Thesis submitted to Agra Univ., Agra, U. P., India. (Not indexed).

Manohar, L and Rao, P. V., 1976. Physiological response to parasitism. I. Changes in carbohydrate reserve of the molluscan host. Southeast Asian Trop. Med. Pub. Hlth., 7: 395-404.

Mishra, G. S., 1985. Observations on larval helminthes in dragonfly, dung beetle and slug. Parasitic infestations of piglets with special reference to lesions encountered. M. V. Sc. Thesis submitted to Agra Univ., Agra, U. P., India. (Not indexed).

Mishra, G. S. and Pande, B. P., 1968. On the metacercaria of *Prosthogonimus putschkowskii* in *Sympatrum decoloratum* (Odonata: Insecta) with observations on its development in chicks. Indian J. Vet. Sci., 37: 156-164.

Muraleedharan, K., 1967. Studies on experimental infections with some of the intestinal flukes in laboratory-raised chicks and laying hens, on normal and deficient rations. M. V. Sc. Thesis submitted to Agra Univ., Agra, U. P., India. (Not indexed).

Muraleedharan, K., 1999. Role of certain insects like ants, lice, grass-hoppers and dragon flies as intermediate hosts of helminth parasites. In: Vectors and Vector-borne Parasitic Diseases. IV Natl. Traing. Progrm., 25-1-1999 to 8-2-1999, Centr. Adv. Stud., Dept. Parasitol., Vet. Coll., Univ. Agri. Sci., Bangalore-560 024. pp.125-131.

Muraleedharan, K., 2000. Biology, economic importance, and control of lice, ants, grass-hoppers and dragon flies. In: Acarines and Insects of Veterinary and Medical Importance. V. Natl. Traing. Progrm., Centr. Adv. Stud., Dept. Parasitol., Vet. Coll., Univ. Agri. Sci., Hebbal, Bangalore-560 024. pp.18-25.

Muraleedharan, K. and Pande, B. P., 1968. Experimental infections with prosthogonimid metacercaria in chicks maintained fed deficient mashes. Indian Vet J., 45: 641-649.

Pande, B. P., Bhatia, B. B. and Chauhan, P. P. S., 1970. Efficacy of carbon tetrachloride in experimental prosthogonimiasis – A preliminary study. Orissa Vet J., 5: 33-37.

Pratap Singh, 1968. Observations on experimental infections of 4-6 month old pullets and laying hens and also pigeon with prosthogonimid metacercariae with a note on eumegacetid form. M. V. Sc. Thesis submitted to Agra Univ., Agra, U. P., India. (Not indexed).

Pratap Singh and Pande, B. P., 1968. Experimental *Prosthogonimus ovatus* infections in the domestic pigeons. Curr. Sci., 37: 500-501.

Pratap Singh and Pande, B. P., 1971. Experimental prosthogonimiasis in 4-6 months pullets and laying hens with special reference to pathological lesions. Indian J. Anim. Sci., 41: 122-136.

Ram Prakash, 1966. Observations on the metacercariae from dragon flies and larval poultry tapeworms in one of the common beetles and studies on helminth parasites of fish eating birds.-kites. M. V. Sc. Thesis submitted to Agra Univ., Agra, U. P., India. (Not indexed).

Ram Prakash and Pande, B. P., 1969. On an experimental infection of two laying white leghorn hens with prosthogonimid metacercaria (a preliminary study). Indian Vet J., 46: 211-215.

Singh, B. P., 1975. A preliminary survey of cercarial fauna of two common snails-*Lymnaea luteola* and *Indoplanorbis exustus* at Mathura: Observations on the post-cercarial developmental stages and host parasite relationships of the two mammalian schistosomes. M. V. Sc. Thesis submitted to Agra Univ., Agra, U. P., India. (Not indexed).

1.7 FAMILY: CYCLOCOELIDAE

Sreekumaran, P., 1966. Studies on the common trematodes of ducks. Thesis (M. Sc.), Vet. Sci. Thesis submitted to the Univ. Kerala, Trivandrum, Kerala.

Sreekumaran, P. and Peter, C. T., 1973a. Studies on the common trematodes of ducks in Kerala II. Life history of *Typhlocoelium cymbium* (Diesing) Kossack, 1911. Indian Vet J., 50: 1060.

Sreekumaran, P. and Peter, C. T., 1973b. Studies on the common trematodes of ducks in Kerala III. 'Cannibalism" among rediae of *Typhlocoelium cymbium*. Indian Vet J., 50: 1061.

1.8 FAMILY: PHILOPHTHALMIDAE

Bhaskar Rao, P and Rao, B. V., 1981. Developmental stages of the *Philophthalmus anatinus* Sugimoto, 1928 in the snail. Indian Vet. J., 58: 525.

Bhatia, B. B., Pathak, K. M. L., and Kumar, D., 1985. Development of philophthalmic fluke in the eye of chickens and their pathogenic effects. Indian J. Parasitol., 9: 285-287.

Madhavi, R., Uma Devi, K. and Swarnakumari, V. G. M., 1997. Community structure of larval trematode fauna of the snail, *Thiara tuberculata* from a freshwater stream at Visakhapatnam, Andhra Pradesh. Curr. Sci., 72: 582-585.

Rezual Karim, 1982. Study on some larval trematodes encountered in the melanoid snail, *Thiara* (*Melanoides*) *tuberculata* (Muller). Indian Vet. J., 59: 562-565.

Rezual Karim, Bhatia, B. B. and Rai, D. N. 1982. Post cercarial development of *Philophthalmus gralli* Mathis and Leger, 1910 in experimental animal. Indian J. Parasit., 6: 275-278.

Satyanarayana Murty, A., 1966. Experimental demonstration of the life cycle of *Philophthalmus* sp. (Trematode: Philophthalmidae). Curr. Sci., 35: 366-367.

Saxena, S. K., 1981. Studies on the life history of *Philophthalmus lucknowensis* Baugh, 1962. II: Hatching and miracidium. Rev. Ibér. Parasitol., 41, 493–518.

Saxena, S. K., 1984. Studies on the life history of *Philophthalmus lucknowensis* Baugh, 1962. III: Rediae and cercariae. Rev. Ibér. Parasitol., 44, 291–307.

Saxena, S. K., 1985. Studies on the life history of *Philophthalmus lucknowensis* Baugh, 1962. IV. Metacercaria to adult. Rev. Ibér. Parasitol., 45, 59–77.

Swarnakumari, V. G. M. and Madhavi, R., 1992a. Growth, development and allometry of *Philophthalmus nocturnus* in the eyes of domestic chicks. J. Helminthol., 66: 100-107.

Swarnakumari, V. G. M. and Madhavi, R., 1992b. The effects of crowding on adults of *Philophthalmus nocturnus* grown in domestic chicks. J. Helminthol., 66: 255-259.

Syed Ismail Koya, M. S., 1988. Ecobiology of helminthes parasites of finfishes and shellfishes of Cochin waters with special reference to digenetic trematodes. Ph.D. Thesis submitted to Cochin Univ. Sci. and Tech., Cochin, Kerala, M. S. 96, pp.302-303.

1.9 FAMILY: LECITHODENDRIIDAE

Dharmendra Nath, 1971. Occurrence of metacercarial cysts of *Eumegacetes singhi* Jaiswal, 1957 and its experimental development in poultry. Indian Vet. J., 48: 1283-1284.

Hanumantha Rao, K. and Madhavi, R., 1961. Metacercaria of *Eumegacetes* sp. (Trematoda: Lecithodendriidae) in dragon fly naiads from a stream of Waltair. Curr. Sci., 30: 303-304.

Ram Prakash and Pande, B. P., 1970. Libellulid dragonflies as second intermediate hosts of flukes Indian J. Helminthol., 21: 150-160.

Singh, P. and Pande, B. P., 1968. Adults of *Eumegacetes artamii* from white leghorn pullets infected with the metacercariae from two dragon flies. Curr. Sci., 37: 563-564.

Swarnakumari, V. G. M., 2001. Comparison of cercarial infection of the snail *Thiara tuberculata* at different habitats. J. Parasitol. Appl. Anim. Biol., 10: 49-54.

Swarnakumari, V. G. M. and Madhavi, R., 1994. The life cycle of *Eumegacetes artamii* Mehra, 1935 (Trematoda, Eumegacetidae). Acta Parasitol., 39: 9–12.

1.10 FAMILY: PARAGONIMIDAE

Anon. Food-borne parasitic zoonosis: Status of metacercarial infection in fishes of Assam. Res. Proj., Reg. Med. Centr, Dibrugarh, Assam. (Research Project, K. Narain: Investigator and J. Mahanta: Co-Investigator, not indexed).

Anon., 2005-2009. Molecular characterization infrapopulation of differentiation of lung flukes in north eastern region of India (Narain, K., Rekha Devi, K. and Mahanta, J.), Res. Proj., Reg. Med. Centr, Dibrugarh, Assam. (Not indexed).

Anon., 2007-2010. Prevalence and molecular diagnosis of lung flukes in north eastern region of India (Narain, K., Rekha Devi, K. and Mahanta, J.), Res. Proj., Reg. Med. Centr, Dibrugarh, Assam. (Not indexed).

Anon., 2008-2009. Annual Report, Reg. Med. Centr., N. E. Region. (ICMR), Dilbrugarh-786 001.(Not indexed).

Anon, 2010. Studies on fish trematode infection in North Eastern Region of India. (An ICMR Proj., No.19. Investigators: J. Mahanta, K. Narain and and K. Rekha Devi, not indexed).

Mahajan, R. C., 2005. Paragonimiasis: An emerging public health problem in India. Indian J. Med. Res., 121: 716-718.

Mukherjee, R. P., 1992. Fauna of India and Adjacent Countries: Larval Trematodes, Part I – Amphistome Cercariae (pp.1-89), Zoological Survey of India, Kolkatta. pp.1-158.

Narain, K., Agatsuma, T. and Blair, D. 2010. Paragonimus. In: Molecular detection of food-borne infection. Dongyou Liu, Ed., CRC Press, Taylor & Francis Group, Boca Raton. Chapter 61, pp. 827-835. (Review, not indexed).

Narain, K., Rekha Devi, K. and Mahanta, J., 2003a. *Paragonimus* and paragonimiasis-A new focus in Arunachal Pradesh, India. Curr. Sci., 84: 985-987.

Narain, K., Rekha Devi, K. and Mahanta, J., 2003b. A rodent model for pulmonary paragonimiasis. Parasitol. Res., 91: 517-519.

Prasad, P. K., Goswami, L.M., Veena Tandon and Chatterjee, A., 2011. PCR-based molecular characterization and insilico analysis of food-borne trematode parasites, *Paragonimus westermanii, Fasciolopsis buski* and *Fasciola gigantica* in northeastern India using ITS 2 r DNA. Asian Pacafico Bioinformatics Network, 9th International Conf. Bioinformatics (Incob), Tokyo, Japan, 26-28th Sept, 2010, Bioinformation, 6 (2): 64-68.

Prasad, P.K., Tandon, V., Biswal, D. K, Goswami, L.M., Chatterjee, A. 2009. Phylogenetic reconstruction using secondary structures and sequence motifs of ITS2 rDNA of *Paragonimus westermanii* (Kerbert, 1878) Braun, 1899 (Digenea: Paragonimidae) and related species. BMC Genomics 10: (Suppl. 3) S25.

Rekha Devi, K., Narain, K., Agatsuma, T., Blair, D., Nagataki, M., Wickramasughe, S., Yatawara, L. and Mahanta, J., 2010. Morphological and molecular characterization of *Paragonimus westermanii* in northeastern India. Acta Trop., 116: 31-38.

Rekha Devi, K., Narain, K., Mahanta, J., Nirmolia, T, Blair, D., Saikia, S.P. and Agatsuma, T., 2013. Presence of three distinct genotypes within the *Paragonimus westermani* complex in northeastern India. Parasitology, 140(1): 76-86.

Singh, T. S., 1996. Studies on *Paragonimus* infections in man and crabs in Manipur, India. Abstr. In: Global Meet Parasit. Dis., New Delhi, India, 18-22 March, 1966. J. Parasitic Dis., 20: 72.

Singh, T. S., 2002. Occurrence of lung fluke, *Paragonimus hueit'ungensis* in Manipur, India. Chinese Med. J. (Taipei), 65: 426-429.

Singh, T. S., 2003. Occurrence of the lung fluke, *Paragonimus hetrotremus*, in Manipur India. Chinese Med. Sci. J. (Beijing), 18: 20-25.

Singh, T. S., 2005. Paragonimus and paragonimiasis: Present status in India. 17th Natl. Congr. Parasitol., 24-26 October, 2005, Reg. Med. Res. Centr., N. E. Region (ICMR), Dibrugarh-786001, Assam, India. Abstr. No. 143-O, p. 67. (Review).

Singh, T. S. and Singh, Y. I., 1997. Three types of Paragonimus metacercariae isolated from Potamiscus manipurensis in Manipur. Indian J. Med. Microbiol., 15: 159-162.

Singh, T. S. and Sugiyama, H., 2008. Paragonimiasis in India: A newly emerging food borne parasitic disease. Clinc. Parasitol., 19: 95- 98. (Review, not indexed)

Singh, T. S., Singh, L. D. and Sugiyama, H., 2006. Possible discovery of Chinese lung fluke, *Paragonimus skrajabini* in Manipur, India. Southeast Asian J. Trop. Med. Publ. Hlth., 37 (Suppl. 3): 53-56.

Singh, T. S., Sugiyama, H. and Rangsiruji, A., 2012. *Paragonimus* & paragonimiasis in India. Indian J. Med. Res., 136: 192-204. (Review article).

Singh, T. S., Sugiyama, H., Rangsiruji, A. and Ranjana Devi, K. H., 2007. Morphological and molecular characterization of *Paragonimus hetrotremus*, the causative agent of human paragonimiasis in India. Southeast Asian J. Trop. Med. Publ. Hlth., 38 (Suppl. 1): 82- 86.

Singh, T. S., Sugiyama, H., Umehara, A., Hiese, S. and Khalo, K., 2009. *Paragonimus hetrotremus* infection in Nagaland. A new focus of paragonimiasis in India. Indian J. Med. Microbiol., 27: 123-127.

Singh, T. S., Sugiyama, H., Ranjana Devi, K. H., Singh, L. D., Binchar, S. and Rangsiruji, A., 2011. Experimental infection of laboratory animals with *Paragonimus hetrotremus*, metacercariae occurring in Manipur India. Southeast Asian J. Trop. Med. Publ. Hlth., 42: 34-38.

Takeda, M., Sugiyama, H. and Singh, T.S., 2012. Some freshwater crabs from northeast India bordered on Myanmar. J. Teikyo Heisei Univ., 23, 199-213.

Veena Tandon, Prasad, P. K., Chatterjee, A. and Bhutia, P.T., 2007. Surface fine topography and PCR-based determination of metacercariae of *Paragonimus* sp. from edible crabs in Arunachal Pradesh, Northeast India. Parasitol. Res., 102: 21-28.

1.11 FAMILY: PARAMPHISTOMIDAE

Agrawal, M. C., 1994a. Devouring amphistome metacercariae by fresh water snails. Sixth Natl. Cong. Vet. Parasitol, 22-24 October, 1994, Dept. Parasitol., Coll. Vet. Sci., JNKVV, Jabalpur-482 001, M. P., India, Abstr. S-2: 12, p.14.

Agrawal, M. C., 1994b. Shedding behavior of amphistome cercariae by fresh water snails. Sixth Natl. Cong. Vet. Parasitol, 22-24 October, 1994, Dept. Parasitol., Coll. Vet. Sci., JNKVV, Jabalpur-482 001, M. P., India, Abstr. S-2: 13, pp.14-15.

Agrawal, M. C., Pandey, S., Sirmour, S. and Deshmukh, P. S., 2000. Endemic form of cercarial dermatitis (Khujlee) in Bastar area of Madhya Pradesh. J. Parasitic Dis., 24: 217-218 (This paper refers to amphistome cercariae too).

Agrawal, M. C., Vohra, V., Gupta, S. and Singh, K. P., 2003. Schistosomiasis and other intestinal helminthes in domestic animals in and around Bhubaneswar, Orissa. Indian J. Anim. Sci., 73: 875-877.

Agarwal, R. D., 1970. Infective stages of two helminth parasites from intermediate/paratenic hosts with notes on their partial life-cycle in experimental mammals. On two amphistome cercariae from common snails, their encystment and development in experimental mammals. M. V. Sc. Thesis, Agra Univ., Agra, U. P., India. vide Agra Univ. J. Res.,20: 137-138.(Not indexed).

Agrawal, R. D. and Pande, B. P., 1971. Cercaria of *Fischoederius elongatus*, its encystment and development in experimental animals. Indian J. Anim. Sci., 41: 1055-1059.

Anantaraman, M. and Balasubramanian, G., 1949. A study of *Cercaria fraseri*, Buckley, 1939 in Madras. Curr. Sci., 18: 124-126.

Arora, R (see Ritu Arora).

Bhalerao, G. D., 1943. The cercarial fauna of the irrigated tract of the Nizam's Dominions with suggestions regarding the relationship to the trematode parasites in man and domesticated and other animals. Indian J. Vet. Sci., 13: 294-296.

Bhalerao, G. D., 1945. The commonest amphistome of domesticated animals of central provinces and its intermediate hosts. Proc. 32nd Indian J. Sci. Cong. (Nagpur), Part III, p.97.

Borkakoty, M. R. and Das, M. R., 1980. Snail and cercarial fauna in Kamrinji district of Assam. J. Res, Assam Agri. Univ., 1: 100-102.

Buckley, J. J. C., 1939. On a new amphistome cercaria (Diplocotylea) from *Planorbis exustus*. J. Helminthol., 17: 25-30.

Chatterji, R. C., 1931. Preliminary observations on the life history of an amphistome cercaria – *Cercaria indicae* XXVI, Sewell, 1922. Zool. Anz., 95: 177-179.

Chaudhuri, S. S and Gupta, R. P., 1985. Viability and infectivity of paramphistome metacercariae stored under different conditions. Indian Vet. J., 62: 470-472.

Chaudhuri, S. S., Gupta, R. P. and Yadav, C. L., 1982. Note on cercarial fauna of aquatic snails of Haryana State. Indian J. Anim. Sci., 52: 1273-1275.

Choubisa, S. L., 1997. Seasonal variation of amphistome cercarial infection in snails of Dungarpur District (Rajasthan). J. Parasit. Dis., 21: 197-198.

Choubisa, S. L., 2002. Focus on seasonal occurrence of larval trematode (cercarial) parasites and their host specificity. J. Parasit. Dis., 26: 72-74.

Choubisa, S. L., 2008. Focus on pathogenic trematode cercariae infecting fresh water snails (Mollusca: Gastropoda) of tribal region of southern Rajasthan. India. J. Parasit. Dis., 32: 47-55.

Dutt, S. C. and Bali, H. S., 1980. Snails of Punjab and their trematode infections. J. Res. Punjab Agri. Univ., 17: 222-228.

Dutt, S. C. and Gill, J. S., 1978. Experimental infections of *Indoplanorbis exustus* (Deshayes) (Gastropoda: Planorbidae) with *Paramphistomum cervi* (Zeder, 1970) Fischoeder, 1901 (Trematoda: Paramphistomidae). Asian Cong. Parasitol., Bombay, 23-26 February, 1978, Abstr., V-29, p. 280.

Dutt, S. C. and Srivastava, H. D., 1966. The life history of *Gastrodiscoides hominis* intermediate host and the cercaria of *Gastrodiscoides hominis* (Trematoda: Gastrodiscoidae) – Preliminary report. J. Helminthol., 40: 45-52.

Dutt, S. C. and Srivastava, H. D., 1972. The intermediate host and the cercaria of *Gastrodiscoides hominis* (Lewis and McConnell, 1876) Leiper, 1913 – the amphistome parasite of man and pigs. J. Helminthol., 46: 35-46.

Dwivedi, P., Prasad, A. and Verma, T. K., 1996a. Pathological changes during experimental paramphistomiasis in lambs and kids induced by *Paramphistomum epiclitum*. Global meet on parasitic diseases, New Delhi, India, 18-22 March, 1996. Abstr., J. Parasit. Dis., 20: 105.

Dwivedi, P., Prasad, A., Malviya, H. C. and Verma, T. K., 1996b. Role of CMI-response in the pathologenesis of immature paraamphistomiasis induced by *Paramphistomum epiclitum* in lambs. Sixth Natl. Cong. Vet. Parasitol, 22-24 October, 1994, Dept. Parasitol., Coll. Vet. Sci., JNKVV, Jabalpur-482 001, M. P., India, Abstr. S-3: 21, p.39.

Gill, J. S., 1987. Studies on epizootoiology, chemotherapy and immunology of *Paramphistomum* infection in sheep. Ph. D. Thesis submitted to Punjab Agri. Univ., Ludhiana -141004. Abstr., J. Vet. Parasitol., 1: 81.

Gill, J. S. and Bali, H.S., 1987a Immunization of sheep with gamma irradiated metacercariae of *Paramphistomum cervi* (Zeder, 1790) Fischoeder, 1901.Cheiron, 17: 79-81.

Gill, J. S. and Bali, H.S., 1987b. Efficacy of four anthelmintics against experimentally induced *Paramphistomum cervi* infection in buffalo calves. Indian J. Anim. Res., 21: 45-47.

Gill, J. S. and Bali, H.S., 1988. Immunization of sheep with gamma irradiated metacercariae of *Paramphistomum cervi* (Zeder, 1790) Fischoeder, 1901.Cheiron, 17: 79-81.

Ghosh, S., Preeti Rawat, Gupta, S. C. and Singh, B. P., 2005a. Comparative diagnostic potentiality of ELISA and dot ELISA in prepatent diagnosis of experimental *Fasciola gigantica* infection in cattle. Indian J. Exp. Biol., 43: 536-541.

Ghosh, S., Preeti Rawat, Veluswamy, R., Joseph David, Gupta, S. C. and Singh, B. P., 2005b. 27 kDa *Fasciola gigantica* glycoprotein for the diagnosis of prepatent fascioliasis in cattle. Vet. Res. Commun., 29: 123-135.

Gupta, A., Dixit, A. K., Pooja Dixit and Mahajan, C., 2012. Prevalence of gastrointestinal parasites in cattle and buffaloes in and around Jablpur, Madhya Pradesh. J. Vet. Parasitol., 26: 186-188.

Hafeez, Md., 1975. Studies on amphistomes and amphistomiasis of sheep and goats in Andhra Pradesh. M. Sc. (Vet.) Thesis submitted to A. P. Agri. Univ., Hyderabad, India.

Hafeez, Md., 1981. Studies on certain immunological aspects of amphistomiasis with special reference to the effect of ionizing radiation on amphistome metacercariae. Ph. D. Thesis submitted to A. P. Agri. Univ., Hyderabad, India.

Hafeez, Md. and Rao, B. V., 1979. Viability of amphistome metacercariae (*C.indicae* XXVI) after irradiation. Indian Vet. J., 56: 622.

Hafeez, Md. and Rao, B. V., 1981a. *Cercariae indicae* XXVI Sewell, 1922 and its adults. Indian J. Parasitol., 5: 233-235.

Hafeez, Md. and Rao, B. V., 1981b. Studies on amphistomiasis in Andhra Pradesh. IV. Immunisation of lambs and kids with gamma irradiated metacercariae of *Cercaria indicae* XXVI. J. Helminthol., 55: 29-32.

Hafeez, Md. and Rao, B. V., 1983a. Immunological studies on amphistomiasis in sheep and goats. II Viability of amphistome cercariae after irradiation. J. Helminthol., 57: 115-116.

Hafeez, Md. and Rao, B. V., 1983b. Antigenic cross-reaction among *Paramphistomum epiclitum*, *Gastrothylax cruminifer* and *Calicophoron cauliorchis*. Indian J. Parasitol., 7: 207-208.

Hafeez, Md. and Rao, B. V., 1984. Immunisation of lambs and kids with gamma irradiated amphistome metacercariae. Indian J. Parasitol., 8: 105-106.

Hafeez, Md. and Rao, B. V., 1986a. Precipitation tests for the detection of amphistomiasis and immune response in lambs and kids vaccinated with irradiated amphistome metacercariae (Cercariae indicae XXVI). Cheiron, 15(1): 1-4.

Hafeez, M. and Rao, B. V., 1986b. Allergic tests for the detection of paramphistomiasis and immune response in lambs vaccinated with irradiated and normal metacecercariae (*Cercariae indicae* XXVI). Cheiron 15(2): 51-55.

Hafeez, Md. and Rao, B. V., 1987. Immune response in lambs vaccinated with gamma irradiated amphistome metacercariae as judged by haematological picture. J. Nucl. Agric. Biol., 16: 154-157.

Hafeez, Md. and Rao, B. V., 1988a. Studies on preservation and *in-vitro* testing of gamma-irradiated amphistome metacercariae. Second Natl. Parasitol., March 3-5, 1988, Dept. Parasitol., Vet. Coll., Bangalore, India, Souvenir, Stream II, p.22.

Hafeez, Md. and Rao, B. V., 1988b. *In-vitro* testing of the viability of amphistome metacercariae (*Cercaria indicae* XXVI). Proc. VIII Natl. Cong. Parasitol., (Calcutta), 10-12 February, 1988, Abstr, p.12.

Hafeez, Md. and Rao, B. V., 1989a. *In-vitro* testing of the viability of amphistome metacercariae (*Cercaria indicae* XXVI). Indian J. Parasitol., 13: 143-144.

Hafeez, Md., Rao, B. V. and Krishna Swamy, S., 1984. Assay for complement mediated cytotoxic antibodies in lambs infected with *Paramphistomum epiclitum*. Indian J. Parasitol., 8: 119-121.

Hafeez, Md., Rao, B. V. and Krishna Swamy, S., 1985. Immunocyto-adherence test for the detection of cell mediated immune response in lambs vaccinated with irradiated amphistome metacercariae (*Cercaria indicae* XXVI). J. Nucl. Agric. Biol., 14: 69-71.

Hafeez, Md., Rao, B. V. and Krishna Swamy, S., 1987. Level of immunogloblins in lambs vaccinated with irradiated and normal amphistome metacercariae (*Cercaria indicae* XXVI). Indian J. Parasitol., 11: 75-77.

Hassan, S. S., Juyal, P. D. and Arvinder Kaur, 2009. Establishing experimental paramphistomosis in lambs. XIX Natl. Cong. Vet. Parasitol. & Natl. Sym., February 3-5, 2009, IAAVP & Dept. Vet. Parasitol., Coll. Vet. Sci., Guru Angad Dev Vet & Anim. Sci. Univ., Ludhiana-141 004, Abstr. OE 44, p.47.

Hassan, S. S. and Juyal, P. D., 2005a. The current epidemiological status paramphistomosis – a parasitic infection in domestic ruminants in Punjab and other adjoining areas. In: the compendium of winter school entitled "Novel approaches for diagnosis and control of parasitic diseases of domestic and wild animals", PAU Ludhiana, pp.151-158. (Review, not indexed).

Hassan, S. S. and Juyal, P. D., 2005b. Establishment of experimental infection of paramphistomosis in sheep: a methodical approach. In: the compendium of winter school entitled "Novel approaches for diagnosis and control of parasitic diseases of domestic and wild animals" PAU, Ludhiana, pp.162-169. (Review, not indexed).

Jain, S. P., 1969. Studies on the biology and comparative morphology of some amphistomatous parasites of domestic animals. Agra. Univ. J. Res. (Sci.), 18: 75-80.

Jain, S. P., 1970. Double infection of larval trematodes in molluscs. Agra. Univ. J. Res. (Sci.), 19: 47-48.

Jain, S. P., 1973. Studies on amphistomes – I. Amphistome cercariae and their life histories. Agra. Univ. J. Res. (Sci.), 22: 63-74. (Review article, mentioned 13 cercariae of pigmentata group and 10 from diplocotylea group,not indexed).

Jain, S. P., 1976. Studies on amphistomes II: A survey of incidence and nature of amphistome infection in aquatic snails. Agra. Univ. J. Res. (Sci.), 25: 81-98.

Jain, S. P., 1977. Studies on amphistomes III: The life history of *Ceylonocotyle dicrancoelium* (Fischoeder, 1901) Nasmark, 1937–an amphistome parasite of ruminants. Zool. Anz. Leipzig., 199: 286-300.

Jain, S. P., 1978. Studies on amphistomes V: On the life history and validity of *Gigantocotyle bathycotyle* (Fischoeder, 1901) Nasmark, 1937–an amphistome parasite of bile duct of bovines. Zool. Anz. Leipzig., 200: 183-218.

Jain, S. P. and Srivatava, H. D., 1969. The life history of *Ceylonocotyle scoliocoelium* (Fischoeder, 1904) Nasmark, 1937– A common amphistome parasite of ruminants in India. Agra. Univ. J. Res. (Sci.) 18: Part III, pp.1-16.

Johnpaul, A., Raghunathan, M. B. and Selvanayagam, M., 2010. Population dynamicas of fresh water molluscan in the lentic eco-systems in and around Chennai. Rec. Res. Sci. Tech., 2: 80-86.

Juyal, P. D., Jyoti, N. K., Singh and S. S.Hassan 2007. An insight into epidemiology and immunodiagnosis of paramphistomosis in domestic ruminants in Punjab. Proc. XVIII Natl. Cong. "Zoonoses and advances in herbal medicine against parasites of veterinary importance", SKUAST, Jammu,7-9 September, 2007, pp.187-190. (Review, not indexed).

Jyothi, 2001. Studies on immunodiagnosis of *Paramphistomum epiclitum* in kids using stage specific antigens. M. V. Sc. Thesis submitted to Deemed Univ., Indian Vet. Res. Inst., Izatnagar, India.

Kumar, A., Deb, A. R., Ansari, M. Z. and Sahai, B. N., 1992. Incidence and biology of snail vectors of paramphistome infection in Bihar. Indian J. Anim. Sci., 62: 946-947.

Kumar, R. R., Prasad, K. D. and Ajit Kumar, 2004. Seasonal incidence of *Paramphistomum* spp. cercariae infection in aquatic snails in some areas of Jharkhand. Indian Vet. Med. J., 28: 242-244.

Kumar, V., 1980. The digenetic trematodes, *Fasciolopsis buski, Gastrodiscoides hominis* and *Artyfechinostomum malayanum* as zoonotic infection in South Asian countries. Ann. Soc. Belge. Med. Trop., 60: 331-339. (not indexed).

Malaviya, H. C., Prasad, A., Varma, T. K. and Dwivdi, P., 1989. On the collection of *Paramphistomum cervi* metacercariae and their infectivity to lambs. J. Vet. Parasitol., 3: 53-55.

Malaviya, H. C., Prasad, A., Varma, T. K. and Dwivedi, P., 1992. Susceptibility of *Gyraulus convexiusculus* to *Paramphistomum lobatum*. J. Vet. Parasitol., 6(1): 53-54.

Malaviya, H. C., Verma, T. K., Prasad, A. and Dwivedi, P., 1994a. Experimental infection of lambs and kids with *Paramphistomum epiclitum*. Sixth Natl. Cong. Vet. Parasitol, 22-24 October, 1994, Dept. Parasitol., Coll. Vet. Sci., JNKVV, Jabalpur-482 001, M. P., Abstr. S-2: 5, p.11.

Malaviya, H. C., Prasad, A., Varma, T. K. and Dwiedi, P., 1994b. Chemotherapy of experimentally induced *Paramphistomum epiclitum* infection in lambs and kids. Sixth Natl. Cong. Vet. Parasitol, 22-24 October, 1994, Dept. Parasitol., Coll. Vet. Sci., JNKVV, Jabalpur-482 001, M. P., Abstr. S-4: 3, p.43-44.

Malaviya, H. C., Prasad, A., Varma, T. K. and Dwiedi, P., 1994c. Chemotherapy of experimentally induced *Paramphistomum epiclitum* infection in lambs. Indian Vet. J., 71: 222-224.

Mohandas, A., 1974. Studies on the freshwater cecariae of Kerala 1. Incidence of infection and seasonal variation. Folia parasit. (Praha), 21: 311-317.

Mohandas, A., 1976. Studies on the freshwater cecariae of Kerala V. Paramphistomatoid and opisthorchioid cercariae. Vest. Cs. Spol. Zool., 40: 196-205.

Mudaliyar, S. V., 1944. Immature forms of *Cotylophoron cotylophorum* causing fatal enteritis in goats. Proc. 31st Indian Sci. Cong. (Delhi). Abstr. 34.

Mudaliyar, S. V., 1945. Fatal enteritis in goats due to immature amphistome probably *Cotylophoron cotylophorum*. Indian J. Vet. Sci., 15: 54-56.

Mudaliar, S. V. and Alwar, V. S, 1947. A check-list of parasites (Class-Trematoda and Cestoda) in the Department of Parasitology, Madras Veterinary College laboratory. Indian Vet. J. 23: 423-434.

Mukherjee, R. P., 1960a. Studies on some amphistomous trematodes of domesticated animals. Ph. D. Thesis, submitted to Agra Univ., Agra, India. (Not indexed).

Mukherjee, R. P., 1960b. Studies on the life history of *Ceylonocotyle scoliocoelium* (Fischoeder, 1904), Nasmark, 1937, an amphistome parasite of sheep and goat. Proc. 47th Indian Sci. Cong. (Bombay), 3: 438-439.

Mukherjee, R. P., 1962. Studies on some amphistomatous trematodes of domesticated animals. Agra Univ. J. Res. (Sci.), 11: 131-135.

Mukherjee, R. P., 1966a. Seasonal variations in cercarial infections in snails. J. Zool. Soc. India., 18: 39-45.

Mukherjee, R. P., 1966b. Studies on the life history of *Fischoederius elongatus* (Poirier, 1883) Stiles and Goldberger, 1910, amphistomatous parasite of cow and buffaloes in India. Indian J. Helminth. 18: 5-14.

Mukherjee, R. P., 1968. Studies on the life history of *Ceylonocotyle indicum* Stiles and Goldberger, 1910 an amphistomatous parasite of ruminants of India. J. Zool. Soc. India. 20: 105-122.

Mukherjee, R. P., 1975. Studies on the life history of *Ceylonocotyle scoliocoelium* (Fischoeder, 1904) Nasmark, 1937, an amphistome parasite of ruminants. Dr. B. S. Chauhan Commemoration Volume, pp.251-226.

Mukherjee, R. P., 1992. Larval Trematodes, Part I – Amphistome cercariae, pp.1-89. In: Fauna of India and Adjacent Countries: Zoological Survey of India, Kolkatta. pp.1-158. (reference book, IH not indexed).

Mukherjee, R. P. and Srivastava, H.D., 1960. Studies on the life history of *Gigantocotyle explanatum* (Creplin, 1847) Nasmark, 1937 a common amphistome parasite in the bile duct and gall bladder of buffaloes. Proc. 47th Indian Sci. Cong. (Bombay), Part III, Sect.VII, p.440.

Muraleedharan, K., 1972. Investigation into the factors governing the epizootology of nasal schistosomiasis in bovines and its control in different field conditions, Ann. Rep. (1971-72), All India co-ordinated Project (ICAR), Bangalore Centre, Dept. Parasitol., Vet. Coll. Univ. Agri. Sci., Hebbal, Bangalore-560024.

Muraleedharan, K., 2000. A case of schistosome cercarial dermatitis in man. J. Parasit. Dis., 24: 231-232.

Muraleedharan, K., Prasanna Kumar, S., Hegde, K. S. and Alwar, V. S., 1977. Larval trematode infections in aquatic snails of Karnataka State – A preliminary study. Mysore J. agric. Sci., 11: 101-104.

Murthy, G. S. S. and D' Souza, P. E., 2008. Fasciolosis and amphistomosis in livestock. In: XI. National Training Programme on Trends and perspectives in the biology, ecology and control of parasitic diseases. 10th to 30th March 2008, Centr. Adv. Stud., Dept.Parasitol., Vet. Coll. Univ., Agri. Sci., Hebbal, Bangalore-560 024. pp.10-18. (Not included in Index list).

Pallavi Devi, Saidul Islam and Manoranjandas, 2003. Ecology and biology of aquatic snails and their control II. Cercarial fauna of fresh water snails from Assam. J. Vet. Parasitol., 17: 131-133.

Pandey, K. C. and Jain, S. P., 1971. On the excretory and reproductive systems of *Cercaria chungathi* Peter and Srivastava, 1955. Indian J. Helminth., 23: 81-85.

Panzoo, G. R., Bali, H. S. and Gill, J. S., 1988. Development of *Gastrothylax crumenifer* (Creplin, 1847) Otto, 1896 in buffalo and goat. Indian J. Parasitol., 12: 301-303.

Panzoo, G. R., Bali, H. S. and Singh Sawai, 1989a. Screening of 8 species of aquatic snails for transmission of *Gastrothylax crumenifer*, a pouched paramphistome of ruminants in Punjab. Proc. VIII Natl. Cong. Parasitol., Calcutta,10-12 February, 1988, Abstr, p.30.

Panzoo, G. R., Bali, H. S. and Singh Sawai, 1989b. Screening of aquatic snails for transmission of *Gastrothylax crumenifer*. J. Vet. Parasitol., 3: 63-65.

Patel, A. I. and Dutt, S. C., 1978. Experimental infection of buffalo with normal and irradiated metacercariae of *Paramphistomum cervi* (Zeder, 1790) Fischoeder (1901). Asian Cong. Parasitol., p.279.

Pati, S. C., Bali, H. S. and Singh, K. S., 1988. On the viability and cercariae producing capacity of *Gyraulus convexiusculus* experimentally infected with miracidia of *Gigantocotyle explanatum*. Indian J. Anim. Res., 22: 111-113.

Pemola Devi, N. and Jauhri, R. K., 2008. Diversity and cercarial shedding of malco fauna collecting from water bodies of Ratnagiri District, Maharashtra. Acta Trop., 105: 249-252.

Peter, C. T., 1949. Studies on fresh water cercarial fauna in Madras. M. Sc.Thesis submitted to Univ. Madras, Madras (Not indexed).

Peter, C. T., 1954. Studies on larval trematodes from fresh water snails in Madras. Proc. 41st Indian Sci. Cong. (Hyderabad), Part III Abstr. pp.221-222.

Peter, C. T., 1955. A note on the precercarial development of *Psuedodiscus collinsi* (Cobbold, 1875) Sonsino, 1895. Proc. 42nd Indian. Sci. Cong. (Baroda), Part III, Abstr. No. 68, pp.352-353.

Peter, C. T., 1956. Studies on the cercarial fauna in Madras IV. The amphistome and gymnocephalus group of cercariae. Indian J. Vet. Sci., 26: 27-30.

Peter, C. T., 1960. Studies on the life history of *Gastrodiscus secundus* Looss, 1907, an amphistome parasite of equines in India. Indian J. Helminth., 12: 18-50.

Peter, C. T. and Mudaliyar, S. V., 1948. On a new cercaria, determined to be the larva of *Gastrodiscus secundus* Looss, 1907. Curr. Sci., 17: 303 -304.

Peter, C. T. and Srivastava, H. D., 1954. Studies on the life history of *Pseudodiscus collinsi* (Cobbold, 1875) Sonsino, 1895, an amphistome parasite of equines in India. Proc. 41st Indian Sci. Cong. (Hyderabad) Part III, p. 221.

Peter, C. T. and Srivastava, H. D., 1955. On five new species of amphistome cercariae from India. Proc. 42nd Indian Sci. Cong. (Baroda). Part III, p.353.

Peter, C. T. and Srivastava, H. D, 1960a. Studies on the life history of *Psuedodiscus collinsi* (Cobbold, 1875) Sonsino, 1895, an amphistome parasite of equines and elephants in India. Indian J. Helminth., 12: 1-17.

Peter, C. T. and Srivastava, H. D., 1960b. On amphistome cercariae in India with a description of some new species. Indian J. Helminth., 12: 51-73.

Peter, C. T. and Srivastava, H. D., 1961. On *Cercaria chungathi* Peter and Srivastava, 1955 and its relationship to *Gastrothylax crumenifer* (Creplin). Parasitol., 51: 111-115.

Pokhriyal, B. P., Jauhari, R. K. and Sudarshana, R., 1998. Trematode cercarial infection in the snail *Thiara* (*M.*) *tuberculata* (Mueller, 1774) in different localities of Deon Valley. J. Exp. Zool., India. 1(2): 107-110.

Pokhriyal, B. P., Mahesh, R. K. and Jauhari, R. K., 1996. Prevalence of trematode cercariae infection in the snail *Lymnaea* (*Pseudosuccinea*) *acuminata* Lamarck, 1822 in different localities of Dehradun-Valley. Global Meet Parasit. Dis., New Delhi, India, 18-22 March, 1996. Abstr., J. Parasit. Dis., 20: 111.

Prasad, A., Malaviya, H. C., Verma, T. K. and Dwivedi, P., 1989. Studies on the interaction between paramphistome and schistosome infection in *Indoplanorbis exustus* and *Lymnaea luteola*. Indian J. Parasitol., 13: 331-334.

Prasad, A., Dwivedi, P., Verma, T. K. and Malaviya, H. C., 1992. Incidence of paramphistome infection in *Indoplanorbis exustus* and *Gyraulus convexiusculus* at Bareilly. Sixth Natl. Cong. Vet. Parasitol, 22-24 October, 1994, Dept. Parasitol., Coll. Vet. Sci., JNKVV, Jabalpur-482 001, M. P., India, Abstr. S-2: 7, p.12.

Prasad, A., Malaviya, H. C., Verma, T. K. and Dwivedi, P., 1994a. Development of *Paramphistomum epiclitum* Fischoeder, 1904 in the snail host. Indian J. Anim. Sci., 64: 568-573.

Prasad, A., Malaviya, H. C., Verma, T. K. and Dwivedi, P., 1994b. On the cercaria of *Paramphistomum epiclitum* Fischoeder, 1904, obtained from *Indoplanorbis exustus* and its comparison to *Cercariae indicae* XXVI, Sewell, 1922. J. Vet. Parasitol., 8: 9-13.

Prasad, K. D., 1973. Studies on the control of amphistomes in goats. M. Sc. (Vet) Thesis submitted to Rajendranagar Agri. Univ., Patna, Bihar.

Prasad, K. D., Sahai, B. N. and Jha, G. J., 1974. Observations on pathogenicity and histochemistry of experimental infections in *Cotylophoron cotylophorum* in goats. Proc. Natl. Acad. Sci. India., Sect. B., 44, 202-208.

Rajkhowa, C., Gogoi, A. R., Borkakoty, M. R., Sharma, P. C. and Das, M. R., 1991. Cercarial fauna and their pattern of seasonal liberation from *Indoplanorbis exustus* in Assam. J. Vet. Parasitol., 5: 49-52.

Ramanujachari, G. and Alwar, V. S., 1955. A check-list of parasites in the Department of Parasitology, Madras Veterinary College. Indian Vet. J., 32: 47.

Rao, M. A. N., 1932. A comparative study of cercarial fauna in Madras. Indian Vet. J., 9: 107-111.

Rao, M. A. N. and Ayyar, L. S. P., 1932. A preliminary report on two amphistome cercariae and three adults. Indian J. Vet. Sci., 2: 402- 405.

Ritu Arora, Singh, N. K, Hassan, S. S. and Juyal, P. D., 2007. Identification of immunodominent antigens of *Paramphistomum epiclitum*. J. Vet. Parasitol, 21: 117-120.

Rita Arora, Singh, N. K. and Juyal, P. D., 2010. Immunoaffinity chromatographic analysis for purification of specific antigens of *Paramphistomum epiclitum*. J. Parasit. Dis., 34: 57-61.

Sahai, B. N., 1969-70. A survey of trematode infection in aquatic snails. J. Vetcol., 10: 9-12.

Sahai, B. N., Singh, R. P. and Prasad, G., 1985. Histochemical alterations in the duodenum of goats experimentally infected with *Paramphistomum cervi*. Vet. Parasitol., 17: 131-138.

Sahay, M. N., 1987. Studies on the epizootoiology and control of paramphistomiasis infection in bovines of Bihar. Ph.D. Thesis submitted to Birsa Agri. Univ., Ranchi-834006. Abstr., J. Vet. Parasitol, 1: 85-87.

Saifullah, M. K., Gul Ahamed and Nizami, M. A., 2000. Effect of parasitism on the biomolecules of *Gyraulus convexiusculus* hepatopancreas. J. Vet. Parasitol, 14: 21-26.

Sardey, M. R., 1982. A study on the prevalence of trematode cercaria in snails of Nagpur region. In: Vector and Vector-borne Diseases. Proc. All India Sym., Trivandrum, Kerala, India, pp.137-141.

Shivjot Kaur, S., Singla, L. D., Hassan, H. S. and Juyal., P. D., 2008. Evaluation of Dot-ELISA for immunodiagnosis of *Paramphistomum epiclitum* in ruminants. J. Parasit. Dis., 32: 118-122.

Shivjot Kaur, S., Singla, L. D., Hassan, H. S. and Juyal., P. D., 2009. Standardization and application of indirect plate ELISA for immunodiagnosis of paramphistomosis in ruminants. J. Parasit. Dis., 33 70-76.

Shrivastav, H. O. P. and Sengar Kamaljeet, 1995. Prevalence of amphistome cercariae in *Indoplanorbis exustus* in and around Jabalpur. Seventh Natl. Cong. Vet. Parasitol., 19-21 August, 1995, Dept. Parasitol., Madras Vet. Coll., Madras -7, Abstr., Sl 29, p. 88.

Sewell, R. B. S., 1922. *Cercariae indicae*. Indian J. Med. Res., 10 (Suppl.): 1-371.

Singh, A., Srivastava, S., Chandra Sehkar and Jaswant Singh, 2009. Prevalence of trematodes in bovines and snails. Indian Vet. J., 86: 206-207.

Singh, K. S., 1958. A redescription and life history of *Giganocotyle explanatum* (Creplin, 1847) Nasmark 1937 (Trematoda: Paramphistomidae) from India. J. Parasitol., 44: 210-224.

Singh, K. S., 1970. On *Srivastavaia indica* n. g., n. sp. (Paramphistomidae), a parasite of ruminants and its life history. H D. Srivastava Commemoration Volume, Div. Parasitol., Indian Vet. Res. Inst., Izatnagar, U.P. (India), pp.117-126.

Singh, M. D. and Pande, B. P., 1972. On experimental infection of guinea pig, rabbits and lambs with amphistome metacercaria. A histopathological study. Indian J. Anim. Sci., 42: 290- 297.

Singh, R. P., Sahai, B. N. and Prasad, K., 1984a. Haematology of goats experimentally infected with *Paramphistomum cervi*. Indian J. Anim. Sci., 54: 132-134.

Singh, R. P., Sahai, B. N. and Jha, G. S., 1984b. Histopathology of the duodenum and rumen of goats during experimental infections with *Paramphistomum cervi*. Vet. Parasitol., 15: 39-46.

Singh, R. P., Prasad, K.D, Anzari, M. Z. and Sahai, B. N., 1983. Observations on intradermal skin test for diagnosis of *Paramphistomum cervi* in goats. Indian J. Anim. Hlth., 22: 67-68.

Sinha, B. B., 1950. Life history of *Cotylophoron cotylophorum*, a trematode parasite from the rumen of cattle, goat and sheep. Indian J. Vet. Sci., 20: 1-11.

Singla, L. D., Kaur, A., Sandhu, B. S. and Chowdhury, N., 1998. Experimental infection of amphistome metacercariae (*Paramphistomum cervi*) in rabbit – A preliminary study. Indian Vet. J., 75: 690-692.

Srivastava, H. D., 1938. A study on the life- history and pathogenecity of *Cotylophoron cotylophorum* (Fischoeder, 1901) Stiles and Goldberger, 1910 of Indian ruminants and a biological control to check the infestation. Indian J. Vet. Sci., 8: 381-385.

Srivastava, H. D., 1944a. A study of the life- history of *Paramphistomum explanatum* of bovines in India. Proc. 31st Indian Sci. Cong. (Delhi), Part III, Sect. VIII, p.113.

Srivastava, H. D., 1944b. A study of the life-history of *Gastrothylax crumenifer* of Indian ruminants. Proc. 31st Indian Sci. Cong. (Delhi), Part III, Sect. VIII, p.113.

Srivastava, H. D., 1947. A study of the life-history of *Gastrothylax crumenifer*, a common pouched amphistome of Indian ruminants. Proc. 34th Indian Sci. Cong. (Delhi), p.142.

Srivastava, H. D. and Tripathi, H.N., 1987. Amphistomes of ruminants 2. Life history of two species of *Palamphistomum* Srivastava and Tripathi (1980) of sheep, goat and buffaloes. Indian J. Anim. Sci., 57: 1043-1053.

Syed Shabih Hassan, Juyal, P. D. and Arvinder Kaur, 2009a. Establishing experimental paramphistomosis in lambs. XIX Natl. Cong. Vet. Parasitol. & Nat. Symp. "National impact of parasitic diseases on livestock health and production". February 3-5, 2009, Dept. Vet. Parasitol., Coll. Vet. Sci., Guru Angad Dev Vet. Anim. Sci. Univ., Ludhiana-141004, Abstr., OE44: p.47.

Syed Shabih Hassan and Juyal, P. D., 2009b. Detection, purification and characterization of paramphistome spp. antigen by immunodiagnostic assays. XIX Natl. Cong. Vet. Parasit., & Nat. Symp. "National impact of parasitic diseases on livestock health and production". February 3-5, 2009, Dept. Vet. Parasitol., Coll. Vet. Sci., Guru Angad Dev Vet. Anim. Sci. Univ., Ludhiana-141004, Abstr., P114: p.124.

Tandon, R. S., 1957. Life history of *Gastrothylax crumenifer* (Creplin 1847). Zeits. Fur Wiss. Zool., 160: 39-41.

Tandon, R. S., 1958. Development and morphology of the cercariae of an amphistome *Fischoderius elongatus* (Stiles and Gold berger, 1910) recovered from a naturally infected *Lymnaea luteola* at Lucknow. Zool. Anz., 161: 200-206.

Thapar, G. S., 1961. Paramphistomiasis infection in bovines of Bihar. Ph.D. Thesis submitted to Birsa Agri. Univ., Ranchi.

Thapar, G. S., 1961. Life-history of *Olevaria indica*, an amphistome parasite from the rumen of Indian cattle. J. Helminth, R.T. Leiper Suppl. (1961), 179-186.

Tripathi, H. N. and Srivastava, H. D. 1987. Amphistomes of ruminants. 2. Life history of two species *Paramphistomum srivastava* and *tripathi* (1980) of sheep, goats and buffaloes. Indian J. Anim. Sci., 57: 1043–1056.

Urvasi, Bali, H. S. and Kaur, A., 1999. Haematological studies in lambs and kids experimentally infected with gamma irradiated paramphistome metacercariae. J. Vet. Parasitol., 13: 139-142.

Urvasi, Bali, H. S. and Kaur, A., 2000. Immunization of lambs and kids with gamma irradiated paramphistome metacercariae. Indian Vet. J., 77: 13-15.

Upadhyay, A. N., 1987. Biochemical, immunological and therapeutic studies of *Gigantocotyle explanatum.* Ph. D. Thesis (1986), Birsa Agri. Univ., Ranchi-834 006. Abst., J. Vet. Parasitol., 1: 82-83.

Upadhyay, A. N. and Sahai, B. N., 1986. Incidence of paramphistome cercariae in fresh water aquatic snails at Ranchi. Indian J. Anim. Hlth., 25: 119-121.

Vaidyanathan, S. N. 1941. Experimental infestation with *Fischoderius elongatus* in a calf at Madras. Indian J. Vet. Sci., 11: 243.

Varma, A. K., 1961. Observations on the biology and pathogenicity of *Cotylophoron cotylophorum* (Fischoeder, 1901). J. Helminthol., 35: 161-168.

Varma, T. K and Prasad, A., 2000. Prevalence of paramphistome cercariae in fresh water aquatic snails in Bareilly, Uttar Pradesh. Intas Polivet, 1: 67-71.

Varma, T. K., Malviya, H. C., Dwivedi, P. and Prasad, A., 1993. Effect of gamma irradiation of *Paramphistomum epiclitum* (Fischoeder, 1904) and its humeral immune response in lambs. Indian J. Anim. Sci., 63: 914-917.

Varma, T. K., Prasad, A., Dwivedi, P. and Malviya, H. C., 1995. Fluke establishment and chemotherapy of experimentally induced *Paramphistomum epiclitum* infection in lambs and kids.7th Natl. Cong. Vet. Parasitol., 19th to 21st August, 1995, Dept. Parasitol, Madras Vet. Coll. Madras-7, Abstr, SIV 13, p.133.

Veluswamy, R., Singh, B. P., Ghosh, S., Chandra, D., Raina, O. K., Gupta, S. C. and Jayraw, A. K., 2006. Prepatent detection of *Fasciola gigantica* infection in bovine calves using metacercarial antigen. Indian J. Exp. Biol., 44: 749-753.

Vohra, R., Rajesh Katoch, Anish Yadav, Khajuria, J. K. and Shilpa Sood., 2009. Prevalence of *cercaria pigmentata* in *Indoplanorbis exustus* in Jammu District from September 2005 to August 2008. XIX Natl. Cong. Vet. Parasit., & Nat. Symp. "National impact of parasitic diseases on livestock health and production". February 3-5, 2009, Dept. Vet. Parasitol., Coll. Vet. Sci., Guru Angad Dev Vet. Anim. Sci. Univ., Ludhiana-141004, Abstr., OE4: p. 27.

Vohra, S., Rajesh Katoch, Yadav, A., Khajuria, J. K. and Ashok Kumar., 2006. Seasonal prevalence of *Cercaria pigmentata* in Jammu District. Proc. XVII Natl. Cong. Vet. Parasitol., November, 15-17, 2006. Dept. Parasitol. Rajiv Gandhi Coll. Vet. & Anim. Sci., Kurumbapet, Puducherry- 605009, Abstr. S-II. 45, p.77.

Yokananth, S., Ghosh, S, Gupta, S. C., Suresh, M. G. and Saravanan, B. C., 2005. Characterization of specific and cross-reacting antigens of *Fasciola gigantica* by immunoblotting. Parasitol. Res., 97: 41-48.

1.12 FAMILY: STRIEGIDAE

Dutt, S. C. and Bali, H. S., 1980. Snails of Punjab State and their trematode infections. J. Res. Punjab agric. Univ., 12: 222-228.

Muraleedharan, K., Prasanna Kumar, S., Hegde, K. S. and Alwar, V. S., 1977. Larval trematode infections in aquatic snails of Karnataka State – A preliminary study. Mysore J. agric. Sci., 11: 101-104.

Sahai, B. N., 1969-1970. A survey of trematode infection in acquatic snails. J. Vet. Coll. Assam Agri. Univ., 10: 9-12.

1.13 FAMILY ISOPARORCHIDAE

Bhalaerao, G. D., 1932. A note on the probability of infection in man and domestic animals by *Isoparorchis hypselobargri*. Indian J. Vet. Sci., 11: 406-407.

Chauhan, B. S., 1947. Note on collection of zoological survey of India. Rec. Indian Mus., 45: 133-137.

Jaiswal, G. P., 1957. Studies on the trematode parasites of fishes and birds found in Hyderabad State. Zool. Jahrb. Abt. Syst., 85: 1-72.

Pandey, K. C., 1967. Studies on the metacercariae of fresh water fishes of India. IV. On the morphology of *Isoparorchis hypselobargri* (Billet, 1898) with a note on its development. Proc. Natl. Acad. Sci., India, B 39, pp.186-190.

Rai, P., 1969. Notes on the histopathology of opisthorchiid, plagiorchiid and isoparaorchiid metacercoidal invasion of some Indian fresh-water fishes. Indian J. Anim. Sci., 39: 177-183.

Rai P. and Pande, B. P., 1965. Isoparorchid infection in some hitherto unrecorded fish species. Curr. Sci., 34: 586- 588.

Southwell, T. and Prasad, B., 1918. Note from Bengal Fisheries Laboratory. No.5. Parasite of Indian fishes with a note on carcinoma in the climbing perches. Rec. Indian Mus., 9: 79-103.

1.14 FAMILY: SCHISTOSOMATIDAE

Agrawal, M. C., 1978. Studies on the heterologous immunity in schistosomiasis. Ph. D. Thesis submitted to Jawaharlal Nehru Krishi Viswa Vidyalaya, Jabalpur, India.Vide Thesis Abstracts 5: 213-214.

Agrawal, M. C., 1981. Recovery of worms from mice infected with cercariae of *Schistosoma incognitum* and *Schistosoma spindale* by different routes. Indian J. Parasitol., 5: 227-228.

Agrawal, M. C., 1996a. Rabbit as a model for *Schistosoma nasale*. J. Parasitol. Appl. Anim. Biol., 5: 25-27.

Agrawal, M. C., 1996b. Habitat of S*chistosoma nasale* in the final host. Abstr. Global Meet on Parasit. Dis., New Delhi, 18-22 March, 1996. J. Parasit. Dis. 20: 105.

Agrawal, M. C., 1998a. New technique for infection and recovery of schistosomes from animals. Indian J. Anim. Sci., 68: 521-523.

Agrawal, M. C., 1999. Schistosomosis: an underestimated problem in animals in South Asia. World Anim. Rev., 92: 55-57. (Not indexed).

Agrawal, M. C., 2000. Final Report on National Fellow Project on "Studies on the strain identification, epidemiology, diagnosis, chemotherapy and zoonotic potentials of Indian schistosomes". ICAR, New Delhi. (Not indexed).

Agrawal, M. C., 2003. Epidimiology of fluke infections. In: Helminthology of India. (M. L. Sood Ed), International Book Distributors, Derhadun, pp.511-542. (Not indexed).

Agrawal, M. C., 2004. Dominant mammalian Schistosome species of in India. Proc. 11th Internatl Cong. Institution for Trop. Vet. Med. and 16th Vet. Assn. Malayasia Cong., Kuala Lumpur, Malaysia, 23-27August, 2004. pp.192-193. (Review, not indexed)

Agrawal, M. C., 2005. Present status of schistosomiasis in India. Proc. Natl. Acad. Sci.India, 50, 75 (B) special issue, pp.184-196. (Review, not indexed).

Agrawal, M. C., 2011. Schistosomes and Schistosomiasis in South East Asia, Springers (India) pvt. Ltd., New Delhi. (Review, not indexed).

Agrawal, M. C. and Alwar, V. S., 1992. Nasal schistosomiasis: A review. Helm. Abst., 61: 373-384. (Review, not listed in index).

Agrawal, M. C. and Anjana Mishra, 1994. Evaluation of host susceptibility against schistosomes infection. Indian J. Parasitol., 18: 229-230.

Agrawal, M. C. and Das, M., 1986. Use of rhesus monkey (*Macaca mulatta*) – *Schistosoma incognitum* model for studying incidental schistosomiasis in man. Seventh Natl. Cong. Parasitol., Ravi Shankar Univ., Raipur, 26-28 December, 1986, Abstr.,25, p.81.

Agrawal, M. C. and Mishra, A., 1996. Comparative susceptibility of guinea pig to *Schistosoma spindale*. Indian Vet. J., 73: 131-135.

Agrawal, M. C. and Rao, V. G., 2011. Indian schistosomes: A need for further investigation. J. Parasitol. Res., 250868, 4 pages, doi.10.1155/2011/200868.

Agrawal, M. C. and Sahasrabudhe, V. K., 1984. Factors affecting heterologous immune response in mice and rats against *Schistosoma incognitum* by immunization with *S. spindale*. Indian Vet. J., 71: 451-457.

Agrawal, M. C. and Sahasrabudhe, V. K. 1988. *Schistosoma bovis* like eggs from *Schistosoma spindale* females in mice. Indian J. Parasitol., 12: 137-138.

Agrawal, M. C. and Singh, K. P., 2000a. Effect of S*chistosoma incognitum* infection on serum biochemistry of rabbits. J. Vet. Parasitol., 14: 31-34.

Agrawal, M. C. and Singh, K. P., 2000b. Trials of oxamaniquine on S*chistosoma incognitum* infected rabbits. J. Vet. Parasitol., 14: 141-143.

Agrawal, M. C. and Shaw, H. L., 1989. A review on S*chistosoma incognitum* Chandler, 1926. Helm. Abstr., 58: 239-251.(Review, not listed in index).

Agrawal, M. C. and Southgate, V. R., 2000. *Schistosoma spindale* and bovine schistosomiasis. J. Vet. Parasitol., 14: 95-107. (Review, not listed in index).

Agrawal, M. C. and Tiwari, A., 1999. Size differences in the adult *Schistosoma spindale* according to host compatibility. Indian Vet. J., 76: 171-173.

Agrawal, M. C., Banerjee, P. S. and Shah, H. L., 1991. Five mammalian schistosome species in an endemic focus in India. Trans. Roy. Soc. Trop. Med. Hyg. 85: 231.

Agrawal, M. C., Bhilegoankar, N. G. Shah, H.L., 1982a. Experimental studies on the fish for biological control of *Schistosoma incognitum*. Indian Vet. J., 59: 850-853.

Agrawal, M. C., Borkakoty, M. R. and Das, M., 1998. Some observations on nasal schistosomiasis (Hur-hurria) in a village of Assam. Indian Vet. J., 75: 80-81.

Agrawal, M. C., George, J. and Gupta, S., 1999. Chemotherapeutic trials using *Schistosoma incognitum* mouse model. J. Parasit. Dis., 23: 137-138.

Agrawal, M. C., Jomini George and Samala Gupta, 2000a. Role of *Lymnaea luteola* in spreading *Schistosoma incognitum* in an endemic area. Indian J. Anim. Sci., 70: 349-352.

Agrawal, M. C., Gupta, S. and George, J., 2000b. Cercarial dermatitis in India. Bull. World Hlth. Org., 72: 278.

Agrawal, M. C., Jain, J. and Rao, K. N. P., 2001. Sudden death of piglets during immature schistosome infection. Indian J. Anim. Sci., 71: 681-682.

Agrawal, M. C., Kumar, J. and Ahlawat, S. P. S., 2003. Helminth infections of livestock in Andaman. J. Vet. Parasitol., 17: 143-146.

Agrawal, M. C., Panesar, N. and Das, M., 1985. Patent infection with S*chistosoma incognitum* to white mouse (*Mus musculus*). Curr. Sci., 54: 640-641.

Agrawal M. C., Panesar, N. and Shah, H. L., 1984. Effect of dry season on sex of recovered schistosomes from experimental animals. Indian J. Anim. Hlth., 23: 103-104.

Agrawal, M. C., Panesar, N. and Shah, H. L., 1986. Diagnosis of experimental schistosomiasis in dogs. Indian Vet. J., 63: 276-280.

Agrawal M. C., Sahasrabudhe, V. K. and Gehlot, K., 1979. Immunization against *Schistosoma incognitum* in mice by administration of cercariae of *Schistosoma indicum*. Indian Vet. J., 56: 682-685.

Agrawal M. C., Sahasrabudhe, V. K. and Kolte, G. N., 1982b. Histopathology heterologous *Schistosoma* Infection in mice. Indian J. Parasitol., 6: 315-317.

Agrawal M. C., Sahasrabudhe V. K. and Shah, H. L., 1983. Immunization against *Schistosoma incognitum* in mice by administration of cercariae of *Schistosoma spindale*. Indian Vet. J., 60: 321-322.

Agrawal M. C., Banerjee, P., Anjana Mishra and Shah, H. L., 1988. Application of CHR-test using blood smear for schistosomiasis survey. Second Natl. Parasitol., March 3-5, 1988, Dept. Parasitol. Vet. Coll., Bangalore, India, Stream II, Abstr., p.18.

Agrawal, M. C., Pandey, S., Sirmour, S. and Deshmukh, P. S., 2000c. Endemic form of cercarial dermatitis (Khujlee) in Bastar area of Madhya Pradesh. J. Parasit. Dis., 24: 217-218.

Agrawal, M. C., Sano, M., Mishra, A. and Solanki, P. K., 1992. Chemotherapeutic trials with praziquantel against *Schistosomia incognitum* and *Hymenolepis* spp. in rats. Indian Vet. J. 69: 279-280.

Agrawal, M. C., Vohra, V., Gupta, S. and Singh, K. P., 2003. Schistosomiasis and other intestinal helminthes in domestic animals in and around Bhubaneswar, Orissa. Indian J. Anim. Sci., 73: 875-877.

Agrawal, M. C., Rao, V. G., Vohra, S., Singh, B. K., Gupta, S., Singh, K. P., Bhondeley, M., Uike, M. and Anvikar, A., 2006. Cercarian hullen reaction for the assessment of human schistosomiasis in India. J. Parasit. Dis., 30: 181-183. (Initial of authors are as given in the paper).

Agrawal, M. C., Rao, V. G., Vohra, S., Bhondeley, M. K., Ukey, M. J., Anvikar, A. R. and Yadav, R., 2007. Is active human schistosomiasis present in India? Current Sci., 92: 889.(Initial of authors are as given in the paper).

Agarwal, R. D., 1989. Studies on parasitic fauna of zoo animals and wild life: studies on the cercarial fauna of Mathura District with special reference to furcocercous cercariae. Ph. D. Thesis submitted to C. S. A. Univ. of Agri. & Tech., Kanpur-208 002. Abstr., J. Vet Parasitol, 3: 83-83.

Ahluwalia, S. S., 1971. Post-cercarial development of *Schistosoma incognitum* in a mammalian host. Indian J. Anim. Sci., 41: 1130-1134.

Ahluwalia, S. S., 1972a. Experimental *Schistosoma incognitum* in pig. Indian J. Anim. Sci., 42: 723-729.

Ahluwalia, S. S., 1972b. Intradermal test in pig infected with *Schistosoma incognitum*. Indian J. Anim. Sci., 42: 729-731.

Ahluwalia, S. S., 1972c. Circumoval precipitin test in pigs infected with *Schistosoma incognitum*. Indian J. Anim. Sci., 42: 955-956.

Ahluwalia, S. S., 1972d. Zoonotic potentials of *Schistosoma incognitum*. Indian J. Anim. Sci., 42: 962-964.

Ahluwalia, S. S., 1972e. "Cercarien-hullen Reaction" (CHR) in pigs with *Schistosoma incognitum*. Indian J. Anim. Sci., 42: 1029-1031.

Ahluwalia, S. S., 1972f. Chemotherapy of *Schistosoma incognitum* in pig I. Antimony potassium tartarate and antimony sodium tartarate. Indian J. Anim. Sci., 42: 1054-1056.

Ahluwalia, S. S., 1973a. Chemotherapy of *Schistosoma incognitum* in pig III. Ambilhar and Miracil-D. Indian J. Anim. Sci., 43: 428-434.

Ahluwalia, S. S., 1973b. A note on chemotherapy of *Schistosoma incognitum* in pigs. II. Stibophen, Triostam and Astiban. Indian J. Anim. Sci., 43: 793-796.

Ahluwalia, S. S., 1975. Blood fluke disease in pig. Food Farmg. & Agri. 7(5): 14-15.

Ahluwalia, S. S. and Dutt, S. C., 1972. Clinical study on *Schistosoma incognitum* infection of the domestic pig: symptomatology, haematology and effects on growth rate. Indian Vet. J. 49: 863-867.

Alwar, V. S., 1957. Schistosomiasis in animals. Subj. 6. VIIth Conf. Vet. Dis. Invest. Officers and Res. Workers in India, November, 1957, Bombay. (Review not index).

Alwar, V. S., 1994. Final report of All India co-ordinated Project "On Investigation into the factors governing the epizootology of nasal schistosomiasis in bovines and its control in different field conditions". ICAR, New Delhi. (Report, not indexed)

Anandan, R. 1986. Studies on *Schistosoma nasale* Rao, 1933. (Trematoda, Schistosomatidae) Ph. D. Thesis submitted to Tamil Nadu Agri. Univ., Coimbatore.

Anandan, R. and Ebenezer Raja, 1988a. Studies on *Schistosoma nasale* Rao, 1933. Population density of *Indoplanorbis exustus* and incidence of *Cercaiae indicae* XXX of *S. nasale* in an enzootic village. Second Natl. Parasitol., March 3-5, 1988, Dept. Parasitol., Vet Coll., Bangalore, India, Stream II, p.9.

Anandan, R. and Ebenezer Raja, 1988b. Studies on *Schistosoma nasale* Rao, 1933. Infectivity of *S. nasale* miracidia of cattle, buffalo, sheep and goat origins to *Indoplanorbis exustus* snails. Second Natl. Parasitol., March 3-5, 1988, Dept. Parasitol., Vet. Coll., Bangalore, India, Stream II, p.10.

Anandan, R. and Ebenezer Raja., 1990. Proc. Nat. workshop on anim. biotech. Dept. Biotech. Govt. of India, New Delhi and Dept. Anim. Biotech., Madras Vet. Coll. TANUVAS, Chennai. (Hetrologous cercariae conferring immunity, *S. incognitum* and *S. nasale* in sheep, not indexed.)

Anandan, R. and Ebenezer Raja., 1990. Post cercarial development of *Schistosoma nasale* in calves. Proc. 1st Asian Cong. Vet. Parasitol. Patna, p.57.

Anantaraman, M., 1958. On schistosome dermatitis. I. Dermatitis in India caused by cercariae of *Schistosoma spindale* Montgomery, 1906; Indian J. Helminthol., 19: 46-52.

Anantaraman, M., 1981. The epizootiology of nasal schistosomiasis in India. Proc. Indian Acad. Sci., (Anim. Sci.). 90: 659-663. (Review, not indexed).

Anjana Mishra, 1991. Development of *Schistosoma spindale* and *S. nasale* in the laboratory and their diagnosis in final hosts. Ph. D. Thesis submitted to Jawaharlal Nehru Krishi Viswa Vidyalaya, Jabalpur, India.

Anjana Mishra and Agrawal, M. C., 1993. Development of *Schistosoma spindale* in albino mouse (*Mus musculus*). J. Vet. Parasitol., 7: 45-50.

Anjana Mishra and Agrawal, M. C., 1994a. Susceptibility of albino rats to *Schistosoma spindale*. Indian J. Parasitol., 18: 99-102.

Anjana Mishra and Agrawal, M. C., 1994b. Maintenance of *Schistosoma spindale* rabbits in laboratory. Indian J. Anim. Sci., 64: 352-354.

Anjana Mishra and Agrawal, M. C., 1998. Role of *Indoplanorbis exustus* snails in spreading schistosomiaisis in endemic area. Indian J. Anim. Sci., 68: 107-110.

Anon, 2007-2009. Morphological, molecular and immunological studies on paddy field dermatitis caused by *Schistosoma spindale* in paddy field workers of Assam (An ICMR Project, not indexed). See Anon, Narain, K. and Mahanta, J., 1999, 2000a, b. Narain *et al.,* 1994;1999.

Anon. (year not specified, Source: Google).Study on paddy field dermatitis (Panikaint) among workers in rice field of upper Assam. Research Project, Reg. Med. Centr., Dibrugash, Assam. (Research Project, not indexed). See also Narain, K. and Mahanta, J., 1999., 2000a,b; Narain *et al.,* 1994; 1999

Avsatthi, B. L., 1976. Studies on immunization of rabbits against *Schistosoma incognitum* using cercariae of different dose levels. Ph. D. Thesis submitted to Jawaharlal Nehru Krishi Viswa Vidyalaya, Jabalpur, India.

Avsatthi, B. L. and Sahasrabudhe, V. K., 1988. Electrophorectic studies of serum proteins in rabbits immunized and subsequently challenged with homologous cercariae at different dose level. Second Nat. Parasitol., March 3-5, 1988, Dept. Parasitol., Vet Coll., Bangalore, India, Souvenir, Stream II, Abstr., pp.18-19.

Banerjee, P. S., 1988. Comparative efficacy of different diagnostic tests in determining the prevalence of schistosmiasis in cattle and buffaloes. M. V. Sc. Thesis, Jawarharlal Nehru Krishi Viswa Vidyalaya, Jabalpur, India.

Banerjee, P. S. and Agrawal, M. C. 1990. Miracidial immobilization test in bovine schistosmiasis. Indian J. Anim. Sci., 60: 628-630.

Banerjee, P. S., Agrawal, M. C. and Shah, H. L., 1990. Diagnosis of bovine schistosomiasis by using ring precipitation test. Indian J. Parasitol., 13: 223-226.

Banerjee, P. S., Agrawal, M. C. and Shah, H. L., 1991. Application of CHR and J-index in bovine schistosomiasis. Indian Vet. J., 68: 1022-1026.

Baugh, S. C., 1978. A century of schistosomiasis in India: human and animal. Rev. Iber. Parasitol., 38: 435-472. (Review, not indexed).

Bhalerao, G. D., 1938. Schistosomes and schistosomiasis in India. Indian J. Vet. Sci. 8: 149-157. (Review, not included in Index).

Bhalerao, G. D., 1943. The cercarial fauna of the irrigated tract of the Nizam's Dominions with suggestions regarding the relationship to the trematode parasites in man and domesticated and other animals. Indian J. Vet. Sci., 13: 294-296.

Bhalerao, G. D., 1948a. Blood fluke problem in India. Sci. and Cult.,13(8) Suppl.(1): 4.

Bhalerao, G. D., 1948b. Blood fluke problem in India. Presidential address, Sec. Med. & Vet. Sci., Proc. 35th Indian Sci. Cong. (Patna), pp.57-71., Patna, (Review) Abstr., In: Curr. Sci.17: 36-37.

Bhalerao, G. D., 1948c. Schistosomiasis in animals. Proc. 4th Internatl. Cong. Trop. Med. Malaria (Washington, May 10-18) 2: 1386-1393. (Review, not indexed).

Bhalerao, G. D., 1948d. Schistosomiasis in animals. Proc. 4th Internatl. Conf. Trop. Med. Malaria. Washington, May 10-18, 1948. 2: 1386-1393. (Review, not indexed and also see under Schistosomes in general, below).

Bhaskar Rao, T., Rao, B. V. and Hafeez, M. D., 1989. Effect of gamma-radiation of cercariae of *Schistosoma incognitum* and *S. spindale*. J. Nucl. Agri. Bio., 18: 126-128.

Bhatia, B. B. and Rai, D. N., 1974. Cercarien-hullen reaction in early phase of experimental infection with *Schistosoma incognitum* in pigs. Indian J. Exp. Biol., 12: 576-578.

Bhatia, B. B. and Rai, D. N., 1974. Distribution and prevalence of *Schistosoma incognitum* in pigs in Uttar Pradesh. Indian J. Anim. Hlth. 15: 109-111.

Bhatia, B. B., Rai, D. N. and Hajela, S. K., 1976. Studies on the morphology of liver of albino mice in experimental *Schistosoma incognitum* in pigs. Indian J. Anim. Sci., 46: 100-104.

Bhilegaonkar, N. G., 1981. Studies on the immunity against *Schistosoma incognitum* in mice by using *in-vitro* derived gamma irradiated schistosomulae. Ph. D. Thesis, Jawaharlal Nehru Krishi Vishwa Vidyalaya, Jabalpur, India. (Not indexed).

Bhilegaonkar, N. G., Babras, M. A. and Sardey, M. R., 1977. Some observations on incidence of nasal schistosomiasis in Maharashtra. Proc. 1st Natl. Cong. Parasitol. (Baroda), pp.43-44.

Bhilegaonkar, N. G., Babras, M. A. and Sardey, M. R., 1978. Some observations on the ecology of the snails, their relation to prevalence oh helminthic diseases in domestic animals particularly nasal schistosomiasis in bovines Maharashtra. Asian Cong. Parasitol., February 23-26, 1978, Bombay, Abstr. III 16, p.139.

Biswas, G. (vide Gita Biswas, infra)

Borakakoty, M. R., 1975. Studies on immunization of albino rats against *Schistosoma incognitum*. M. V. Sc. Thesis submitted to Jawaharlal Nehru Krishi Viswa Vidyalaya, Jabalpur, India.

Borkakoty, M. R., and Das, M. R., 1980. Snail and cercarial fauna in Kamrinji District of Assam. J. Res, Assam Agri. Univ., 1: 100-102.

Choubisa, S. L., 2002. Focus on seasonal occurrence of larval trematode (cercarial) parasites and their host specificity. J. Parasit. Dis., 26: 72-74.

Choubisa, S. L., 2008. Focus on pathogenic trematode cercaria infecting fresh water snails (Mollusca: Gastropoda) of tribal region of southern Rajasthan. India. J. Parasit. Dis., 32: 47-55.

Chaudhuri, S. S., Gupta, R. P. and Yadav, C. L., 1982. Note on cercarial fauna of aquatic snails of Haryana State. Indian J. Anim. Sci., 52: 1273-1275.

Chauhan, B. S., 1959. Schistosomiasis in India. Proc. 1st All-India Cong. Zool., Zool. Soc. India, 34, Chittaranjan Avennue, Calcutta-12.(Review, not indexed).

Daniel, K. and Joseph, A., 2010. Swimming and crawling behahiour of *Schistosoma spindale* (Montgomery, 1906). The Bioscsan, 5: 219-223.

Daniel, K. and Joseph, A., 2011. Penetration behahiour of cercariae of *Schistosoma spindale* (Montgomery, 1906). J. Vet. Parasitol., 25: 185-186.

Das, M., 1984. Studies on experimental schistosomiasis in rhesus monkeys (*Macaca mulatta*). Ph. D. Thesis submitted to Jawaharlal Nehru Krishi Viswa Vidyalaya, Jabalpur, India.

Das, M. and Agrawal, M. C., 1986. Experimental infection of rhesus monkeys with *Schistosoma incognitum* and *Orientobilharzia dattai*. Vet. Parasitol., 22: 151-155.

Das, M. and Joshi, S. C., 1988a. Haematological observations in *Schistosoma incognitum* infection in rhesus monkeys, *Macaca mulatta* Proc. VIII Natl. Cong. Parasitol., Calcutta,10-12 February, 1988, Abstr, p.49.

Das, M. and Joshi, S. C., 1988b. Clinical observations in *Schistosoma incognitum* infection in rhesus monkeys, *Macaca mulatta*. Proc. VIII Natl. Cong. Parasitol., Calcutta,10-12 February, 1988, Abstr, p.77.

Das, M. and Joshi, S. C., 1988c. Serological tests in *Schistosoma incognitum*.infection Rhesus monkeys (*Macaca mulatta*) with Proc. VIII Natl. Cong. Parasitol., Calcutta,10-12 February, 1988, Abstr., p.104.

Das, M. and Joshi, S. C., 1988d. Infection of monkeys, *Macaca mulatta* with *Schistosoma incognitum*. Proc. VIII Natl. Cong. Parasitol., Calcutta, 10-12 February, 1988, Abstr., pp.113-114.

Das, M., Panesar, N. and Agrawal, M. C., 1987a. Cercarian hullen reaction in experimental *Schistosoma incognitum* infection in rhesus monkeys. J. Vet. Parasitol., 1: 63-65.

Das, M., Agrawal, M. C., Joshi, S. C. and Katiyar, A. K., 1987b. Studies on pathology of experimental *Schistosoma incognitum* infection in rhesus monkey (*Macaca mulatta*). Indian J. Parasitol., 11: 71-74.

Das, M., Agrawal, M. C., Joshi, S. C. and Shah, H. L., 1986. Histochemical studies in experimental *Schistosoma incognitum* infection in rhesus monkeys (*Macaca mulatta*). Indian J. Anim. Sci., 56: 840-843.

Das, M., Amita Kushwaha, A., Agrawal, M. C. and Joshi, S. C.,1988a. Some biochemical changes in *Schistosoma incognitum* infected rhesus monkey (*Macaca mulatta*). Proc. VIII Natl. Cong. Parasitol., Calcutta,10-12 February, 1988, Abstr., p.66.

Das, M., Kushwaha, A., Agrawal, M.C., Joshi, S. C. and Shah, H. L., 1988b. Biochemical determination of pathogenesis of *Schistosoma incognitum* infection in *Macaca mulatta*. J. Vet. Parasitol., 2: 133-135.

Deka, J. K., 1999. Epidemiological studies on *Schistosoma nasale* Rao, 1993, in and around Azara, near Guwahati. M. V. Sc Thesis submitted to Assam Agri. Univ., Khanapara, Guwahati-781 022.

De Bont, J. and Vercruysee, J., 1998. Schistosomiasis in cattle. Adv. Parasitol., 41: 285-365. (Reviewed the Indian scenerio, not indexed).

Dutt, S. C., 1957a. Development of female *Schistosoma spindale* in guinea pig. Nature, 179: 1359.

Dutt, S. C., 1957b. On cercarial dermatitis in India with description of the causal parasite, *Cercaria srivastavai* n. sp., Proc. 44th Indian Sci. Cong. (Delhi), Part III, p.78.

Dutt, S. C., 1962. Studies on the susceptibility of the guinea pig to infection with *Schistosoma spindale* Montgomery, 1906. Parasitol., 52: 199-206.

Dutt, S. C., 1965. On the susceptibility of *Macaca mulatta* to infection with *Schistosoma incognitum*. Curr. Sci., 34: 49-50.

Dutt, S. C., 1967a. On the susceptibility of *Macaca mulatta* to *Schistosoma incognitum* with observations on pathology of infection. Indian J. Med. Res., 55: 1173- 1180.

Dutt, S. C., 1967b. Studies on *Schistosoma nasale* Rao, 1933. I Morphology of the adults, egg and larval stages. Indian J. Vet. Sci., 37: 249-262.

Dutt, S. C., 1970. On the two dermatitis producing cercariae *Cercaria srivastavai*, Dutt, 1957 and *C. hardayali* n. sp. with a note on schistosoma dermatitis in India. In: H. D. Srivastava Commemoration Volume. Singh, K. S. and Tandon, B. K. (Eds.), Division of Parasitology, Indian Vet. Res. Inst., Izatnagar, pp.311-318.

Dutt, S. C. and Bali, H. S., 1980. Snails of Punjab State and their trematode infections. J. Res. Punjab agric. Univ. 12: 22-228.

Dutt, S. C. and Srivastava, H. D., 1952. On the morphology and the life history of mammalian blood-fluke – *Ornithobilharzia dattai* n. sp. (Preliminary report). Parasitol., 42: 144- 150.

Dutt, S. C. and Srivastava, H. D., 1955a. Studies on the life history of *Orientobilharzia dattai* (Dutt and Srivastava, 1952) n. comb –a blood-fluke of domestic mammals. Proc. 42nd Indian Sci. Cong. (Baroda), Part III, Abstr. pp.350-351.

Dutt, S. C. and Srivastava, H. D., 1955b. Biological studies on *Orientobilharzia dattai* (Dutt and Srivastava, 1952) – a blood-fluke of domestic mammals. Proc. 42nd Indian Sci. Cong. (Baroda), Part III, Abstr. 64, 351.

Dutt, S. C. and Srivastava, H. D., 1955c. Biological studies *Schistosoma indicum* Montgomery, 1906 – a blood-fluke of Indian ungulates. Proc. 42nd Indian Sci. Cong. (Baroda), Part III, Abstr. 64, p.353.

Dutt, S. C. and Srivastava, H. D., 1955d. Life history of *Schistosoma indicum* Montgomery, 1906) a common blood-fluke of Indian ungulates. Proc. 42nd Indian Sci. Cong. (Baroda), Part III, Sect VII, p.354.

Dutt, S. C. and Srivastava, H. D., 1961. Morphology and life history of the mammalian blood-fluke *Orientobilharzia dattai* (Dutt and Srivastava). I. Morphology of adult, egg and miracidium. Indian J. Vet. Sci., 31: 288-303.

Dutt, S. C. and Srivastava, H. D., 1962a. Studies on the morphology and life history of the mammalian blood-fluke *Orientobilharzia dattai* (Dutt and Srivastava) Dutt and Srivastava. II. The molluscan phase of the life cycle and the intermediate host specificity. Indian J. Vet. Sci., 32: 33-43.

Dutt, S. C. and Srivastava, H. D., 1962b. Biological studies on *Orientobilharzia dattai* (Dutt and Srivastava, 1952) Dutt and Srivastava, 1955, a blood-fluke of domestic mammals. Indian J. Vet. Sci., 32: 216-229.

Dutt, S. C. and Srivastava, H. D., 1962c. Studies on the morphology and life history of the mammalian blood-fluke *Orientobilharzia dattai* (Dutt and Srivastava, 1952)

Dutt and Srivastava, 1955. III Definitive host specificity. Indian J. Vet. Sci., 32: 260-268.

Dutt, S. C. and Srivastava, H. D., 1964. Studies on the life history of *Orientobilharzia turkestanicum* (Skrjabin, 1913) Dutt and Srivastava, 1955 (Preliminary report). Curr. Sci., 33: 752-753.

Dutt, S. C. and Srivastava, H. D., 1968. Studies on *Schistosoma nasale* Rao, 1933 II. Molluscan and mammalian hosts of the blood fluke. Indian J. Vet. Sci., 38: 210-216.

Fairley, N. H. 1927.The early spontaneous cure of bilharziosis (*S. spindalis*) in monkeys (*Macacus sinicus*) and its bearing on species immunity. Indian J. Med. Res., 14: 685-700.

Fairley, N. H. and Jesudasan, F., 1927. The experimental transmission of *Schistosoma spindale* in the Indian water buffaloes. Indian J. Med. Res., 14: 701-706.

Fairley, N. H. and Jasudasan, F., 1930a. Studies on *Schistosoma spindale*. Part I. The infection rate, seasonal incidence and source of supply of infected mollusks. Indian Med. Res. Mem., 17: 1-10.

Fairley, N. H. and Jesudasan, F., 1930b. Studies on *Schistosoma spindale*. Part II.The definitive hosts of *S. spindale* with reference to alimentary infection in ruminants. Indian Med. Res. Mem., 17: 11-15.

Fairley, N. H. and Jesudasan, F., 1930c. Studies on *Schistosoma spindale*. Part IV. The complement fixation reaction with cercarial antigen – A study in experimental serology. Indian Med. Res. Mem., 17: 73-180.

Fairley, N. H., Mackie F. F. and Jesudasan, F., 1930a. Studies on *Schistosoma spindale*. Part IV. Further observations on spontaneous cure in *Macacus sinicus*. Indian Med. Res. Mem., 17: 53-60.

Fairley, N. H., Mackie F. F. and Jesudasan, F., 1930b. Studies on *Schistosoma spindale*. Part V. The guinea pig as a host for male schistosomes. Indian Med. Res. Mem., 17: 61-72.

Gadgil, R. K., 1961. Schistosomes and schistosomosis in India with special reference to recent researches. Subject No. 1. Schistosomiasis in India. In: Conference on parasites and parasitic diseases. ICAR, New Delhi. (Review, not indexed).

Gita Biswas., 1975. Studies on *Schistosoma nasale* Rao, 1933 including host parasite relationship. Ph. D. Thesis submitted to Agra Univ., Agra.

Gita Biswas and Subramanian, G., 1978a Host parasite relationship of *Schistosoma nasale* and *Indoplanorbis exustus*. Asian Cong. Parasitol., Feb. 23-26, 1978, Bombay, Abstr. V 26, p. 277.

Gita Biswas and Subramanian, G., 1978b. A note on the incidence of nasal schistosomiasis in Bareilly District, Uttar Pradesh. Indian J. Anim. Sci., 48: 544-545.

Gita Biswas and Subramanian, G., 1980. Sequential clinicopathological response of cattle and buffaloes to experimental infection with *Schistosoma nasale*. Proc. 3rd Natl. Cong. Parasitol., Haryana Agri. Univ., Hissar, Abst.23B.

Gita Biswas and Subramanian, G., 1988. Development pattern of *S. nasale* in experimentally reared *I. exustus*. Indian J. Anim. Hlth., 27: 123-126. (Note: *S. nasale* and *I. exustus*, not expanded).

Gita Biswas and Subramanian, G., 1990. Experimental infection of common laboratory animals with a parasite of bovines. Indian J. Anim. Sci., 60: 149-150.

Gita Biswas and Tewari, H. C. 1988. Efficacy of some chemotherapeutic agents against experimental infection of *Schistosoma incognitum* in mice. Indian Vet. J., 65: 304-305.

Gowrisankar, D., 1968. Nasal schistosomiasis in buffaloes in Madras- its incidence, aetiology and pathology. M. V. Sc. Dissertation submitted to Univ. Madras, Madras.

Gupta, S., 2002. Clinical, biochemical and parasitological studies and prevalence of caprine schistosomiasis in and around Jabalpur. Ph. D. Thesis submitted to Rani Durgawati Univ., Jabalpur. (Not indexed).

Gupta, S. and Agrawal, M. C., 2005. Excretion of eggs and miracidia of schistosomes in experimentally infected goats. J. Parasit. Dis., 29: 23-28.

Gupta, S. and Agrawal, M. C., 2007. Anaemia in experimental caprine schistosomiasis. J. Parasit. Dis., 31: 68-71.

Gupta, S., Agrawal, M. C. and Khanna, A., 2006a. Heterologous experimental schistosome infection and fluke recovery from the goats. Indian J. Anim. Sci., 76: 882-885.

Gupta, S., Agrawal, M. C. and Vohra, S., 2006b. Clinical observations in experimental caprine schistosomosis. J. Vet. Parasitol., 20: 29-31.

Hajela, S. K., Bhatia, B. B., and Rai, D. N., 1975. A note on complement fixation test in the diagnosis of *Schistosoma incognitum* infection in the pigs. Indian J. Anim. Sci., 45: 799-801.

Hajela, S. K., Bhatia, B. B., and Rai, D. N., 1976. Haemagglutination test (HAT) in experimental porcine schistosomiasis. Indian J. Anim. Res., 10: 45-46.

Hassan, S. S. and Juyal, P. D., 2009. Schistosomiasis epidemiology in domestic ruminants in Punjab, India. XIX Natl. Cong. Vet. Parasitol. & Natl. Sym. Natl., Impact on parasitic diseases of livestock health and production, 3-5 February, 2009, IAAVP & Dept. Vet. Parasitol., Coll. Vet. Sci., Guru Angad Dev Vet. & Anim. Sci. Univ., Ludhiana-141 004, Abstr. OE 43, p.46.

Jain, J. (vide, Jyothi Jain, infra).

Jain, S. P., 1970. Double infection of larval trematodes in molluscs. Agra. Univ. J. Res. (Sci.) 19: 47-48.

Johnpaul, A., Raghunathan, M. B. and Selvanayagam, M., 2010. Population dynamics of freshwater molluscan in the lentic ecosystems in and around Chennai. Rec. Res. Sci. & Tech., 2: 80-86.

Jyothi Jain, Agrawal, M. C. and Rao, K. N. P., 2000. Effect of praziquantel and closantel treatment on haematology and faecal egg count in experimental porcine schistosomiasis. Indian J. Anim. Sci., 70: 823-825.

Jyothi Jain, Agrawal, M. C., Rao, K. N. P and Jain, N. K., 2001. Effect of praziquantel on histopathology in experimental porcine schistosomiasis. Indian J. Anim. Sci., 71: 631-634.

Jyothi Jain, Gupta, S. and Rao, K. N. P., 1999. Cercarial hullen reaction in experimental porcine schistosomiasis. JNKVV Res. J., 33: 85-87.

Kaur, N., 1985. Comparative studies on serodiagnosis and faecal examination in experimental schistosomiasis. Ph. D. Thesis submitted to Rani Durgavati Univ., Jabalpur. (Not indexed).

Kohli, S., 1991. Studies on different techniques for separating schistosome cercariae prevalent in and around Jabalpur. Ph. D. Thesis, Rani Durgavati Viswa Vidyalaya, Jabalpur. (Not indexed).

Kohli, S and Agrawal, M. C., 1994a. Identification of four mammalian schistosome cercariae on the basis of their argenophilic papillae. J. Parasitol. Appl. Anim. Biol., 3: 63-70.

Kohli, S. and Agrawal, M. C., 1994b. Differentiation of schistosome cercariae with pH sensitive stain. Indian Vet. J., 71: 1049-1050.

Kohli, S. and Agrawal, M. C., 1995. Differentiation of mammalian schistosome cercariae on basis of their furcal tip. Indian J. Anim. Sci., 65: 41-43.

Kohli, S. and Agrawal, M. C., 1996. Alkaline phosphatase activity in schistosome cercariae. Indian J. Anim. Sci., 66: 545-548.

Kolte, S. W., Kurkure, N. V., Maske, D. K. and Khatoon, S., 2012. Prevalence of *Schistosoma nasale* infection in bovines of eastern Vidharbha (Maharashtra) vis-à-vis infection in *Indoplanorbis exustus*. J. Vet. Parasitol., 26: 140-143.

Koshy, T. J. and Alwar, V. S., 1974. Electrophoretic studies in nasal schistosomiasis in bovines. Cheiron 3: 114-118.

Koshy, T. J., Achutan, H. N. and Alwar, V. S., 1975. Cross transmissibility of *Schistosoma nasale* (Rao, 1933) infection between cattle and buffaloes. Indian Vet. J., 52: 216-218.

Kumar, V., 1973. Studies on snail hosts of *Orientobilharzia turkestanicum* (Skrjabin, 1913) Dutt and Srivastava, 1955 (Schistosomatidae: Trematoda) in India. Ann. Soc. Belg. Med. Trop., 53: 17-23.

Kumar, V. and de Burbure, A., 1986. Schistosomiasis in animals and man. Helm. Abstr., 55: 469-480. (Review paper, not indexed).

Liston, W. G. and Soparkar, M. B., 1918. Bilharziais among animals in India. The life cycle of *Schistosoma spindalis*. Indian J. Med. Res., 5: 567-569.

Madhavi, R., 1977. Argentophilic papillae of the cercaria of *Schistosoma spindale*, Montgomery, 1966. Abstr. 1st Natl. Cong. Parasitol., Baroda, 24-26 February, 1977.

Mahajan, M. R., 1933. Bovine nasal granuloma (snoring disease of cattle) in Hyderabad State. Indian J. Vet. Sci., 3: 346-349.

Mahapatra, D., 1997. Epidemiology, clinical pathology and chemotherapy of schistosomiasis in bovines. M.V.Sc. Thesis. Orissa Univ. Agri. Tech., Bhubaneswar.

Manohar, L., Venketeswara Rao, P and Swami, K. S. 1972. Variation in aminotransferase activity and total free aminoacid level in the body fluid of the snail, *Lymnaea luteola* during different larval trematode infections. Invert. Pathol., 19: 36-41.

Mahanta, J., Narain, K., and Srivastava, V. K., 1996. Paddy field dermatitis in Assam −a focus of concern for north-east India. Global meet on parasitic diseases, New Delhi, India 18-22 March, 1996. Abstr., J. Parasit. Dis., 20: 97-98.

Meitram Ratnadevi, Swamy, M. and Agrawal, M.C., 2007. Detection of schistosome antigens in tissue sections by immunoperoxidase technique. Indian J. Vet. Sci., 77: 58-59.

Mishra, A. (Vide Anjana Mishra, above).

Mohandas, A., 1974a. Studies on fresh water cecariae of Kerala 1. Incidence of infection and seasonal variation. Folia Parasitol. (Praha), 21: 311-317.

Mudaliar, S. V. and Alwar, V. S, 1947. A check-list of parasites (Class-Trematoda and Cestoda) in the Department of Parasitology, Madras Veterinary College Laboratory. Indian Vet. J., 23: 423-434.

Mukherjee, R. P., 2007. Fauna of India and Adjacent Countries: Larval Trematodes, Part III – Distome furcocercus cercariae. Zoological Survey of India, Kolkatta. p.1-158.(Reference book, not indexed).

Muraleedharan, K., 2000. A case of schistosome cercarial dermatitis in man. J. Parasit. Dis., 24: 231-232.

Muraleedharan, K., 2007. Snails and snail-borne diseases and their control. *In*: Focus with Special Reference to Emerging Diseases. X. Natl. Trng. Prog. Vet. Parasitol. 10th to 30th March 2007. Dept. Parasitol. Vet. Coll., Univ. Agri. Sci., Hebbal, Bangalore-560 024, pp.61-81.(Review paper, not included in Index list).

Muraleedharan, K., 1972. Investigation into the factors governing the epizootology of nasal schistosomiasis in bovines and its control in different field conditions. Annual Report (1971-72) of All India co-ordinated Project (ICAR), Bangalore Centr., Dept. Parasitol., Vet. Coll. Univ. Agri. Sci., Hebbal, Bangalore-560024.

Muraleedharan, K., 1973. Investigation into the factors governing the epizootology of nasal schistosomiasis in bovines and its control in different field conditions. Annual Report (1972-73) of All India co-ordinated Project (ICAR), Bangalore Centr., Dept. Parasitol., Vet. Coll. Univ. Agri. Sci., Hebbal, Bangalore-560024.

Muraleedharan, K., 2009. Schistosomosis of domestic animals in India- current status. In: Emerging and Re-emerging Parasitic Diseases of Veterinary and Public Health Importance, XII. Natl. Trng. Prog., 2nd to 22nd March 2009, Centr. Adv. Stud.,

Dept. Parasitol., Vet. Coll., Univ. Agri. Sci., Hebbal, Bangalore-560 024, pp. 61-81. (Review paper, not included in Index list).

Muraleedharan, K. and Prasanna Kumar, S., 1974. Investigation into the factors governing the ewpizootology of nasal schistosomiasis in bovines and its control in different field conditions. Final Report (1971-74) of All India co-ordinated Project (ICAR), Bangalore Centr., Dept. Parasitol. Vet. Coll., Univ. Agri. Sci., Hebbal, Bangalore-560 024.

Muraleedharan, K., Prasanna Kumar, S. and Hegde, K. S., 1975a. Predatory activity of the guppy, *Lebistes recticulatus* (Peters, 1859) on cercaria and miracidia of *Schistosoma nasale*, Rao, 1933. Indian Vet. J., 52: 763-776.

Muraleedharan, K., Prasanna Kumar, S. and Hegde, K. S., 1975b. On the longevity of cercaria of *Schistosoma nasale* Rao, 1933. Curr. Res. (UAS, Bangalore), 4: 140-141.

Muraleedharan, K., Prasanna Kumar, S., Hegde, K. S.and Alwar, V. S., 1975c. Experimental infection of *Indoplanorbis exustus* (Deshayes) with miracidia of *Schistosoma nasale* Rao, 1933. Curr. Res. (UAS, Bangalore), 4: 116-117.

Muraleedharan, K., Prasanna Kumar, S., Hegde, K. S. and Alwar, V. S., 1976a. A note on the intermediate host of *Schistosoma nasale* Rao, 1933 and seasonal prevalence of its cercariae. Indian Vet J., 53: 819-820.

Muraleedharan, K., Seshadri, S. J., Jagadish Babu, Prasanna Kumar, S., Hegde, K. S. and Alwar, V. S., 1976b. Experimental infection of buffalo calves with the cercaria of *Schistosoma nasale* Rao, 1933 of cattle origin and a study of lesions encountered. Mysore J. Agric. Sci., 10: 673-680.

Muraleedharan, K., Prasanna Kumar, S., Hegde, K. S. and Alwar, V. S., 1977. Larval trematode infections in aquatic snails of Karnataka State – A preliminary study. Mysore J. Agric. Sci., 11: 101-104.

Narain, K. and Mahanta, J., (year unknown, Source: Google). Studies on paddy field dermatitis (Panikaint) among workers in rice fields of upper Assam. Reg. Med. Res. Centre, Dibrugarh, Assam.(Not Indexed).

Narain, K. and Mahanta, J., 1999. Scanning electron microscopy of the integumental surface of *Schistosoma spindale* isolated during an outbreak of cercarial dermatitis in Assam. India. Indian J. Vet. Parasitol., 13: 103-106.(see Anon).

Narain, K. and Mahanta, J., 2000a. Cercarial dermatitis of animal schistosomes origin: In: Proceedings of young scientists' workshop on Identification and Diagnosis of Parasites of Biomedical Importance (A. K. Yadav, Ed.). North East Hill Univ., Shillong, pp. 29-34. (Review, not indexed). (see Anon).

Narain, K. and Mahanta, J., 2000b. Dermatitis associated with paddy field environment in Assam- A review- In: Man- Environment Relationship (M. K. Bhasim and Veena Bhasim, Eds), Kamala Raj Enterprises, Delhi. pp.213-220. (Review, not indexed). (see Anon).

Narain, K., Rajaguru, S. K. and Mahanta, J., 1998. Incrimination of *Schistosoma spindale* as a causative agent of farmer's dermatitis in Assam with a note on liver pathology in mice. J. Comm. Dis., 30: 1-6. (see Anon).

Narain, K., Mahanta, J., Dutta, R and Dutta, P., 1994. Paddy field itch in Assam: A cercarial dermatitis. J. Comm. Dis., 26: 26-30. (see also Anon).

Nikhale, S. G., 1972. Studies on experimental *Schistosoma incognitum* infection in mice with particular reference to its immunology. M. V. Sc. Thesis submitted to Jawaharlal Nehru Krishi Viswa Vidyalaya, Jabalpur, India.

Niphadkar, S. M. and Rao, S. R., 1967. Natural occurrence of *Schistosoma spindale* in *Bandcota bengalensis* in Bombay together with a note on their possible role in the dessemination of *Schistosoma spindale* in domestic animals.Indian Vet. J., 44: 563-565.

Panesar, N and Agrawal, M. C., 1986. Comparative efficacy of faecal examination and serological methods in experimental schistosomiasis in mouse. Indian J. Med. Res. 84: 366-373.

Panesar, N., Agrawal, M.C., and Das, M., 1987. Application of precipitin tests in *Schistosoma incognitum* infected rhesus monkeys. Indian J. Parasitol., 11: 59-61.

Panesar, N., Agrawal, M. C., Arora, S. and Shaw, H. L., 1986. Evaluation of diagnostic techniques in experimental *Schistosoma incognitum* in rabbits. Indian Vet. J., 63: 799-803.

Pemola Devi, N. and Jauhri, R. K. 2008. Diversity and cercarial shedding of malco fauna collecting from water bodies of Ratnagiri District, Maharashtra. Acta Trop., 105: 249-252.

Peter, C. T., 1949. Studies on freshwater cercarial fauna in Madras. M. Sc.Thesis submitted to Univ. Madras, Madras. (not indexed).

Peter, C. T., 1954. Studies on larval trematodes from fresh water snails in Madras. Proc. 41st Indian Sci. Cong. (Hyderabad), Part III, Abstr., pp. 221-222.

Peter, C. T., 1955. Studies on cercarial fauna in Madras. Part. I. The furcocercariae cercariae. Indian J. Vet. Sci., 25: 121-127.

Peter, C. T., 1955. Studies on the cercarial fauna in Madras. II. A new species of schistosome cercaria. Indian J. Vet. Sci., 25: 219-234.

Pokhriyal, B. P., Jauhari, R. K. and Sudarshana, R., 1998. Trematode cercariae infection in the snail, *Thiara* (M.) *tuberculata* (Mueller, 1774) in different localities of Doon Valley. J. Expt. Zool., India, 1(2): 107-110.

Pokhriyal, B. P., Mahesh, R. K. and Jauhari, R. K., 1996. Prevalence of trematode cercariae infection in the snail *Lymnaea* (*Pseudosuccinea*) *acuminata* Lamarck, 1822 in different localities of Dehradun-Valley. Global Meet on Parasit. Dis., New Delhi, India 18-22 March, 1996. Abstr., J. Parasit. Dis., 20: 111. (reported cercaria of liver fluke, amphistome and schistosome)

Prasad, A., Malaviya, H. C., Verma, T. K. and Dwivedi, P., 1989. Studies on the interaction between paramphistome and schistosome infection in *Indoplanorbis exustus* and *Lymnaea luteola*. Indian J. Parasitol., 13: 331-334.

Rai, D. N., Bhatia, B. B. and Hajela, S. K., 1975. Efficacy of Ambilhar (Ciba) in experimental *Schistosoma incognitum* in pigs. Indian Vet. J., 52: 860-862.

Rajamohanan, K., 1972. Studies on nasal schistosomiasis in cattle and buffaloes. M.V.Sc. Thesis, University of Kerala, Trivandrum, Kerala.

Rajamohanan, K. and Peter. C. T., 1972a. Studies on nasal schistosomiasis in cattle and buffaloes. Indian Vet. J., 49: 1063-1065.

Rajamohanan, K. and Peter, C. T., 1972b. Pathology of nasal schistosomiasis in buffaloes. Kerala J. Vet. Sci., 6: 94-100.

Rajamohanan, K., Sunderam, R. K. and Peter. C. T., 1972. On "Cercarien-Hullen Recktion" of Vogel & Minning 1949 in *Schistosoma nasale* infection. Kerala J. Vet. Sci., 3: 76-77.

Rajkhowa, C., 1989. Studies on schistosomiasis in domestic animals in Assam with special reference to *Schistosoma indicum*. Ph. D. Thesis submitted to Assam Agri. Univ., Khanapara, Guwahati, Assam.

Rajkhowa, C., Gogoi, A. R. and Borkakoty, M. R., 1997. Cercarien-Hullen Reaction in the serum of bull calves experimentally infected with *Schistosoma indicum*. Indian Vet. J., 74: 176-177.

Rajkhowa, C., Gogoi, A. R., Borkakoty, M. R. and Sharma, B. C., 1992. Some observations on haematological changes in experimental *Schistosoma indicum* infection in cattle. Indian J. Anim. Sci., 62: 125-126.

Rajkhowa, C., Gogoi, A. R., Borkakoty, M. R. and Sharma, B. C., 1996. Changes in serum protein fraction in cattle experimentally infected with *Schistosoma indicum*. J. Vet. Parasitol., 10: 185-189.

Rajkhowa, C., Gogoi, A. R., Borkakoty, M. R. and Sharma, B. C., 1997. Some biochemical changes in the blood of bull calves experimentally infected with *Schistosoma indicum*. J.Vet. Parasitol., 11: 169-173.

Rajakhowa, C., Gogoi, A. R., Bhattacharya, M., Mukit, A. and Borkakoty, M. R., 1999. Histoemnzymic changes in liver of bull calves and mice experimentally infected with *Schistosoma indicum*. J. Vet. Parasitol., 13: 30-32.

Rajkhowa, C., Gogoi, A. R., Borkakoty, M. R., Sharma, P. C. and Das M. R., 1991. Cercarial fauna and their pattern of seasonal liberation from *Indoplanorbis exustus* in Assam. J. Vet. Parasitol., 5: 49-52.

Ramanujachari, G. and Alwar, V. S., 1954. A check-list of parasites (Classes Trematoda, Cestoda and Nematoda) in the Department of Parasitology, Madras Veterinary College (additions since 1947). Indian Vet. J., 31: 46-56

Rao, K. H. and Murty, A. S., 1968. A note on the occurrence of schistosoma dermatitis in Kondakarla, Andhra Pradesh. Curr. Sci. 37: 407-408.

Rao, M. A.N., 1932. A comparative study of cercarial fauna in Madras. Indian Vet. J., 9: 107-111.

Rao, M. A.N., 1933. A preliminary report on the successful infection with nasal schistosomiasis in experimental calves. Indian J. Vet. Sci., 3: 161-162.

Rao, M. A.N., 1934. A comparative study of *Schistosoma spindalis,* Montgomery, 1906 and *Schistosoma nasalis,* n. sp. Indian J. Vet. Sci., 4: 1-28.

Rao, M. A. N., 1935. Further observations on bovine nasal schistosomiasis in experimental calves. Indian J. Vet. Sci., 5: 266-273.

Rao, V. G., Dash A. P., Agrawal M. C., Yadav, R. S., Anvikar, A. R., Vohra, S., Bhondeley, M. K., Ukey, M. J., Das, S. K., Minocha, R. K. and Tiwari, B. K., 2007. Cercarial dermatitis in central India: an emerging health problem among tribal communities. Ann. Trop. Med. Parasitol., 101: 409-413.

Raut, S. K., 1983. Factors influencing cercarial liberation from their snail hosts. Proc. Sym. Host Environ. Zool. Surv. India, pp.55-65.

Reddy, R. K., 1964. Development of *Schistosoma spindale* (Montgomery, 1907) in the molluscan host as well as white mice with particular reference to histopathology of various organs. P. G. Thesis, Bombay Vet. Coll., Bombay. (not indexed).

Sahai, B. N., 1969-70. A survey of trematode infection in aquatic snails. J. Vetcol., 10: 9-12.

Sahasrabudhe, V. K., 1977.ICAR Scheme on investigation into the immunologyof schistosome infection with a view to evolve a vaccine. Annual Report, 1976-77. Dept. Parasitol.,Coll. Vet. Sci. & Anim. Husb., J. N. K.V.V., Jabalpur.

Sahay, M. N., 1975. Studies on experimental infection of *Schistosoma nasale* in laboratory animals. M. V. Sc. Thesis, Bihar Vet. Coll., Patna.

Sahay, M. N. and Sahai, B. N., 1976. Histopathology of experimental nasal schistosomiasis in laboratory animals, kids and lambs. Indian J. Anim. Hlth., 15: 93-95.

Sahay, M. N. and Sahai, B. N., 1978. Studies on susceptibility of the laboratory animals, kids and lambs to experimental infection with *Schistosoma nasale*. J. Parasitol., 64: 1135-1136.

Sahay, M. N., Sahai, B. N. and Prasad, G., 1977. Histochemical observations on liver, lung and heart of laboratory animals, kids and lambs in experimental nasal schistosomiasis. Indian J. Anim. Sci., 47: 814-818.

Sapate, P. P., Bhilegoankar, N. G. and Maske, D. K., 2000. Golden hamster as a model for immunological studies in *Schistosoma nasale*. J. Vet. Parasitol., 14: 85-86.

Sapate, P. P., Bhilegoankar, N. G. and Maske, D. K., 2001. Development of *Schistosoma nasale* in hamsters and mastomyes and their pathogenicity. Indian Vet. J., 78: 14-17.

Sardey, M. R., 1982. A study on the prevalence of trematode cercaria in snails of Nagpur region. In: Vector and Vector-borne Diseases. Proc. All India Sym., Trivandrum, Kerala, India, pp.137-141.

Sewell, R. B. S., 1922. *Cercariae indicae.* Indian J. Med. Res., 10 (Suppl): 1-371.

Shah, H.L. and Agrawal, M.C., 1990. Schistosomiasis: In: A Review on Parasitic Zoonosis (Ed. S. C. Parija). ATIBS Publishers, Delhi-143-172. (Not indexed).

Shameem, K. and Radhika, J., 2001. Rhythmicity in the emergence of the cercariae of *Schistosoma incognitum* and *S. spindale* from the snail hosts. J. Parasitol. Appl. Anim. Biol., 10: 21-32.

Shames, N., 1998. Chemotherapeutic studies on experimental porcine schistosomiasis. M. V. Sc., Thesis, Rani Durgavati Viswa Vidyalaya, Jabalpur. (Not indexed).

Shames, N., Jain, J., Agrawal, M. C. and Rao, K. N. P., 2000a. Body weight loss in porcine schistosomiasis. Indian Vet. J., 78: 154-156.

Shames, N., Agrawal, M. C. and Rao, K. N. P., 2000b. Chemotherapeutic efficacy of praziquantel and closantel in experimental porcine schistosomiasis. Indian J. Anim. Sci., 70: 797-800.

Shrivastava, H. O. P., 1977. Studies on immunity to *Schistosoma incognitum* using irradiated cercariae, worm homogenates and excretions and secretions of the adult fluke. Ph. D. Thesis submitted to Jawaharlal Nehru Krishi Viswa Vidyalaya, Jabalpur, India.

Shrivastava, H. O. P., Dhawedkar, R, G. and Shah, H. L., 1981a. Indian Vet. Med. J., 6: 48.

Shrivastava, H. O. P., Dutt, S. C., Shah, H. L. and Sahasrabudhe, V. K., 1981b. Studies on immunity to *Schistosoma incognitum* in mice using irradiated cercariae. Indian Vet. J., 58: 865-868.

Shrivastava, H. O. P., Dhavedkar, R. G. and Shah, H. L., 1983. Allergic intradermal test for diagnosis of schistosomiasis in pigs. Indian J. comp. Microbiol. Immunol. infect. Dis., 4: 191-193.

Shrivastava, M. K., 1997. Chemotherapeutic trials in laboratory animals infected with *Schistosoma incognitum* and *S. spindale*. M. V. Sc. Thesis, Jawaharlal Nehru Krishi Vishwa Vidhalaya, Jabalpur, India. (Not indexed).

Shrivastava, M. K. and Agrawal, M. C., 1999a. Effect of triclabendazole and oxyclozanide on experimental *Schistosoma incognitum* in mouse. Indian Vet. J., 76: 493-496.

Shrivastava, M. K. and Agrawal, M. C., 1999b. Effect of rafoanide and CDRI compound 81-470 against *Schistosoma incognitum* in albino mice. Indian J. Anim. Sci., 69: 97-99.

Shrivastava, M. K. and Agrawal, M. C., 1999c. Doubtful efficacy of lithium antimony tartarate (Anthiomaline) in schistosomiasis. J. Vet. Parsitol., 13: 67-68.

Singh, A., Srivastava, S., Chandra Sehkar and Jaswant Singh, 2009. Prevalence of trematodes in bovines and snails. Indian Vet. J., 86: 206-207.

Singh, B. K. and Agrawal, M. C., 2005. Triclabendazole on bovine hepatic schistosomosis. Indian J. Anim. Sci., 75: 654-655.

Singh, B. P., 1975. A preliminary survey of cercarial fauna of two common snails-*Lymnaea luteola* and *Indoplanorbis exustus* at Mathura: Observations on the post-cercarial developmental stages and host parasite relationships of the two mammalian schistosomes. M. V. Sc. Thesis, Agra Univ., Agra, U. P., India. (Not indexed, refer Singh and Ahluwalia, 1977a, b).

Singh, B. P. and Ahluwalia, S. S., 1977a. Post-cercarial development of *Orientobilharzia dattai* (Trematoda: Schistosomatidae). Indian Vet. J., 54: 207-212.

Singh, B. P. and Ahluwalia, S. S., 1977b. Efficacy of Neguvon (Bayer) against *Orientobilharzia dattai* in sheep, goat and white mice. Indian Vet. J., 54: 859-861.

Sinha, P. K. and Srivastava, H. D., 1954. Life history of S*chistosoma incognitum*, Chandler, 1926 Syn. *S. suis* Rao and Ayyar, 1938) and its new mammalian hosts. Proc. Indian Sci. Cong.(Hyderabad), 41: 222.

Sinha, P. K. and Srivastava, H. D., 1960. Studies on S*chistosoma incognitum*, Chandler, 1926.II. On the life history of blood-fluke. J. Parasitol., 46: 629- 641.

Sinha, P. K. and Srivastava, H. D., 1965. Studies on S*chistosoma incognitum*, Chandler, 1926 on the host specificity of the blood-fluke. Indian Vet. J., 42: 335-341

Sinha, P. K. and Srivastava, H. D., 1960. Studies on S*chistosoma indicum* Chandler, 1926. II. On the life history of the fluke. J. Parasitol., 46: 629-641.

Singh, R. N., 1959. Seasonal infestation of *Indoplanorbis exustus* (Deshayes) with furcocercus cercariae. Proc. Nat. Acad. Sci. India. Section B., 29: 62-72.

Sivaseelan, S., Kathiresan, D. and Anna, T., 2004. Persistant nasal schistosomosis in a village. Indian Vet. J., 81: 454-455.

Soparkar, M. B., 1921. The cercaria of S*chistosoma spindalis* (Montogomery). Indian J. Med. Res., 9: 1-22.

Sreekumar, P., 1966. Studies on the common trematodes of ducks. M. Sc. Thesis. Univ. of Kerala.

Srivastava, H. D., 1957. Schistosomiasis in animals. Subject 6. VII Conf.Vet. Dis. Invest.Officers and Res. Workers, India, Bombay, November, 1957.

Srivastava, H. D., 1960. Blood flukes. Presidential address. Proc. 47th Sci. Congr. Part II, Bombay, pp. 105-121.

Srivastava, H. D. and Dutt, S. C., 1951. Life history of S*chistosoma indicum* Montogomery, 1906 – a common blood fluke of Indian ungulates. Curr. Sci., 20: 273-275.

Srivastava, H. D. and Dutt, S. C., 1955. Life history of S*chistosoma indicum* Montogomery, 1906 – a common blood fluke of Indian ungulates. Proc. 42nd Indian Sci. Cong.(Baroda), III, p.354.

Srivastava, H. D. and Dutt, S. C., 1961. Schistosomes and schistosomosis in India with special reference to recent researches. Subject No. 1. Schistosomiasis in India. In: Conf. on Parasites and Parasitic diseases. Indian Council of Agricutural Research. (Review, not indexed).

Srivastava, H. D. and Dutt, S. C., 1962. Studies on *Schistosoma indicum*. Indian Council of Agricultural Research, Series No.34, New Delhi, pp.1-91.

Srivastava, H. D., Muralidharam, S. R. G. and Dutt, S. C., 1963. Pathogenecity of experimental infection of *Schistosoma indicum* Montogomery, 1906 to young sheep. Proc. 50th Sci. Indian Cong.(Delhi), III, p.583.

Srivastava, H. D., Muralidharam, S. R. G. and Dutt, S. C., 1964. Pathogenicity of experimental infection of *Schistosoma indicum* Montogomery, 1906 to young sheep. Indian J. Vet. Sci., 34: 35-40.

Srivastava, M. K. and Agrawal, M. C., 1999a. Doubtful efficacy of lithium antimony tartarate (Anthiomaline) in schistosomiasis. J. Vet. Parsitol., 13: 67-68.

Shrivastava, M. K. and Agrawal, M. C., 1999b. Efficacy of rafoxanide and CDRI compound 81-470 against *Schistosoma incognitum* in albino mice. Indian J. Anim. Sci., 69: 97-99.

Shrivastava, M. K. and Agrawal, M. C., 1999c. Effect of triclabendazole and oxyclozanide on experimental S*chistosoma incognitum* in mouse. Indian Vet. J. 76: 493-496.

Subramanian, G. Verma, J. C., Verma, T. K. and Singh, K. S., 1973. Time course of antibody response in experimental S*chistosoma incognitum* infection in pigs. Indian J. Anim. Sci., 43: 223-225.

Sumanth, S., D'Souza, P. E. and Jagannath, M. S., 2003. Immunodiagnosis of nasal and visceral schistosomosis in cattle by Dot-Elisa. Indian Vet. J., 80: 495-498.

Sumanth, S., D'Souza, P. E. and Jagannath, M. S., 2004a. Serodiagnosis of nasal and visceral schistosomosis in cattle by counter current immuno electrophoresis. Vet. Arh., 74: 427-433.

Sunil Wasudeo Kolte, 2002. Studies on immunology and pathology of *Schistosoma nasale* in bovines. Ph. D. (Veterinary Parasitology) submitted to Deemed Univ., Indian Vet. Res. Inst., Izatnagar-243 122, India. (vide thesis abstr., J. Vet. Parasitol., 16: 198., 2002), vide, Kolte, S. W., supra

Syed Shabih Hassan and Juyal, P. D., 2009. Schistosomiasis epidemiology in domestic ruminants in Punjab. XIX Natl. Cong. Vet. Parasit., & Natl. Symp. on national impact of parasitic diseases on livestock health and production. February 3-5, 2009, Dept. Vet. Parasitol., Coll. Vet. Sci., Guru Angad Dev Vet. Anim. Sci. Univ., Ludhiana- 141004, Abstr., OE46: p.43.

Taakre, M. D., 1996. Studies on experimental infection of *Schistosoma nasale* in goats. M. V. Sc. Thesis, Punjabrao Deshmukh Krishi Viswapeeth, Angola, India. (indexed).

Tewari, H. C. and Biswas, G., 1972. Experimental studies on the immunology of *Schistosoma incognitum* Chandler, 1926 by vaccination with gamma irradiated cercariae and passive transfer. Z. Parasitenkd., 38: 48-53.

Tewari, H. C. and Singh, K. S., 1979. Pathogenesis of S*chistosoma incognitum* in mice with special reference to the mechanism of anaemia. Indian J. Anim. Sci., 49: 380-383.

Tewari, H. C., Dutt, S. C. and Iyer, P. K. R., 1963. Observations on the pathogenecity of experimental infection of *Schistosoma incognitum* Chandler, 1926 in dogs. Proc. 50th Sci. Indian Cong. III, p. 584.

Tewari, H. C., Dutt, S. C. and Iyer, P. K. R., 1966. Observations on the pathogenecity of experimental infection of S*chistosoma incognitum* Chandler, 1926 in dogs. Indian J. Vet. Sci., 36: 227-231.

Thakre, M. D., 1996. Studies on experimental infection of *Schistosoma nasale* in goats. M. V. Sc. Thesis, Punjabrao Deshmukh Krishi Viswapeeth, Angola, India. (indexed).

Thakre, M. D. and Bhilegaonkar, N. G., 1998. Incidence of *Indoplanorbis exustus* snails and *Schistosoma nasale* infection in Bhandra District (Maharashtra). J.Vet. Parsitol., 12: 54-55.

Varma, A. K., 1954. Studies on the nature, incidence, distribution and control of nasal schistosomiasis and fasciloiasis in Bihar. Indian J.Vet. Sci., 24: 11-34.

Varma, A. K., 1957. Schistosomiasis in animals. Subject 6. VII Conf. Vet. Dis. Invest. Officers and Res. Workers in India, November, 1957, Bombay.

Varma, T.K., Singh, B. P. and Tiwari, H. C., 1983. Immunity to S*chistosoma incognitum* in mice by previous exposure to *Schistosoma spindale*. J. Helminthol., 57: 37-38.

Vinai Kumar, 1973. Studies on snail hosts of *Orientobilharzia turkestanicum* (Skrjabin, 1913) Dutt and Srivastava, 1955(Schistosomatidae: Trematoda) in India. Ann. Soc., belge. Med. Trop., 53: 17-23.

Vohra, S., 2005. Development of immunological methods for diagnosis of schistosomosis in small ruminants. Ph. D. Thesis (Abstracts). Indian J. Vet. Med., 22: 82-84.

Vohra, S. and Agrawal, M. C. and Malik Y. P. S., 2006. Dot-ELISA in diagnosis of caprine schistosomosis. Indian J. Anim. Sci. 76: 988-991.

1.15 CERCARIAE CAUSING HUMAN DERMATITIS

Agarwal, M. C., 2000. Final Report on National Fellow Project on Studies on the strain identification, epidemiology, diagnosis, chemotherapy and zoonotic potentials of Indian schistosomes. ICAR, New Delhi. (not indexed).

Agrawal, M. C., Gupta, S. and George, J., 2000a. Cercarial dermatitis in India. Bull. Wld. Hlth. Org., 72: 278.

Agrawal, M. C., Pandey, S., Sirmour, S. and Deshmukh, P. S., 2000b. Endemic form of cercarial dermatitis (Khujlee) in Bastar area of Madhya Pradesh. J. Parasit. Dis., 24: 217-218.

Anantaraman, M., 1958. On schistosome dermatitis. I. Dermatitis in India caused by cercariae of *Schistosoma spindale* Montgomery, 1906; Indian J. Helminth., 19: 46-52.

Anandan, R., 1985. Studies on *Schistosoma nasale* Rao, 1933. (Trematoda, Schistosomatidae) Ph. D. Thesis, Tamil Nadu Agri. Univ., Coimbatore.

Anon., 2007-2009. Morphological, molecular and immunological studies on paddy field dermatitis caused by *Schistosoma spindale* in paddy field workers of Assam. (Investigators: Rekha Devi, K. Mahanta, J.and Narain, K.). (Res. Project, not indexed).

Bhalerao, G. D., 1948a. Blood fluke problem in India (Presidential address, Soc. Med. & Vet. Sci.), 35th Indian Sci. Cong. (Patna), pp.57-71. (Abstr. In: Curr. Sci. 17: 36-34.)

Dutt, S. C., 1970. On two dermatitis producing cercariae – *Cercaria srivastavai*, Dutt, 1957 and *C. hardayali* n. sp. with a note on schistosoma dermatitis in India. H. D. Srivastava Commemoration Volume. (Singh, K. S. and Tandon, B. K. Eds.), Div. Parasitol., Indian Vet. Res. Inst., Izatnagar, pp.311-318.

Hafeez, Md, 1999. Cercaria and cercarial dermatitis In: Vectors and Vector-borne Parasitic Diseases. IV Natl Trng. Prog. 25-1-1999 to 8-2-1999. Centr. Adv. Stud., Dept. Parasitol., Vet. Coll., Univ. Agri. Sci., Bangalore-560 024. pp.85-87. (Review, not indexed).

Mahanta, J., Narain, K., and Srivastava, V. K., 1996. Paddy field dermatitis in Assam – a focus of concern for north-east India. Abstracts: Global meet on parasitic diseases, New Delhi, India, 18-22 March, 1996, J. Parasit. Dis., 20: 97-98.

Muraleedharam, S. R. G., 1965. Studies on cercaria from Andhra Pradesh and Madras State. M. V. Sc. Dissertation submitted to Univ. of Madras, Madras.

Muraleedharan, K., 2000. A case of schistosome cercarial dermatitis in man. J. Parasit. Dis., 24: 231-232.

Narain, K. and Mahanta, J., (year not specified, Source: Google). Studies on paddy field dermatitis (Panikaint) among workers in rice fields of upper Assam. Reg. Med. Res. Centr. Dibrugarh, Assam.

Narain, K. and Mahanta, J., 1999. Scanning electron microscopy of the integumental surface of *Schistosoma spindale* isolated during an outbreak of cercarial dermatitis in Assam. India. Indian J. Vet. Parasitol., 13: 103-106.(see Anon).

Narain, K. and Mahanta, J., 2000a. Cercarial dermatitis of animal schistosomes origin: In: Proc.Young scientists workshop on identification and diagnosis of parasites of biomedical importance. (A. K. Yadav, Ed). North East Hill Univ., Shillong, pp. 29-34. (Review, not indexed). (see Anon).

Narain, K. and Mahanta, J., 2000b. Dermatitis associated with paddy field environment in Assam- A review- In: (M. K. Bhasim and Veena Bhasim, Eds.) Man- Environment Relationship. Kamala Raj Enterprises, Delhi. pp.213-220.(Review, not indexed). (see Anon).

Narain, K., Rajaguru, S. K. and Mahanta, J., 1998. Incrimination of *Schistosoma spindale* as a causative agent of farmer's dermatitis in Assam with a note on liver pathology in mice. J. Comm Dis., 30: 1-6. (see Anon).

Narain, K., Mahanta, J., Dutta, R and Dutta, P., 1994. Paddy field itch in Assam: A cercarial dermatitis. J. Comm Dis., 26: 26-30. (see Anon).

Narain, K., Rajguru, S. K. and Mahanta, J., 1998.Incrimination of *Schistosoma spindale* as a causative agent of farmer's dermatitis in Assam with a note n liver pathology in mice. J. Comm., Dis., 30: 1-6.

Rao, K. H. and Murty, A. S., 1968. A note on the occurrence of *Schistosoma dermatitis* in Kondakarla, Andhra Pradesh. Curr. Sci. 37: 407-408.

Rao V. G., Dash A. P., Agrawal M. C., Yadav, R. S., Anvikar, A. R., Vohra, S., Bhondeley, M. K., Ukey, M. J., Das, S. K., Minocha, R. K. and Tiwari, B. K., 2007. Cercarial dermatitis in central India: an emerging health problem among tribal communities. Ann. Trop. Med. Parasitol., 101: 409-413.

1.16 CERCARIAE FAUNA (IN GENERAL)

(not included in index table)

Agarwal, R. D., 1985. Studies on parasitic fauna of zoo and wild life: Studies on the cercarial fauna of Mathura District with special reference to furcocercous cercariae. Ph.D. Thesis submitted to C. S. A. Univ. of Agri. & Tech., Kanpur-208 002.(Vide abstract. J. Vet. Parasitol.,3: 83).

Ammanulla, B. and Shahul Hameed, P., 1996. Studies on molluscan diversity I. Kaveri river system (Tiruchirapalli, India) with special reference to vector snails of trematode parasites. Curr Sci., 71: 473-475. (Snails positive for unspecified cercariae were *Indoplanorbis exustus, Lymnaea ovalis, Thiara tuberculata* and *T. scabra*).

Bhalaerao, G. D., 1943.The cercarial fauna of the the irrigated tract of Nizam Dominians, with suggestions regarding their relationship to the trematode parasites in man and domestic and other animals. Indian J.Vet. Sci., 13: 294-296.

Borkakoty, M. R., and Das, M. R., 1980. Snail and cercarial fauna in Kamrinji District of Assam. J. Res, Assam Agri. Univ., 1: 100-102.

Brinesh, R. and Janardhanan, K. P., 2011. Three new species of xiphidiocercariae from thiarid snail *Thiara tuberculata* in Palakkad, Kerala, India. J. Parasit. Dis., 35: 42-49.

Chaudhuri, S. S., Gupta, R. P. and Yadav, C. L., 1982. Note on cercarial fauna of some aquatic snails of Hariana State. Indian J. Anim. Sci., 52: 1273-1275.

Choubisa, S. L., 1986a. The biology of certain larval trematodes infecting freshwater snails of lakes of Udaipur. Ph. D. Thesis submitted to M. L. Sukhadia Univ., Udaipur, Rajasthan.

Choubisa, S. L., 1986b. Incidence of larval trematode infection and their seasonal variation in the fresh water molluscs of southern Rajasthan. Rec. Zool. Surv. India. 83: 69-80.

Choubisa, S. L., 1991. Comparative study of cercarial behavior and their host specificity. Indian J. Parasitol., 15: 125-128.

Choubisa, S. L., 1997. Seasonal variation of amphistome cercarial infection in snails of Dungarpur district (Rajasthan). J. Parasit. Dis., 21: 197-198.

Choubisa, S. L., 2002. Focus on seasonal occurrence of larval trematode (cercarial) parasites and their host specificity. J. Parasit. Dis., 26: 72-74. (amphistome, echinostome, furcocercous, gymnocephalous).

Choubisa, S. L. and Sharma, P. N., 1983. Sesonal variations of cercarial infection in snails of Fateh Nagar Lake of Udaipur. Indian J. Parasitol., 7: 111-113. (Snails involved: *Indoplanorbis exustus, Gyraulus convexiusculus* and *Lymnaea luteola*).

Choubisa, S. L. and Sharma, P. N., 1986. Incidence of larval trematode infection and their seasonal variation in the fresh water molluscs, Southern Rajasthan. Rec. Zool. Surv. India, 83 (1 &2): 69-90.

Chaudhuri, S. S., Gupta, R. P. and Yadav, C. L., 1982. Note on cercarial fauna of aquatic snails of Haryana State. Indian J. Anim. Sci., 52: 1273-1275.

Chowdhury, N., 2001. Helminths of domestic animals in Indian subcontinent. In: Perspectives of Helminthology (Chowdhury, N and Tada, I Eds.), Oxford & IDI I Publishing Co. Pvt. Ltd., New Delhi. (Refer page 98 of the book, Table 4. for information on hosts and intermediate hosts).

Das Gupta, B., 1973. Biology of some freshwater gastropods and their trematode parasites. Ph.D. Thesis, Univ. Calcutta, pp.170

Dhar, D. N., Bansal, G. C. and Sharma, R. H., 1985.Studies on aquatic snails of Kashmir Valley with particular reference to *Lymnaea sensu stricto*. Indian J. Parasitol., 9: 241-244.

Dharmendra Nath, 1974. A note on metacercarial fauna encountered in Indian fresh-water fishes. Indian Vet. J., 51: 481-483.

Dutt, S. C., 1958. *Cercaria sibi*, n. sp. a new furcocercous cercaria from *Lymnaea acuminata*. Proc. 45th Indian Sci. Cong. (Madras), Part III, p.342.

Dutt, S. C. and Bali, H. S., 1980. Snails of Punjab State and their trematode infections. J. Res. Punjab agric. Univ. 12: 222-228.

Gupta, N. K. and Taneja, S. K., 1969. Two furcocercous cercariae from fresh water snails of Chandigarh. Res. Bull. Punjab Univ., 20 (I-II): 223-228.

Gupta, P. K., Sinha, S. R. P., Sucheta Sinha, Suchita Kumari and Mandal, K. G., 2010.Prevalence of ruminant digenetic trematode transmitting snails in Bihar. XX NCVP Parasitology Today-Ecology to Molecular Biology, 18-20 February, 2010, Dept. Vet. Parasitol., CVS, CCS, Haryana Agri. Univ., Hissar-125004, p.80.

Halij, M. T., 2005. Studies on larval trematode from *Melania tuberculata* a fresh water snail from Nashik District, *Geobios*, 321: 25-32.

Jithendran K. P. and Krishna Lal, 1990. Fresh water snails with larval trematodes in Kangra valley, Himachal Pradesh. J. Vet. Parasitol., 7: 111-113.

Khatri, A. K. and Saxena, M. M., 2008. Trematode infection and its seasonal trends in the gastropod fauna in somewaters in the arid region of Rajasthan. 20th Natl. Cong. Parasitol., Nov. 3-5, 2008, Dept. Parasitol. and Indian Soc. Parasitol., North Eastern Hill Univ., Shillong 793 022, Abstr. PP-44, p.135.

Malaki, A and Singh, K. S., 1962. Parasitological survey of Kumaun region. Part XVI.Three cercariae from Kumaun. J. Helminthol., 14: 133-153. (Not indexed).

Majumdar, S. S., Das Gupta, B. and Chowdhury, A., 1983. *Indoplanorbis exustus* (Deshayes) as the intermediate host of furcocercous cercariae in West Bengal. Proc. Sym. Host as Environment, from March 27-28, 1983, Zool. Surv. India, Calcutta. pp.17-22.

Manohar, L and Rao, P. V., 1976. Physiological response to parasitism I. Changes in carbohydrate reserve of the molluscan host. Southeast Asian Trop. Med. Pub. Hlth., 7: 395-404. (Not Indexed).

Matta, S. C., 1965. Studies on larval trematodes in local snails with special reference to metacercariae and giant eimerian schizonts in abomasum and small intestine of local sheep – a histological study. M. V. Sc. Thesis, Agra Univ., Agra, U. P., India.

Matta, S. C. and Pande, B. P., 1966. Studies on some metacercariae in local snails. Indian J. Helminth., 18: 128-141.

Matta, S. C. and Rai, D. N., 1971. A note on a metacercarial form the snail, *Indoplanorbis exustus* (Deshayes). Indian J. Anim. Res., 5: 55-58.

Mayura, A. K., Jadav, J. P. and Prajapati, R. K., 1996. A new echinostome cercaria from Jaipur., U. P., India. Global meet on parasitic Diseases, New Delhi, India 18-22 March, 1996. Abstr., J. Parasit. Dis., 20: 110.

Mohandas, A., 1974a. Contribution to the cercarial fauna of Kerala. Ph. D. Thesis, Univ. Kerala, Trivandrum, Kerala. (Not indexed).

Mohandas, A., 1974b. Studies on fresh water cercariae of Kerala 1. Incidence of infection and seasonal variation. Folia Parasitol. (Praha), 21: 311-317. (Indexed under echinostomes).

Mukherji, R. P., 1966. Seasonal variations of cercarial infections in snails. J. Zool. Soc. India, 18: 39-45.

Mukherjee, R. P., 1966a. Seasonal variations in cercarial infections in snails. J. Zool. Soc. India., 18: 39-45.

Mukherji, R. P., 1980. Collection and study of larval trematodes (Platyhelminthes). Proc. Workshop on Techniques for Parasites. Zool. Surv, India, 23-26.

Muralidharam, S. R. G., 2005. Studies on cercariae from Andhra Pradesh and Madras states. M. V. Sc. Desssertation, Madras Univ., Madras.

Muraleedharan, K., 2007. Snails and snail-borne diseases and their control. In: X. National Training Programme on Veterinary Parasitology in Focus with Special Reference to Emerging Diseases.10th to 30th March 2007, Centr. Adv. Stud., Dept. Parasitol. Vet. Coll., Univ. Agri. Sci., Hebbal, Bangalore-560 024. pp. 61-81. (Snail intermediaries of trematodes and final hosts recorded from India have been tabulated in this review paper, but not included in Index table).

Muraleedharan, K., 2010. Snails and snail-borne parasitic diseases. In: XIII. National Training Programme on Trends and Perspectives in the Biology, Eecology and Control of Parasitic Diseases. 11th to 30th March 2010. Centr. Adv. Stud., Dept. Parasitol., Vet. Coll. Univ. Agri. Sci., Hebbal, Bangalore-560 024. pp. 10-18. (Snail intermediaries of trematodes and final hosts recorded from India has been tabulated in this review paper, not included in Index list).

Muraleedharan, K., Prasanna Kumar, S., Hegde, K. S. and Alwar, V. S., 1977. Larval trematode infections in aquatic snails of Karnataka State– A preliminary study. Mysore J. Agric. Sci., 11: 101-104.

Pallavi Devi, Saidul Islam and Manoranjandas, 2003. Ecology and biology of aquatic snails and their control II. Cercarial fauna of fresh water snails from Assam. J. Vet. Parasitol., 17: 131-133.

Pandey, K. C. and Nirupama Agrawal, 1977. Studies on cercarial fauna of Kathauta Tal, Lucknow. Indian J. Zool., 18: 1-50.

Pandey, K. C. and Nirupama Agrawal, 1978. Larval trematodes and their seasonal variations in snails of Kathauta Tal, Lucknow. Indian J. Parasitol., 2: 139-143.

Pemola Devi, N. and Jauhari, R. K., 2008. Diversity and cercarial shedding of malaco fauna collecting from water bodies of Ratnagiri District, Maharashtra. Acta Trop., 105: 249-252.

Peter, C. T., 1949. Studies on the fresh water cercarial fauna in Madras. M. Sc.Thesis, submitted to the Univ. Madras.

Peter, C. T., 1954. Studies on larval trematodes from fresh water snails in Madras. Proc. 41st Indian Sci. Cong., Hyderbad, Part III, Abstr., pp.221-222.

Pokhriyal B. P., Jauhari, R. K. and Sudarshana, R., 1998. Trematode cercariae infection in the snail, *Thiara (M.) tuberculata* (Mueller, 1774) in different localities of Doon Valley. J. Exptl. Zool., India, 1(2): 107-110.

Pokhriyal, B. P., Mahesh, R. K. and Jauhari, R. K., 1996. Prevalence of trematode cercariae infection in the snail *Lymnaea (Pseudosuccinea) acuminata* Lamarck, 1822 in different localities of Dehradun-Valley. Global Meet on Parasit.Dis., New Delhi, India 18-22 March, 1996. Abstr., J. Parasit. Dis., 20: 111.

Poonam Devi, 2011. Trematode cercariae infection in the snail, *Thiara (Melanoides) tuberculata* (Muller, 1774) in different localities of Doon Valley, India. Deccan Curr. Sci. 4: 221-223.

Preveen Kr. Gupta, 2009. Studies on cercarial infected snails affecting ruminants with special reference to their control. M. V. Sc., Thesis submitted to Bihar Vet. Coll., Patna, Bihar, India.

Rai, D. N., 1966. Observations on metacercariae encountered in snails and studies on helminth parasites of fish-eating birds, snake bird and little cormorant. M. V. Sc. Thesis submitted to Agra University, Agra, U. P., India.

Rajkhowa, C., 1982. Studies on snail and cercarial fauna from East Khasi Hills District of Meghalaya. Indian Council of Agricultural Research, Annual Report, ICAR. pp.292-297.

Rajkhowa, C.,Gogoi,A. R., Borkakoty, M. R. Sharma, P. C. and das, 1991. Cercarial fauna and their pattern of seasonal liberation from *Indoplanorbis exutus* in Assam. J. Vet. Parasitol., 5: 49-52.

Ratnaparkhi, M. R., Rajurkar, S. R., Deshpande and Mandakhalikar, M. V., 1993. Studies on the incidence of cercariae from various species of snails from Parbhani District (Maharashtra State). Livestock Adviser., 18: (10).

Rao, M. A. N., 1932. A comparative study of cercarial fauna in Madras. Indian Vet. J., 9: 107-111.

Rao, M. R., Reddy, M. M., Saheb, S. K. and Raveendra, K. 1988. Effect of mixed larval trematode infections on carbohydrate pool of the snail, *Lymnaea luteola*. Environ. Ecol., 6: 594-596.

Raut, S. K., 1983. Factors influencing cercarial liberation from their snail hosts. Proc. Sym. Host Environ. Zool. Surv. India, pp.55-64.

Sardey, M. R., 1982. A study on the prevalence of trematode cercaria in snails of Nagpur region. In: Vectors and vector-borne diseases. Proc. All India Sym., Trivandrum, Kerala, India. February 26-28, 1982. pp.137-141.

Sewell, R. B. S., 1922. *Cercariae indicae.* Indian J. Med. Res., 10 (Suppl): 1-371. (Not indexed).

Singh, A., Srivastava, S., Chandra Sekhar and Singh, J., 2009. Prevalence of trematodes in bovines and snails. Indian Vet. J. 86: 206-207.

Singh, B. P., 1975. A preliminary survey of cercarial fauna of two common snails-*Lymnaea luteola* and *Indoplanorbis exustus* at Mathura: Observations on the post-cercarial developmental stages and host parasite relationships of the two mammalian schistosomes. M. V. Sc. Thesis submitted to Agra University, Agra, U. P., India.

Singh, R. N., 1959. Seasonal infestation *Indoplanorbis exustus* (Deshayes) with furcocercus cercariae. Proc. Natl. Acad. Sci. India, 29: 62-72.

Soparkar, M. B., 1921. Notes on some furcocercous cercariae from Bombay. Indian J. Med. Res., 9: 23-32.

Srivastava, N. and Jauhari, R. K., 1990. Influence of some abiotic factor on the occurrences of larval trematodes in *Lymnaea acuminata* (Lamarck) in Doon Valley. Rev. Parassitol., 6: 335-341.

Swarnakumari, V. G. M., 2001.Comparison of cercarial infection of the snail, *Thiara tuberculata* at different habitats. J. Parasitol. Appl. Anim. Biol., 10: 49-54.

Swarna Kumari, V. G. M., 2002. Studies on the cercarial found of *Thiara (M.) tuberculata* and the relationship between the snail size and its infection status. Uttar Pradesh. J. Zool., 22(2): 125-130.

Tamloorkar, S. L. Narladkar, B. W. and Deshpande, P. D., 2001. Prevalence of snail species in Marathwada region and their cercariae carrier status. Twelfth Natl. Cong. Vet. Parasitol., Aug 26-27, 2001, Dept. Parasitol., Coll. Vet. Sci., Acharya Ranga Agri. Univ., Tirupati-517 502, India. Abstr. S-2: 68, p.83.

Thapar, G. S., 1970. Studies on the life histories of trematode parasites. II. Some new and little known cercariae from Lucknow and its environments. Indian J. Helminth., 21 (2): 119-146

Varma, T. K. and Prasad, A., 2000. Prevalence of paramphistome cercariae in fresh water aquatic snails in Bareilly, Uttar Pradesh. Intas Polivet, 1: 67-71

Upadhyay, C., 2007. Ecological conditions in propagation of *Indoplanorbis exustus* and *Lymnaea luteola* and their potentials in transmitting trematode infections. Ph. D.Thesis, Rani Durgavati Vishwa Vidyalaya, Jabalpur, India.

Vasanda Kumar, M. V. and Jha, K. P., 2005. Two new species of furcocercous cercariae from fresh water snails in Malabar, Kerala. Uttar Pradesh J. Zool., 25: 93-96.

Vasanda Kumar, M. V. and Jha, K. P., 2006. Two new species of cercariae VII and VIII n.sp. from fresh water snails in Malabar, Kerala. Uttar Pradesh J. Zool., 26: 61-64.

Venkata Rama Krishna, G., 1983. The effect of density of snails on the incidence of larval trematode infection. Geobios, 10: 40-41.

Vohra, R., Rajesh Katoch, Anish Yadav, Khajuria, J. K. and Shilpa Sood., 2009. Prevalence of *Cercaria pigmentata* in *Indoplanorbis exustus* in Jammu District from September 2005 to August 2008. XIX Natl. Cong. Vet. Parasit., & Natl. Symp. "National impact of parasitic diseases on livestock health and production". February 3-5, 2009, Dept. Vet. Parasitol., Coll. Vet. Sci., Guru Angad Dev Vet. Anim. Sci. Univ., Ludhiana-141004, Abstr., OE4: p.27.

2. CLASS: CESTODA

2.1 FAMILY: MESOCESTOIDIDAE

Srivastava, H. D., 1939. A study of the life history of a common tapeworm, *Mesocestoides lineatus* of Indian dogs and cats. Indian J. Vet. Sci. 9: 187: -190.

2.2 FAMILY: ANOPLOCEPHALIDAE

Anantaraman, M., 1951. The development of *Moniezia,* the large tapeworm of domestic ruminants. Sci. and Cult., 17: 155-157.

Balakrishnan, M. M. and Haq, M. A., 1984. New oribatid vectors of cestodes from India.In: 15. Medical and Veterinary Acarology. (Eds. D. A. Griffiths and C. E. Bowman), Ellis Harwood Ltd., England. Acarology VI, 2: 1161-1164.

Balakrishnan, M. M. and Haq, M. A., 1985. Three new oribatids of anoplocephaline cestodes of Kerala. Indian J. Acarol., 10: 9-13.

Baruah, N., 1994. Studies on moneziasis in goats in Assam. Ph. D. Thesis in Assam Agri. Univ., Assam.

Baruah, N., Gogoi, A. R., Baruah, G. K., Borkakoty, M. R. and Lakhar, B. C., 2002. Moneziasis in goats.: A histopathological study. J. Vet. Parasitol., 16: 59-60.

Deshpande, P. D. Shastri, U. V. and Gafoor, M. A. 1980a. Role of *Hypozetes* sp., an oribatid mite from India as intermediate host of *Moniezia expansa.*, Proc. 3rd Natl. Cong. Parasitol, Abstr., Indian J. Parasitol., 4 (Suppl.): p.75.

Deshpande, P. D., Anantwar, L. G., Shastri, U. V. and Gafoor, M. A. 1980b. Effect of development of *Moniezia expansa* in kids on some haematological and biochemical values. Proc. 3rd Natl. Cong. Parasitol, Abstr., Indian J. Parasitol., 4 (Suppl.): p. 96.

Haq, M. A., 1984. New natural oribatid mites as vectors from India: Progress in Acarology. Ed: Channabasavanna, G. P. and Virakatamath, C. A., Oxford & IBH Delhi.

Haq, M. A., 1988. In: Bicovas Proceedings 1. Anathasubramaniyan, K. S. Venkatesan, P and Sivaraman, S., (Eds). New Metro Printers, India, 93-108.

Haq, M. A., 1990. Mites in disease transmission. In: Proc. II Kerala Sci. Cong., Thiruananthapuram, Balakrishnan Nair (Ed.). pp.239-240.

Haq, M. A., 1991. Harmful soil fauna and their management. 3.15. New natural oribatid mites as vectors from Kerala. In: Advances in Management and Conservation of Soil Fauna. Veeresh G. K., Rajagopal, D. and Virakatamath, C. A. (Eds), Oxford & IBH Publishing, New Delhi, pp.259-264.

Haq, M. A., 1999. On oribatid mites as vectors of cestodes. In: Vector and Vector-borne Diseases, Centr. Adv. Stud., Dept. Parasitol., Vet., Coll., Bangalore, pp.169-181.

Haq, M. A., 2001. Potential oribatid mite infection of cestode parasites of Kerala. In: Acarology: Proc. 10th International Cong. held at Canberra Australia, Halliday, R. B., Walker, D. E., Proctor, H. C., Norton, R. A. and Colloff, M. J. (Eds.), CISRO Publishing Collingwood,150, Oxford Street, P. O. Box No. 1151,Victoria 3066, Australia. pp.569-575.

Haq, M. A., Ramani, M. and Prakashan, K., 1999. On oribatid mites as transmitting agents of cestodes in Kerala. J. Acarol., 15: 68-72.

Jeyathilakan, S., Raman, M., Abdul Basith and Lalitha John, 2010. Incidence of oribatid mites in a livestock farm. Indian Vet. J. 87: 13-14.

Kaur, A., 1995. Epizootology and biology of common anoplocephaline cestodes in sheep along with histopathology and haematology of the definitive host. Ph. D. (Zoology, Parasitology) Thesis, Punjab Univ., Chandigarh-160 014. Abstr., J. Vet. Parasitol., 9: 153-154.

Kaur, A., 1996. Observation on the life cycle of common anoplocephaline cestodes of sheep in India. Global meet on parasitic diseases, New Delhi, India, 18-22 March, 1996, Abstr., J. Parasit. Dis., 20: 110.

Kaur, A. and Bali, H. S., 1994. Development of *Moniezia expansa* Rudolphi, 1810 in intermediate and definitive hosts. Sixth Natl. Cong. Vet. Parasitol, 22-24 October, 1994, Dept. Parasitol., Coll. Vet. Sci., JNKVV, Jabalpur-482 001, M. P., India, Abstr. S-2: 15, p.15.

Kaur, A., Bali, H. S. and Duggal, C. L., 1993a. New potential of intermediate host of *Moniezia expansa* (Rudolphi, 1810) in India. J. Vet. Parasitol., 7: 55-58.

Kaur, A., Bali, H. S. and Gill, J.S., 1993b. A note on the vectors of *Thysaniezia giardi* (Moniez, 1879) in India. J. Vet. Parasitol., 7: 23-28.

Kaur, A., Bali, H. S. and Gill, J.S., 1995. Laboratory infections of *Scheloribates curvialatus* (Hammer, 1961) with common anoplocephaline cestodes. Abstr. 12th Nat. Cong. Parasitol. Panaji, Goa, 23-25 Jan., 1995, J. Parasit. Dis., 19: 103.

Kaur, A., Bali, H. S., Duggal, C. L. and Singh, H., 1997. Laboratory infections of *Scheloribates curvialatus* (Hammer, 1961) with common anoplocephaline cestodes. J. Parasit. Dis., 21: 203-204.

Kaur, A., Bali, H. S. and Rai, H. S., 2000. Laboratory infection of *Scheloribates curvialatus* (Hammer, 1961) with common anoplocephaline cestodes. Indian Vet. J., 77: 385-388.

Kaur, A., Rai, H. S. and Singh, N. K., 2007. Prevalence of oribatid mites and its correlation with cestodes of veterinary importance in Ludhiana. J. Vet. Parasitol., 21: 149-152.

McAloom, F. M., 2004. Oribatid mites as intermediate hosts of *Anoplocephala manubriata*, cestode of asian elephant in India. Exp. Appl. Acarol., 32: 181-185.

Mehra, K. N. and Srivastava, H. D., 1955a. Studies on the life cycle of *Moniezia expansa* (Rudolphi, 1810), a broad tapeworm of ruminants. Proc. 42nd Sci. Cong (Baroda)., Part III, Abstr. 67, p.352.

Mehra, K. N. and Srivastava, H. D., 1955b. Studies on the life cycle of *Moniezia benedini* (Moniez, 1879), tapeworm of ruminants. Proc. 42nd Sci. Cong. (Baroda), Part III, p.352.

Nadakal, A. M., 1960a. Observations on the life cycle of *Avitellina centripunctata* (Rivolta, 1874) an anoplocephaline cestode from sheep zand goat. J. Parasitol., 46: 12.

Nadakal, A. M., 1960b. *Protoschelorbates* sp. An orbatid mite from India, as a potential vector of the sheep tapeworms, *Moniezia expansa* and *M. benedeni*. J. Parasitol., 46: 817.

Narsapur, V.S., 1971. Ecological and biological studies on the oribatid fauna of India (Bombay region) together with some observations on the life cycle of common anoplocephalid tapeworms. Ph. D. Thesis Mahatma Phule Krishi Vidyapeedh, Rahuri, Maharashtra, India.

Narsapur, V.S., 1974a. A note on the vectors of *Avitellina lahorea* (Woodland, 1927) in India. Indian Vet. J., 51: 54-56.

Narsapur, V.S., 1974b. Ecological and biological studies on the oribatid fauna of India (Bombay region) together with observations on the life cycle of common anoplocephaline tapeworms. Indian Vet. J., 51: 165-166.

Narsapur, V. S., 1976a. Observation on the biology of sheep tapeworm, *Moniezia expansa* (Rudolphi, 1810) in India. Indian J. Anim. Sci. 45: 603-609.

Narsapur, V.S., 1976b. Intermediate hosts and larval development of *Moniezia benedini* and *M. expansa* (Moniez, 1879) in India. J. Parasitol., 62: 720.

Narsapur, V. S., 1976c. Laboratory infections of *Scheloribates* spp. (oribatid mites) with *Moniezia expansa* and *M. benedini*. J. Helminthol. 50: 153-156.

Narsapur, V. S., 1988. Pathogenesis and biology anoplocephalid cestodes of domesticated animals (a review article). Ann. Rech.Vet., 19: 1-17. (Review article, not indexed).

Pythal, C., 1974. Studies on cestodes of ruminants. Thesis (M. Sc.) submitted to University of Kerala, Trivandrum.

2.3 FAMILY: THYSANOSOMIDAE

Kaur, A., 1995. Epizootology and biology of common anoplocephaline cestodes in sheep along with histopathology and haematology of the definitive host. Ph. D. (Zoology, Parasitology) Thesis, Punjab University, Chandigarh-160 014. Abstr., J. Vet. Parasitol., 9: 153-154.

Kaur, A., 1996. Observation on the life cycle of common anoplocephaline cestodes of sheep in India, Global Meet on Parasitic Diseases, New Delhi, India, 18-22 March, 1996, Abstr, J. Parasit. Dis., 20: 110.

Kaur, A., Bali, H. S. and Gill, J. S., 1993. A note on the vectors of *Thysaniezia giardi* (Moniez, 1879) in India. J. Vet. Parasitol., 7: 23-28.

Kaur, A., Bali, H. S. and Gill, J. S., 1995. Laboratory infections of *Scheloribates curvialatus* (Hammer, 1961) with common anoplocephaline cestodes. Abstr. 12th Natl. Cong. Parasit. Panaji, Goa, 23-25 Jan., 1995, J. Parasit. Dis., 19: 103.

Kaur, A., Bali, H. S., Duggal, C. L. and Singh, H., 1997. Laboratory infections of *Scheloribates curvialatus* (Hammer, 1961) with common anoplocephaline cestodes. J. Parasit. Dis., 21: 203-204.

Kaur, A., Bali, H. S. and Gill, G. S., 1993. A note on the vectors of *Thysanezia giardi* (Moniez, 1879). J. Vet. Parasitol., 7: 23-28.

Kaur, A., Bali, H. S. and Rai, H. S., 2000. Laboratory infections of *Scheloribates curvialatus* (Hammer, 1961) with common anoplocephaline cestodes. Indian Vet. J., 77: 385-388.

Nadakal, A. M., 1960a. Observations on the life cycle of *Avitellina centripunctata* (Rivolta, 1874) on anoplocephaline cestode from sheep and goats. J. Parasitol., 46: No. 5. Sec. 2, 12.

Nadakal, A. M., 1960b. *Protoschelorbates* sp. An orbatid mite from India, as a potential vector of the sheep tapeworms, *Moniezia expansa* and *M. benedeni*. J. Parasitol., 46: 817.

Narsapur, V. S., 1974a. A note on the vector of *Avitellina lahorea* (Woodland, 1927) in India. Indian Vet. J., 51: 54-56.

Tandon, R. S., 1963. Obervations on the life cycle of the anoplocephaline cestodes, *Stilesia globipunctata* (Rivolta, 1974) subfamily Thysanosominae, a common parasite of ruminants in India. Parasitologia, 5: 183-188.

2.4 FAMILY: DAVAINEIDAE

Bhowmik, M. K., 1981. Studies on taeniasis in poutry with special reference to pathogenesis and immunological response to *Raillietina cesticillus* (Molin) infection. Ph. D. Thesis, Bidhan Chandra Krishi Viswavidyalaya, West Bengal, India.(Not indexed).

Bhowmik, M. K. and Sinha, P. K., 1983. Studies on the protective immune response to *Raillietina cesticillus* infection in fowl. Indian J. Anim. Hlth., 22: 25-28.

Bhowmik, M. K., Sinha, P. K. and Chakraborthy, A. K., 1982. Studies on the pathobiology of chicks experimentally infected with *Raillietina cesticillus* (cestoda). Indian J. Poult. Sci., 17: 207-213.

Bhowmik, M. K., Sinha, P. K. and Sarkar, P. B., 1983. Cellular and humoral responses to *Raillietina cesticillus* infection in fowl. Indian J. Anim. Sci., 53: 688-691.

Bhowmik, M. K., Sinha, P. K. and Chakraborthy, A. K., 1985. Effect of vitamin-A deficient diet on fowl cestode, *Raillietina cesticillus*. Indian J. Anim. Hlth., 24: 63-66.

Chand, K., 1961. Studies on taenlasis in poultry. M. V. Sc. Thesis, submitted to Punjab Agri. Univ., Ludhiana.

Chand, K., 1964a. Taeniasis in poultry. Indian Poult. Gaz., 46: 47-49.

Chand, K., 1964b. Life cycle of *Cotugnia digonopora* (Pasquale, 1890) Diamere, 1893 and role of ants in taeniasis in poultry. J. Res. Punjab Agri. Univ. 1: 93-100.

Chand, K., 1969. Studies on the biology, host-parasite relationship, treatment and control of some common tapeworms of poultry. Ph. D. Thesis, submitted to Agra Univ., Agra.

Chand, K., 1970a. Lesion caused by *Railletina tetragona* (Molin, 1958) Furhmann, 1920, in the domestic fowl. In: H. D. Srivastava Commemoration Volume, Div. Parasitol., Indian Vet. Res. Inst. Izatnagar, U. P., pp.295-298.

Chand, K., 1970b. Preliminary studies on common tapeworms of the fowl. Punjab Vet. 9: 31-35.

Chellappa, D. J., Sunder, N. and Gomathinayagam, S., 1993.The role of ants as intermediate host for poultry tapeworms. Paper presented at III Indo-Pacific and VI All India Sym. on Invertebrate Reproduction at Ayya Nadar Janaki Ammal College, Sivakasi, 10-12 Dec., 1993.

Dutt, S. C., 1961. Effects of light infection of the fowl cestode, *Raillietina cesticillus* (Molin, 1958) on young chicken. Indian J. Vet. Sci., 31: 252-256.

Dutt, S. C. and Mehra, K. N., 1962. Studies on the life history of *Hymenolepis farciminosa* (Goeze, 1782) Cestoda: Hymenolepididae. Parasitol., 52: 397-400.

Dutt, S. C., Sinha, P. K. and Mehra, K. N., 1950. Studies on the life history and biology of fowl cestode, *Raillietina cesticillus* (Molin). Proc., 37th Indian Natl. Cong. (Pune), III, 242-243.

Dutt, S. C., Sinha, P. K, and Mehra, K. N., 1961. Studies on the life history and biology of fowl cestode, *Raillietina cesticillus* (Molin, 1958) Furhmann, 1920. Indian J. Vet. Sci., 31: 108-120.

Harikrishnan, T. G. and Ponnudurai, G., 2010. Occurrence of *Cotugnia digonopora* in young broilers. Indian Vet. J., 87: 831-832.

Gogoi, A. R. and Chaudhuri, R. P., 1982a. Infectivity of newly formed cysticercoids of *Raillietina tetragona, R. echinobothrida* and *R. cesticillus* to chicks. Indian J. Parasitol., 6: 51-52.

Gogoi, A. R. and Chaudhuri, R. P., 1982b. Observation on the larval and adult forms of *Raillietina echinobothrida* (Megnin, 1881) with a note about the validity of *Raillietina pseudoechinobothrida* (Meggitt, 1926). J. Assam Sci. Soc., 25: 61-67.

Gogoi, A. R. and Chaudhuri, R. P., 1982c. Efficacy of febendazole (Panacur, Hoechst A G) against and mixed infection of fowl cestodes fowl cestodes *Raillietina tetragona, R. cesticillus and R. echinobothrida*. Indian J. Anim. Sci., 52: 62-66.

Gogoi, A. R. and Chaudhuri, R. P., 1982d. Contribution to the biology of fowl cestodes *Raillietina tetragona, R. echinobothrida* and *R. cesticillus*. Indian J. Anim. Sci., 52: 246-253.

Gogoi, A. R. and Chaudhuri, R. P., 1982e. Effect of bunamidine hydrochloride against fowl cestodes *Raillietina tetragona, R. cesticillus* and *R. echinobothrida*. Indian J. Anim. Sci., 52: 556-560.

Gogoi, A. R. and Hazarika, R. N., 1975a. Effect of *Raillietina echinobothrida* on poultry: An experimental study. Indian Poult. Rev., 7: 13-16.

Gogoi, A. R. and Hazarika, R. N., 1975b. *Teramorium tortosum* (Rog), an intermediate host of poultry tapeworms, *Raillietina tetragona, R. echinobothrida* and *R. tetragona*. Indian Vet. J., 52: 939-941.

Gogoi, A. R. and Hazarika, R. N., 1975c. Efficacy of dichlorophen against natural and artificial infection of *Raillietina echinobothrida*. Orissa Vet. J., 10: 55-59.

Joseph, S. A. and Karunamoorthy, G. and Lalitha, C. M. 1987. *Pheidole* sp., a new intermediate host for fowl tapeworm, *Cotugnia digonopora*. Cheiron, 16: 275.

Lalitha John, 1999. Beetle as vectors of parasitic diseases. In: Vectors and Vector-borne Parasitic Diseases. IV Natl. Trng Prog., 25-1-1999 to 8-2-1999. Centr. Adv. Stud., Dept. Parasitol., Vet. Coll., Univ. Agri. Sci., Bangalore-560 024. pp.31-39. (Review, not index

Malaviya, H. C. and Dutt, S. C., 1970. Morphology and life-history of *Cotugnia srivastavai* sp. N. (Cestoda: Davaieidae) from domestic pigeon. H. D. Srivastava Commemoration Volume, pp.103-106.

Malaviya, H. C. and Dutt, S. C., 1971a Redescription and life-history of *Cotugnia meggitti* Yamaguti, 1935. Indian J. Helminth., 23: 104-114.

Malaviya, H. C. and Dutt, S. C., 1971b. Morphology and life-history of *Raillietina (Raillietina) nagapurensis mehrai*, sp. nov. (Cestoda: Davaineidae). Indian J. Anim. Sci., 41: 1003-1007.

Malaviya, H.,C. and Dutt,S. C., 1971c, Proc.Nat. Acad. Sci.,41 (B): 1V, 357-362. (*Raillietina torquata*).

Mathur, S. C. and Pande, B. P., 1969. *Raillietina cesticillus* and *R. tetragona* infection in chicks reared on normal and deficient feeds – an experimental study. Indian J. Anim. Sci., 39: 115-134.

Muraleedharan, K., 1988. Insect vectors of poultry cestodes of India – A short review. General article section, Second Natl. Cong.Vet. Parasitol., March 3rd-5th, 1988, Bangalore, India. pp.1-3. (Review, not indexed).

Muraleedharan, K., 1996. Larval cestodes of poultry. In: Recent Trends in Diagnosis and Control of Parasitic Diseases. Natl. Trng. Prog. Teachers/Scientists. 25th-30th March, 1996, Centr. Adv. Stud., Dept. Parasitol., Vet. Coll., Univ. Agri. Sci., Bangalore-560 024., Bangalore. pp.23-28. (Review, not indexed).

Muraleedharan, K., 1999. Role of certain insects like ants, lice, grass-hoppers and dragon flies as intermediate hosts of helminth parasites. In: Vectors and Vector-borne Parasitic Diseases. IV Natl. Trng. Prog., 25-1-1999 to 8-2-1999. Centr. Adv. Stud., Dept. Parasitol., Vet. Coll., Univ. Agri. Sci., Bangalore-560 024. pp.125-131. (Review, not indexed)

Muraleedharan, K., 2000. Biology, economic importance, and control of lice, ants, grasshoppers and dragon flies. In: Acarines and Insects of Veterinary and Medical Importance. V. Natl. Trng. Prog. Centr. Adv. Stud., Dept. Parasitol., Vet. Coll., Univ. Agri. Sci., Bangalore-560 024. pp.18-25. (Review, not indexed).

Nadakal, A. M., 1968-73. Resistance potential of certain breeds of domestic fowl exposed to *Railletina tetragona* infections. Final Tech. Rept. on PL-480 Project, A7-ADP-25, Mar Ivanios College, Trivandrum-695015, Kerala. (Not indexed).

Nadakal, A. M. and Vijayakumaran Nair, K., 1979. Studies on metabolic disturbances caused by *Raillietina tetragona* (Cestoda) infection in domestic fowl: Effect of infection on certain aspects of carbohydrate metabolism. Indian J. Exp. Biol., 17: 310-311.

Nadakal, A. M. and Vijaykumaran Nair, K., 1982. A comparative study on the mineral composition of the poultry cestode *Railletina tetragona* (Molin, 1858) and certain tissues of its host. Proc. Indian Acad. Sci. (Anim. Sci.) 91: 153-158.

Nadakal, A. M., John, K. O., Muraleedharan, K. and Mohandas, A. 1970a. Resistance of certain breeds of domestic fowl exposed to *Railletina tetragona* infections. I. Contribution to the biology of *Railletina tetragona* (Molin, 1858). Proc. Helm. Soc. Wash., 37: 141-143.

Nadakal, A. M., John, K. O., Mohandas, A. and Muraleedharan, K., 1970b. Resistance of certain breeds of domestic fowl exposed to *Railletina tetragona* infections. II. Studies on the periodicity of segment discharge by the domestic fowl infected with *Railletina tetragona* (Molin, 1858). Proc. Helm. Soc. Wash., 37: 144-146.

Nadakal, A. M., Muraleedharan, K., Mohandas, A. and John, K. O., 1970c. Observations on certain aspects of biology of the fowl tapeworm, *Cotugnia digonopora* (Pasquale, 1890) Diamere, 1893. Jap. J. Parasit., 19: 196-198.

Nadakal, A. M., Mohandas, A., John, K. O. and Muraleedharan, K., 1971a. Resistance of certain breeds of domestic fowl exposed to *Railletina tetragona* infections. 3. Species of ants intermediate hosts for certain fowl cestodes. Poult. Sci., 50: 115-118.

Nadakal, A. M., Muraleedharan, K., John, K. O. and Mohandas, A., 1971b. Resistance of certain breeds of domestic fowl exposed to *Railletina tetragona* infections. V. Pathogenic effects of cestode on growing chickens. Jap. J. Parasit., 20: 433-438.

Nadakal, A. M., John, K. O. and Mohandas, A. and Muraleedharan, K., 1972. Resistance of certain breeds of domestic fowl exposed to *Railletina tetragona* infections. 6. Effects of starvation of the host chicken on the tapeworm, *Railletina tetragona*. Poultry Sci., 50: 1027-1031.

Nadakal, A. M., John, K. O., Mohandas, A. and Muraleedharan, K., 1973a. Resistance of certain breeds of domestic fowl exposed to *Railletina tetragona* infections. 7. Effects of cortisone on *R. tetragona*. Poultry Sci., 52: 682-687.

Nadakal, A. M., John, K. O., Mohandas, A. and Simon, M., 1973b. Resistance of certain breeds of domestic fowl exposed to *Railletina tetragona* infections. IX. Effect of protein deficient diet on R. *Railletina tetragona* infection. Riv. Parasitol., 34: 185-191.

Nadakal, A. M., Mohandas, A. John, K. O. and Muraleedharan, K., 1973c. Resistance of certain breeds of domestic fowl exposed to *Railletina tetragona* infections. 4. Effects of light on *R. tetragona* infection. Indian J. Anim. Hlth., 12: 121-125.

Nadakal, A. M., Mohandas, A. John, K. O. and Simon, M., 1973d. Resistance of certain breeds of domestic fowl exposed to *Railletina tetragona* infections.VIII. Effect of thiouracil on R. *Railletina tetragona* infections. Poultry Sci., 52: 1069-1074.

Nadakal, A. M., Mohandas, A. John, K. O. and Simon, M., 1973e. Resistance of certain breeds of domestic fowl exposed to *Railletina tetragona* infections. X. Influence of sex hormones on *Railletina tetragona* infections. Z. Parasitenk., 41: 147-156.

Nadakal, A. M., Mohandas, A. John, K. O. and Muraleedharan, K., 1973f. Contribution to the biology of fowl cestode, *Railletina echinobothrida* with notes on its pathogenicity. Trans. Amer. Micros. Soc., 92: 273-276.

Nadakal, A. M., Mohandas, A. John, K. O. and Simon, M., 1974a. Resistance of certain breeds of domestic fowl exposed to *Railletina tetragona* infections. XI. Effects of *Railletina tetragona* infections on egg laying birds. Arch. Fur Geflugelk., 38: 138-142.

Nadakal, A. M., Mohandas, A. John, K. O. and Simon, M., 1974b. Resistance of certain breeds of domestic fowl exposed to *Railletina tetragona* infections. XIII. Effects of Vitamin A deficient diet on the host on *Railletina tetragona* infection. Arch. Fur Geflugelk., 38: 174-177.

Nadakal, A. M., Mohandas, A. John, K. O. and Simon, M., 1975a. Resistance of certain breeds of domestic fowl exposed to *Railletina tetragona* infections. XII. Effects of calcium deficient diet on the host on *Railletina tetragona* infections. Riv. Parasitol., 36: 41-46.

Nadakal, A. M., Mohandas, A. John, K. O. and Simon, M., 1975b. Resistance of certain breeds of domestic fowl exposed to *Railletina tetragona* infections. XIV. Effects of host dietary carbohydrate deficiency on *Railletina tetragona* infection. Arch. Fur Geflugelk. 39: 15-20.

Nadakal, A. M. and Vijayakumaran Nair, K., 1979. Metabolic disturbances caused by *Raillietina tetragona* (Cestoda) infection in domestic fowl: Effect of infection on certain aspects of carbohydrate metabolism. Indian J. Exp. Biol., 17: 310-311.

Nadakal, A. M. and Vijayakumaran Nair, K., 1982. A comparative study on the mineral composition poultry cestode *Raillietina tetragona* (Molin, 1858) and certain tissues of its host. Proc. Indian Acad. Sci. (Anim. Sci.), 91: 153-158.

Narasapur, V. S., 1996. Anoplocephaline cestodes of domestic animals. In: Recent Trends in Diagnosis and Control of Parasitic Diseases. Natl. Trng. Prog. Teachers/ Scientists.25th-30th March, 1996. Centr. Adv. Stud., Dept. Parasitol., Vet. Coll., Univ. Agri. Sci., Bangalore-560 024. pp.11-21. (Review, not indexed).

Ponnudurai, G. and Chellappa, D. J., 2001a. Observation on the biology of *Cotugnia digonopora* and their natural host in Tamilnadu. Indian J. Poult. Sci., 36: 113-114.

Ponnudurai, G. and Chellappa, D. J., 2001b. Observation on the biology of *Choanotaenia infundibulum* and *Raillietina cesticillus* and their natural host in Tamilnadu, India. J. Vet. Parasitol., 15: 51-53.

Rajendran, M. and Nadakal, A. M., 1984. Effect of Yomesan (niclosamide) on developmental and adult stages of fowl cestode, *Railletina tetragona* (Molin, 1958) in experimental infection. Indian J. Anim. Sci., 54: 93-95.

Rajendran, M. and Nadakal, A. M., 1988. The efficiency of praziquantel (Droncit R) against *Railletina tetragona* (Molin, 1958) in domestic fowl. Vet. Parasitol., 26: 253-260.

Rajendran, M., Vijayakumaran Nair, K. and Nadakal, A.M., 1981. A note on an abnormal cysticercoids of *Railletina echinobothrida* (Megnin, 1881) (fowl cestode). Proc. Helm. Soc. Wash., 48: 104.

Sinha, A. K., Chitra Sinha and Nikkil, R., 1986. On the mode of attachment of *Railletina tetragona* (Molin, 1858) Fuhrmann 1920 and related tissue changes in the intestine of domestic fowl. Indian J. Parasitol., 10: 237-238.

Sinha, P. K. and Srivastava, H. D., 1958. Studies on the age resistance to superinfection of poultry against *Raillietina cesticillus* (Molin) with some observations on the host specificity of the parasites. Indian Vet. J. 35: 288-291.

Srivastava, J. S. and Pande, B. P., 1967a. On the effects of *Raillietina* (*Raillietina*) *tetragona* and *R.* (*Skrjabina*) *cesticillus* infections in weight gains of laboratory raised chicks – a preliminary study. Indian J. Anim. Hlth., 6: 93-100.

Sunder, N. and Chellappa, D. J., 2001. Life cycle of *Cotugnia digonopora* in chicken. J. Vet. Parasitol., 15: 153-154.

Vijayakumaran Nair, K. and Nadakal, A. M., 1980. Metabolic disturbances caused by *Raillietina tetragona* (Cestoda) infection in domestic fowl: Effect of infection on certain aspects of protein metabolism. Proc. Indian Acad. Sci. (Anim. Sci.), 89: 543-549.

Vijayakumaran Nair, K. and Nadakal, A. M., 1981. Haematological changes in domestic fowl experimentally infected with the cestode *Raillietina tetragona* (Molin, 1858). Vet. Parasitol., 8: 49-56.

Vijayakumaran Nair K and Nadakal A.M., 1982. A comparative study on the mineral composition of the poultry cestode, *Raillietina tetragona* Molin 1858 and certain tissues of its host. Proc. Indian Acad. Sci. (Anim. Sci.), 91: 153-158.

Vijayakumaran Nair, K. and Nadakal A.M., 1986. Site selection by *Raillietina tetragona*. All-India Sym. Animal behaviour, 15th Ann. Conf. Ethological Soc. India, August 7-9, 1986, Annamalai Univ., Annamalai Nagar.

Vijayakumaran Nair, K., Rajendran, M and Nadakal, A. M., 1982. Certain aspects of the ecology and host parasite relations of the larval *Raillietina tetragona* (Cestoda) infection in ant vectors. *In*: Vectors and Vector-borne Diseases. Proc. All-India Sym., Trivandrum, Kerala State, India. February 26-28, 1982, pp.159-164.

2.5 FAMILY: DILEPIDIDAE

Chand, K., 1970. Preliminary studies on some common tapeworms of the fowl. Punjab Vet., 9: 31-35.

Chandra, R. and Sinha, K. S., 1972. Histopathological studies on the lesions caused by *Amoebotaenia sphenoides* (Railliet,1892) in chicks. Indian J. Anim. Sci., 42: 45-50.

Dutt, S.C. and Sinha, P. K., 1950. Studies on the life history of the fowl tapeworm, *Choanotaenia infundibulum* (Bloch). Proc. 37th Indian Sci. Cong.(Pune), 3: 242.

Dutt, S. C. and Sinha, P. K., 1961. Studies on the life history of the fowl tapeworm, *Choanotaenia indundibulum* (Bloch). Indian J. Vet. Sci., 31: 121-131.

Madhavan Pillai, K. M., 1968. Studies on tapeworms commonly encountered in fowls. Thesis (M. Sc.) submitted to Univ. Kerala.

Madhavan Pillai, K. M. and Peter, C. T., 1971. Studies on tapeworms commonly encountered in fowls. Indian Vet. J., 48: 430- 431.

Ponnudurai, G. and Chellappa, D. J. 2001. Observation on the biology of *Choanotaenia infundibulum* and *Raillietina cesticillus* and their natural host in Tamilnadu. J. Vet. Parasitol., 15: 51-53.

Ponnudurai, G., Harikrishnan, T. G. and Anna, T., 2003. Incidence of cysticercoids of poultry tapeworm, *Choanotaenia infundibulum* in *Musca domestica* in Namakkal, Tamil Nadu. *Indian J. Poult. Sci.,* 38: 308-310.

Srivastava, J. S. and Pande, B. P., 1967. The tenebrionid beetle, *Gonocephalum depressum*, F., from poultry units, as the vector of some of the larval helminths. Indian J. Ent., 29: 205-212.

Sunderam, R. K., and Radhakrishnan, C. V., 1962. Occurrence of *Amoebotaenia sphenoides* (Railliet, 1892) in desi fowls of Kerala State and some observations on its life cycle. Kerala Vet., 1: 98-101.

2.6 FAMILY: DIPYLIDIIDAE

Agarwal, R. D., 1970. Infective stages of two helminth parasites from intermediate/paratenic hosts with notes on their partial life-cycle in experimental mammals. On two amphistome cercariae from common snails, their encystment and development in experimental mammals. M. V. Sc. Thesis, Agra Univ., Agra, U. P., India. vide Abst. Thesis, 1971, Agra Univ. J. Res. 20: 137- 138.

Devi, S., Deka, D. K., Neog, R., Islam, S., Upadhyaya, T. N. and Das, M., 2011. *Dipylidium caninum* in dogs and screening of fleas as possible vectors in Greater Guwahati. J. Vet. Parasitol., 25: 76-78.(see Sarika Dev, *et al.,* infra).

Gadre, D. V., Kumar, A. and Mathew, M., 1993. Infection by *Dipylidium caninum* through pet cats. Indian J. Pediatr., 60: 151-152.

Gupta, V. P., 1970. A dilepid cysticercoid from *Uromastix hardwickii* and its development in pup. Curr. Sci., 39: 137-138

Joseph, S. A., 1974. *Ctenocephalides felis orientis,* Jordan, 1925 as an intermediate host of the dog tapeworm, *Dipylidium caninum* (Linn, 1758) Railliet, 1892. Cheiron, 3: 70-74.

Sangarika Devi, Deka, D. K., Neog, R., Islam, S., Upadhayaya, T. N. and Das, M., 2009. Occurrence of fleas and other incriminating vectors towards transmission of *Dipylidium caninum* in dogs in Guwahati. XIX Natl. Cong. Vet. Parasit. & Natl. Symp. "National Impact of Parasitic Diseases on Livestock Health and Production". February, 3-5, 2009, Dept. Vet. Parasitol., Coll. Vet. Sci., Guru Angad Dev Vet. Anim. Sci. Univ., Ludhiana-141004, Abst. PE 17, p.54.

Sangarika Devi, Deka, D. K., Neog, R. Islam, S., Upadhayaya, T. N. and Das, M., 2010. Experimental trial for development cycle of *Ctenocephalides felis felis* and metacestodes of *Dipylidium caninum* in laboratory conditions. XX NCVP Parasitology Today-Ecology to Molecular Biology, 18-20 February, 2010, Dept. Vet. Parasitol., CVS, CCS, Haryana Agri. Univ., Hisar-125004, p.32.

Sangarika Devi, Deka, D. K., Neog, R, Islam,S., Upadhyaya, T. N. and Das, M., 2011. Studies on the life cycle of *Ctenocephalides felis felis vis-à-vis* development of *Dipylidium caninum.* J. Vet. Parasitol., 25: 177-178.

Singh, M. D. and Pande, B. P., 1972. Experimental development of cysticercoids of *Joyuxiella echinorhynohoides* in a kitten and pups with a note on histopathology of larval lesions. Indian J. Anim. Sci., 42: 207-213.

2.7 FAMILY: HYMENOLEPIDIDAE

Baruah, N. and Gogoi, A. R., 1983. Studies on the incidence of cestodes in ducks in Kamrup District of Assam with special reference to the life-cycle of *Microsomacanthus collaris.* 5th Natl. Cong. Parasitol., June 25-27, 1983, Coll. Vet. Sci., A. P. Agri. Univ., Tirupati, A. P. and Indian Soc. Parasitol., Abstr. D-13, pp.84-85.

Baruah, N. and Gogoi, A. R., 1985. On the biology of duck cestode *Hymenolepis, (Microsomacanthus) collaris.* Indian J. Parasitol., 9: 43.

Mehra, K. N., 1950. Studies on the life history of *Hymenolepis carioca,* a cestode of poultry. Proc. 37th Indian Sci. Cong. (Pune), 3: 243.

Srivastava, J. S. and Pande, B. P., 1968.The cysticercoid of *Hymenolepis cantaniana* (Polonio, 1860) in *Orthophagus quadridentatus* (F) (Scarabaeidae: Coleoptra). Indian J. Anim. Hlth. 7: 289-291.

2.8 FAMILY: TAENIIDAE

Abdul Basith, S., Jeyathilakan, N. and Lalitha John, 2005. Western blot analysis of fluid antigens of hydatid cyst and other metacestodes of sheep. Proc. XVI Natl. Cong. Vet. Parasitol., December. 6-8, 2005, Ab-33, pp.69-70.

Abdul Basith, S., Jeyathilakan, N., Lalitha John Daniel Joy Chandran, N and Dhinakar Raj., 2009. Evaluation of SkDa hydatid antigen based on latex agglutination test in immumunodiagnosis of cystic echinococcosis in sheep. XIX Natl. Cong. Vet. Parasitol. & Natl. Sym. Natl Impact on Parasit. Dis, Livestock Hlth. Prod., Feb, 3-5, 2009, IAAVP & Dept. Vet. Parasitol., Coll. Vet. Sci., Guru Angad Dev Vet & Anim. Sci. Univ., Ludhiana-141 004, OI 36,p.115.

Abraham, J., 1979. Studies on the larval cestodes of zoonotic importance in Kerala with special reference to hydatid. M. V. Sc. Thesis.

Abraham, J., Pillai, K. M. and Iyer, R. P., 1980a. Fertility rate in hydatid cysts in domestic animals. Kerala J. Vet. Sci., 11: 155-158.

Abraham, J., Pillai, K. M. and Iyer, R. P., 1980b. Incidence of hydatidosis in animals slaughtered in Kerala. Kerala J. Vet. Sci., 11: 247-251.

Ahamad, S. M., 1948. Meat inspection in India. J. Army Vet. Corps., 19: 72-75.

Ahluwalia, S. S., 1960. Some larval helminthic infections in pigs. Indian J. Vet Sci., 30: 235-239.

Alam, A., Prasad, M. C. and Mohan, K., 1971. *Coenurus cerebralis* in lateral ventricle of the brain of goat. Indian J. Anim. Hlth., 10: 123-124.

Ali Afsar and Nizami, W. A., 1995. Comparative biochemical profile of surface plasma membranes of pulmonary and hepatic protoscolices of *Echinococcus granulosus*. J. Vet. Parasitol., 9: 63-71.

Alwar, V. S. and Lalitha, C. M., 1961. A check- list of helminth parasites in Department of Parasitology, Madras Veterinary College. Indian Vet. J., 38: 142-148.

Alwar, V. S., 1958. Parasites of pigs (*Sus scorfa domestica*) in Madras. Indian Vet. J. 35: 112-116.

Anjana Parihar and Nama, H. S., 1978. Helminthic fauna of *Funanmbulus pennanti* (Rodentia: Sciuridae) in Jodhpur, Rajasthan. Indian J. Parasitol., 2: 35-36.

Arora, B. M., 1991. Some diseases encountered in wild and captive mammals. In: Semr. Vet. Med. on wild and captive animals, November 8 to 10, 1991, Bangalore, India, Abstr., 8.4, p. 26.

Arora, B. M., Verma, T. K., Tewari, H. C. and Mandal, C. K., 1984. Cysticercosis caused by *Cysticercus tenuicollis* in a *Beisa oryx*.Vet Rec., 114: 149.

Arora, R. G. and Dixit, S. N., 1970. Generalised hydatidosis in a bullock. Haryana Vet., 9: 33-36.

Arora, R. G., Kalra, D. S. and Iyer, P. K. R., 1971. Studies on the pathology of caprine and ovine hepatic cysticercosis. Haryana Agri. Univ. J. Res., 1: 27-33.

Aruljyothi, N., Balagopalan, T. P. and Ramesh Kumar, B., 2013. Surgical management of coenurus cyst at the croup regionof a Malabari kid. Indian Vet. J., 90(4): 88-89.

Arvind Sharma and Tirth Singh, 1994. Hydatid cyst in myocardium of a buffalo and liver of its foetus. Livestock Adviser, 19, (4): 22-23.

Asim Krishna Pal and Sinha, P. K., 1970. Multilocular hydatidosis in a bullock. Indian Vet. J., 47: 910-912.

Aslam, M. R. and Razmi Gh., R., 2001. Muscular coenurosis in sheep. J. Vet. Parasitol., 15: 157-158.

Avapal, R. S., Sharma, S. K. and Juyal, P. D., 2003. Occurrence of *Cysticercus cellulosae* in pigs in and around Ludhiana City. J. Vet. Parasitol., 17: 69-70.

Awachat, K. G. and Iyer, G. R., 1971. Rumeno-reticular dysfunction due to hydatid cyst of spleen in in a bullock. Indian Vet. J., 48: 967-969.

Ayub, M. A., 1998. Investigation on hydatidosis in sheep, goats and human beings in Bikanir. M. V. Sc. Thesis submitted to Rajasthan Agri. Univ., Bikaner.

Ayyar, V. V. V., 1944. Indian Vet. J., 18: 27-28.

Bajpai, A. K., 1988. Biochemical changes in pigs infected with metacestodes of *Taenia solium* Second Natl. Cong. Parasitol., March 3-5, 1988, Dept. Parasitol., Vet. Coll., Bangalore, India, Souvenir, Stream II, Abstr., p.19.

Balamurugan, T., Senthil Selvakumar, S., Anna, T., Harikrihnan, T. J. and Ponnudurai, G., 2003. Prevalence of hydatidosis in food animals. XIV Natl. Cong. and Natl. Sym. on Milestones in immunological research and control of animal and poultry parasitism in new millennium. Maharashtra Anim. and Fish. Sci. Univ. and Indian Asso. Adv. Vet. Parasitol., Nagpur, 15th to 17th October, 2003, p.75.

Bali, H. S., Kwatra, M. S. and Fotedar, D. N., 1976. Studies on the pathology on some helminth borne diseases in sheep of Kashmir. Punjab Agri. Univ. J. Res., 11: 114-119.

Bali, H. S. and Chabbra, R. C., 1978. A study on secondary hydatids from a buffalo and their experimental infection in pups. Indian J. Anim. Sci., 48: 432-435.

Bali, H. S., 1972. On the development of *Multiceps multiceps* and its effects in pups infected with *Coenurus* recovered from a sheep. J. Res., PAU, (Suppl.), 9: 201-205.

Bali, H. S., Dutt, S.C., Rathor, S. S. and Gupta, P. P., 1978. A note on subcutaneous hydatidosis in a bullock. Indian Vet. J., 55: 735-736.

Bali, H. S., Gupta, M. P., Sood, N. K., Dua, K. and Arvinder Miglani, 1990. Coenuriasis in a buffalo heifer – A case report. Indian Vet. J., 67: 663-665.

Bam, J., Deori, S., Paul.,V., Bhattacharya, V., Bera, A.K., Bora, L. and Baruah, K. K., 2012. Sesonal prevalence of parasitic infection of yak in Arunachal Pradesh, Assam. Pacafic J. Trop. Med. 2(4): 264-267.

Bandyopadhyay, S. and Basu, A. K., 1987. Chemical composition of sterile and fertile hydatid cysts of cattle. J. Vet. Parasitol., 11: 73-76.

Bandyopadhyay, S. and Basu, A. K., 1996. Serological survey of hydatid disease in cattle in Calcutta. J. Vet. Parasitol., 10: 75-78.

Bandyopadhyay, S. and Singh, B. P., 1987. Antibody reponse in secondary hydatidosis using partially purified antigen. J. Vet. Parasitol., 11: 23-30.

Bandyopadhyay, S. and Singh, B. P., 2000. Antibody reponse in secondary hydatidosis by ELISA using affinity purified buffalo hydatid cyst fluid antigen. J. Vet. Parasitol., 14: 17-20.

Bandyopadhyay, S., 1994. Isolation, purification and characterisation of hydatid antigen. Ph. D. Thesis (Vet. Parasitol.). Deemed Univ., Indian Vet. Res. Inst., Izatnagar. Abstr., J. Vet. Parasitol., 11: 23-30.

Banerjee, D. and Sinha, K. S., 1969a. Studies on *Cysticercus fasciolaris*. I. Studies on early stages of infection in cysticerciasis in rats. Indian J. Anim. Sci., 39: 149-154.

Banerjee, D. and Sinha, K. S., 1969b. Studies on *Cysticercus fasciolaris*. II. Histological studies on *Taenia taeniformis*: Changes in rat intestine and oncosphere during penetration. Indian J. Anim. Sci., 39: 155-163.

Banerjee, D. and Sinha, K. S., 1969c. Studies on *Cysticercus fasciolaris*. III. Histopathology and histochemistry of rat liver in cysticerciasis. Indian J. Anim. Sci., 39: 242-249.

Banerjee, D. and Sinha, K. S., 1969d. Studies on *Cysticercus fasciolaris*. IV. Immunity to *Cysticercus fasciolaris* in rats. Indian J. Anim. Sci., 39: 250-256.

Banerjee, P. S., Bhatia, B. B. and Pandit, B. A., 1994. *Sarcocystis suihominis* in human beings in India. J. Vet. Parasitol., 8: 57-58.

Bari, N. S., Bhangale, G. N., Jumde, P. D., Maske, D. K., Narkhede and Wagahode, H. J., 2009. Incidence of hydatidosis in food animals at Nagpur. XIX Natl. Cong. Vet. Parasitol. & Natl. Sym. Natl. Impact on Parasit. Dis, Livestock Hlth. and Prod., February, 3-5, 2009, IAAVP & Dept. Vet. Parasitol., Coll. Vet. Sci., Guru Angad Dev Vet. & Anim. Sci. Univ., Ludhiana-141 004, PZ13, p.174.

Baruah, T. P., Datta, B. M. and Rahman, T., 1985. Studies on the pathology of porcine pneumonia (parasitic pneumonia). Indian Vet. J., 62: 1013-1016.

Bhadrige, V.V., Zende, R. J., Paturkar, A. M., Bhandare, S. G., Sajjan, S. A. and Katkar, S. M., 2011. Characterisation of whole cyst, scolex and host antigens of *Cysticercus cellulosae* by sodium dodecyl sulphate-polyacylamide electrophoresis (SDS-PAGE). XII Natl. Cong. Vet. Parasitol., 5-7th January, 2011, IAAAP and Dept. Vet. Parasitol., & Bombay Vet. Coll., Mumbai, S3-P6, p.119.

Bhalerao, G. D., 1939. A few unusual helminthes of some domestic animals in India. Indian J. Vet. Sci., 9: 371-374.

Bhalla, N. P. and Nagi, M.S., 1962. Occurrence of larval *Multiceps multiceps* over the heart of a goat. Indian Vet. J., 39: 55-56.

Bhangale, G. N., Maske, D. N., Pratibha Jumde, Priya Gawande and Reena Kadukar, 2010. Studies on hydatidosis in cattle and buffaloes of Nagpur region. XX NCVP Parasitolology Today-Ecology to Molecular Biology, 18-20 February, 2010, Dept. Vet. Parasitol., CVS, CCS, Haryana Agri. Univ., Hisar-125004, p.9.

Bhangale, G. N., Maske, D. N., Pratibha Jumde, Priya Gawande and Reena Kadukar, 2010. Biochemical characteristics of hydatid fluid in cattle and buffaloes. XX NCVP Parasitology Today-Ecology to Molecular Biology, 18-20 February, 2010, Dept. Vet. Parasitol., CVS, CCS, Haryana Agri. Univ., Hisar-125004, p.31.

Bhaskara Rao, T., Vara Prasad, P. V. and Hafeez Md., 2003. Prevalence of *Cysticercus tenuicollis* infection in slaughtered sheep and goats at Kakinada, Andhra Pradesh. J. Parasit. Dis., 27: 126-127.

Bhatia, B. B. and Pathak, K. M. L., 1990. Echinococcosis. In: Parija, S. C. (Ed.), Review of Parasitic Zoonosis. AITBS Publishers, New Delhi, India. p.268. (Not indexed).

Bhatia S. N., 1936. Parasitic cystitis in a goat. *Indian Vet. J.*, 12: 340.

Bhattacharya, D., Sikadar, A., Chattopadhyay, U. K. and Das, S. C., 1999. Utilisation of indirect enzyme linked immunosorbant assay (I-ELISA) for detection of hydatidosis in cattle and buffaloes. Indian J. Anim. Hlth., 38: 105-106.

Bhattacharya, D., Sikadar, A., Das, S. C. and Chattopadhyay, U. K., 2001. Rapid antibody detection of buffalo hydatidosis by coagglutination test. Indian J. Anim. Sci., 71: 40-42.

Bhattacharya, D., Bera, A. K., Bera, B. C., Maily, A. and Das, S. K., 2007. Genotypic characterization of Indian cattle, buffalo and sheep isolates of *Echinococcus granulosus*. Vet Parasitol., 143: 371-374.

Bhattacharya, D., Pan, D., Bera, A. K., Konar, A. and Das, S. K., 2008a. Mutation scan screening appraisal of *Ecchinococcus granulosus* isolates of Indian origin. Vet. Res. Commun., 32: 427-432.

Bhattacharya, D., Bera, A. K., Bera, B. C., Pan, D. and Das, S. K., 2008b. Molecular appraisal of Indian isolates of *Echinococcus granulosus*. Indian J. Med. Res., 127: 383-387.

Bhattacharya, D., Sikadar, A., Sarma, U., Ghosh, A. K. and Biswas, G., 1998. Concurrent infection of *Capillaria hepatica* and *Cysticercus fasciolaris* in rat (*Rattus ratus*) – A preliminary note. Indian Vet. J., 75: 486.

Bhattacharya, D., Laha, R., Das, S. C., Ramakrishna, C., Roy, R. N., Bandyopadhay and Sikdar, A., 2000. Hydatidosis in cattle and buffaloes with special emphasis on viability of protoscolisces- a pilot study. J. Parasitol. Appl. Anim. Biol., 9: 25-30.

Bhattacharya, D., Pan, D., Das, S., Bera, A.K., Bandyopadhyay, S. and Das, S. K., 2009. An evaluation of antigen B family of *Echinococcus granulosus*, its conformational propensity and elucidation of the agretope. J. Helminthol. 83: 219-224.

Binjola, S. K. and Gour, S. N. S., 1994. Evaluation of some immunodiagnostic tests for hydatidosis in buffaloes. J. App. Anim. Res., 6: 183-186.

Bishar, P., 1991. Studies on the incidence of cysticercosis in pigs and cattle with special reference to taeniasis in human beings in Assam and Meghalaya. M.V.Sc. Thesis submitted to Assam Agri. Univ., Guwahati, Assam, India.

Biswas, G., *et al.*, 1989. (Vide Gita Biswas *et al.*, 1989, below).

Borkataki, S., Islam, S., Borkakoty, M. R. and Goswami, P., 2009. Experimental porcine cysticercosis and its haematological study. XIX Natl. Cong. Vet. Parasitol. & Nat. Sym. National Impact on Parasitic Diseases of Livestock Health and Production, Feb, 3-5, 2009, IAAVP & Dept. Vet. Parasitol., Coll. Vet. Sci., Guru Angad Dev Vet. & Anim. Sci. Univ., Ludhiana-141 004, PZ2, p.171.

Borkataki, S., Islam, S., Borkakoty, M. R., and Goswami, P., 2010. Purification of *Cysticercus cellulosae* antigen by FPLC. XX NCVP Parasitology Today-Ecology to Molecular Biology, 18-20 February, 2010, Dept. Vet. Parasitol., CVS, CCS, Haryana Agri. Univ., Hisar-125004, p.40.

Borkataki, S., Islam, S., Borkakoty, M. R. and Goswami, P., 2010. Distribution and dimension of *Cysticercus cellulosae* in different organs of naturally infected pigs. XX NCVP Parasitology Today-Ecology to Molecular Biology, 18-20 February, 2010, Dept. Vet. Parasitol., CVS, CCS, Haryana Agri. Univ., Hisar-125004, p.79.

Borkataki, S., Islam, S., Borkakati, M. R., Goswami, P. and Deka, D. K., 2012. Prevalence of porcine cysticercosis in Nagaon, Morigaon and Karbianglong district of Assam, India. Vet. World, 5: 86-90.

Borkataki, S., Islam, S., Sarma, D. K., Borkakoty, M. R. and Goswami, P., 2009. Comparison study of AGPT, CCIEP and ELISA using homologous hyperimmune sera of *Cysticercus cellulosae*. XIX Natl. Cong. Vet. Parasitol. & Natl. Sym. National Impact on Parasitic Diseases of Livestock Health and Production, 3-5 February, 2009, IAAVP & Dept. Vet. Parasitol., Coll. Vet. Sci., Guru Angad Dev Vet & Anim. Sci. Univ., Ludhiana-141 004. O14, XIX, p.99

Borkataki, S., Pandit, B. A., Shahardar, R. A., and Goswami, P., 2009. Prevalence of hydatidosis in ruminants of Kashmir Valley. XIX Natl. Cong. Vet. Parasitol. & Natl. Sym. National Impact on Parasitic Diseases of Livestock Health and Production, 3-5 February, 2009, IAAVP & Dept. Vet. Parasitol., Coll. Vet. Sci., Guru Angad Dev Vet & Anim. Sci. Univ., Ludhiana-141 004. Stream 1, OE1, p.26.

Chakraborty, A., Gogoi, A. R. and Datta, B. M., 1984. Occurrence of *Cysticercus ovis* in tunica vaginalis and urinary bladder of as goat in Assam. Indian J. Parasitol., 8: 215-216.

Chakraborty, A., 1999. Mortality in different species of animals of Assam Zoo, Guwahati. In: Proc. Workshop on "Health and Management of Zoo Animals for Zoo Veterinarians", held at Guwahati, 28-30 October, 1999.

Chakraborty, T., Bose, A. K., Barat, S. and Bhowmik, M. K., 1998. Spontaneous coenuriosis in Himalayan Tahr and Markhor. Indian J. Anim. Hlth., 37: 33-34.

Chandrapal Reddy, M. and Radhakrishna Reddy, K., 1989. A study on incidence of hydatidosis in food animals in and around Hyderabad, A. P. Livestock Adviser, 14 (6): 35-39.

Chataopadhaya, Singh, B. P. and Chattopadhyay, S. K., 1988. Parasitic diseases in exotic and crossbred sheep farms of Jammu and Kashmir with particular reference to echinococcosis / hydatidosis. Proc. Second Natl. Sem. Sheep and Goat Dis., CSWRI, Avikanagar, India, p.181.

Chaudhri, S. S., Satyavir Singh and Singh, S., 2000. Helminth parasites of domestic animals in Haryana. Haryana Vet., 39: 1-12.

Chauhan, H. V. S., and Rao, U. R. K., 1972. Pathology of some common parasitic diseases in indigenous pigs. Indian J. Anim. Hlth., 11: 75 -79.

Chhabra, R. C. 1983. *In vitro* study on viability of hydatid cysts from buffalo. Indian Vet. J., 60: 939-940.

Choudhari, V R , Garg, U. K., Das, G., Supriya Shukla and Gupta, R. S., 2005. Prevalence of pulmonary hydatidosis in buffaloes (*Bubalis bubalis*) in the Malwa region of Madya Pradesh. Proc. XVI Natl. Cong. Vet. Parasitol., December 6-8, 2005 Dept. Parasitol., Coll.Vet. Sci. & Ani. Husb., Indira Gandhi Agri. Univ., Anjora, Durg, 491 001, Ab-11, p.116.

Chowdary, Ch., Narayanaswamy, B., Shivasankar, V., Ramamohan Rao M. R. and Hararam Das, J., 1987. Hydatidosis in lungs of an American bison (*Bison bison*). Indian Vet. J., 64: 713 -714.

Chowdhury, N. and Rajvir Singh, 1993. Distribution of some elements in hydatid cysts of *Echinococcus granulosus* from buffaloes (*Bubalis bubalis*). J. Helminthol., 67: 112-114.

Chowdhury, N. and Rajvir Singh, 1995. Distribution of cobalt in parasitic helminthes. J. Helminthol., 69: 259-261.

Chowdhury, N., Kinger, S. and Ahuja, S. P., 1986. The chemical composition of secondary hydatid cysts of buffalo origin. Acta. Trop. Med. Parsitol., 80: 469-471.

Darzi, M. M., Pandit, B. A., Shahardar, R. A. and Mir, M. S., 2002. Pathology of *Taenia hydatigena* cysticercosis in a naturally infected Corriedale lamb. J. Vet. Parasitol., 16: 173-174.

Das, U. and Das, A. K., 1998. Cystic hydatidosis of food animals in greater Calcutta. Indian Vet. J.75: 387-388.

Das, P. K., Dey, P. C., Tripathy, S. B. and Nayak, B. C., 1976. Hydatidosis in a bullock. Indian J. Anim. Hlth., 171-172.

Das, S. S. and Sreekumar, R., 1998. Prevalence of hydatidosis in sheep and goats in Pondicherry. J. Vet. Parasitol., 3: 145-146.

Das, S. S., Kumar, D. and Sreekrishnan, R., 2003. Hydatidosis in animals and man. In: Helminthology in India. M. L. Sood (Ed.) International Book Distributors, Dehradun, India. pp.425-451. (review).

Dash, A. K. Pande, M. R. and Pande, D. N., 2000. Prevalence of metacestodes in ruminants in and around Bhubaneswar. XI Natl. Cong. Vet. Parasitol., Bhubaneswar, 4-6 February, 2000, Abstr.p.67.

Dey (1909). Cited by Nagaty, H. F.1946. Helmin Soc. 13: 3: 33-46.

Debasis Bhattacharya, 2003. Phenotypic and molecular characteristic of bladder worm of *Echinococcus granulosus* and its serodiagnosis by Enzyme-Immuno Transfer Blot (EITB). Ph.D. Thesis submitted to Univ.Agri. Sci., Bangalore-560 065, India.

Debasis Jana and Mousumi Jana, 2006. A note on successful surgical management of gid (sturdy) in a black Bengal doe. Intas Polivet, 7: 75-76.

Deka, D. K., 1989. Epidemiology and immunodiagnosis of certain common larval cestodes of domestic animals. Ph. D. Thesis submitted to G.B. Pant Univ. Agri. and Tech., Pantnagar, India. (not indexed).

Deka, D. K., 1992. Occurrence of metacestode infection in domestic animals of Lakhimpur district of Assam. Lakhimpur Vet. Coll., 4: 1-2.

Deka, D. K. and Gaur, S. N. S., 1990a. Epidemiology of hydatidosis in buffaloes of western parts of Uttar Pradesh. J. Vet. Parasitol., 4: 49-53.

Deka, D. K. and Gaur, S. N. S., 1990b. *Taenia solium* cysticercosis in pigs in western parts of Uttar Pradesh. J. Vet. Parasitol., 4: 59-63.

Deka, D. K. and Gaur, S. N. S., 1990c. Counter current immunoelectrophoresis in the diagnosis of *Taenia hydatigena* cysticercosis in goats. Vet. Parasitol., 37: 223-228.

Deka, D. K. and Gaur, S. N. S., 1991a. Studies on immunoelectrophoresis patterns of hydatidosis. J. Vet. Parasitol., 5: 19-23.

Deka, D. K. and Gaur, S. N. S., 1991b. Diagnosis of *Taenia hydatigena* cysticercosis in goats by immunoelectrophoroesis. J. Vet. Parasitol., 5: 35-37.

Deka, D. K. and Gaur, S. N. S., 1993a. Epidemiological study on *Taenia hydatigena* cysticercosis in certain domestic animals of Uttar Pradesh. J. Vet. Parasitol., 7: 29-34.

Deka, D. K. and Gaur, S. N. S., 1993b. Immunoelectrophoresis test against porcine cysticercosis. J. Vet. Parasitol., 7: 35-43.

Deka, D. K. and Gaur, S. N. S., 1993c. Serodiagnosis of *Taenia solium* cysticercosis in naturally infected pigs by ELISA. J. Vet Parasitol., 7: 86-92.

Deka, D. K. and Gaur, S. N. S., 1993d. Counter-current immunoelectrophoresis: A test of choice for serodiagnosis of porcine cysticercosis. Indian Vet. Med. J., 17: 66-69.

Deka, D. K. and Gaur, S. N. S., 1993e. ELISA in the diagnosis of hydatidosis in buffaloes. Indian J. Anim. Sci., 63: 1254-1255.

Deka, D. K. and Gaur, S. N. S., 1994a. Parasitological, pathological and biochemical observations in kids immunized against *Taenia hydatigena* infection. Sixth Natl. Cong. Vet. Parasitol, 22-24, October, 1994, Dept. Parasitol., Coll. Vet. Sci., JNKVV, Jabalpur-482 001, M. P., India, Abstr. S-3-22, p. 39.

Deka, D. K. and Gaur, S. N. S., 1994b. Alterations in serum proteins and enzyme levels in kids infected with *Cysticercus tenuicollis*. Indian Vet. J., 71: 1032-1034.

Deka, D. K. and Gaur, S. N. S., 1994c. Immunity trial in pups against *Taenia hydatigena* infection. Indian Vet. J., 71: 1070-1072.

Deka, D. K. and Gaur, S. N. S., 1994d. Pathological alterations in liver of kids immuned against larval stages of *Taenia hydatigena*. Indian J. Vet. Path., 18: 165.

Deka, D. K. and Gaur, S. N. S., 1998. Some studies on the occurrence of hydatidosis in Western Uttar Pradesh. J. Vet. Parasitol. 12: 43-45.

Deka, D. K, Borkakoty, M. R. and Lahkar, S. C., 1985. Cysticercosis in domestic animals in north eastern region of India. Indian J. Parasitol., 9: 83-85.

Deka, D. K., Choudhury, S. and Chakraborty, A., 1995. Parasites of domestic animals and birds in Lakhimpur (Assam). J. Vet. Parasitol., 9: 21-25.

Deka, D. K., Gaur, S. N. S. and Singh, G. K., 1990 Tissue reaction in pigs naturally infected with cysticerci of *Taenia solium*. Indian J. Anim. Sci., 60: 651-653.

Deka, D. K., Srivastava, G. C. and Chhabra, R. C., 1982. Fertility of hydatid cysts from ruminants. Indian J. Parasitol., 3: 243-244.

Deka, D. K., Srivastava, G. C. and Chhabra, R. C., 1983a. Incidence of hydatidosis in ruminants. Indian J. Anim. Sci., 53: 200-202.

Deka, D. K., Srivastava, G. C. and Chhabra, R. C., 1983b. Potentiality of refrigerated *Echinococcus granulosus* cyst of buffalo origin and experimental infection in dogs. Indian Vet. J., 60: 407-408.

Deka, D. K., Srivastava, G. C. and Chhabra, R. C., 1983c. *In vitro* study on hydatid cyst from buffalo. Indian Vet. J., 60: 939-940.

Deka, D. K., Saidul Islam, Borkakoty, M.R., Abdul Saleque, Isfaqul Hussain and Krishna Nath, 2008. Prevalence of *Echinococcus granulosus* in dogs and hydatidosis in herbivores of certain north eastern states of India. J. Vet. Parasitol., 22: 27-30.

Deka, D. K., Borkakoty, M. R., Sarma, S., Rahman, M., Islam, S., Hussain, I. and Krishna Nath, 2009. Biochemical profile of cystic fluid of some major metacestodes naturally occurring in domestic animals. XIX Natl. Cong. Vet. Parasitol. & Nat. Sym. Natl. Impact on Parasit. Dis, Livestock Hlth. Prod., February, 3-5, 2009, IAAVP & Dept. Vet. Parasitol., Coll. Vet. Sci., Guru Angad Dev Vet. & Anim. Sci. Univ., Ludhiana-141 004, Abstr. OE 29, p.39.

Deodhar, N. S. and Narasapur, V. S., 1968. "Pneumonitis cysticercosis" – a new disease condition of goats by migrating *Cysticercus tenuicollis* in the lungs. Indian Vet. J. 45: 202-204.

Deshpande, M. S., 1970. Incidence of hydatidosis in animals slaughtered at Bombay abattoirs. M.V.Sc. Thesis, Bombay Vet. Coll., Bombay. (not indexed).

Deshpande, N. S., 1977a. Incidence of hydatid cysts in cattle. Mahavet, 3: 35-37.

Deshpande, N. S., 1977b. Hydatidosis in *Sus scrofa domestica* (Swine). Mahavet, 3: 91-93.

Deshpande, N. S., Niphadkar, S. M. and Narasapur, V. S., 1984. Incidence of hydatidosis in animals slaughtered at Bombay abattoirs. Abstr. IV. Natl. Semi. Zoonosis in India and their control. 5: 42.

Dev Sarma, M. K., 1999. Some aspects of hydatidosis and other common cestodes of domestic animals. M. V. Sc. Thesis, submitted to A. A. U., Khanapara, Guwahati-22, Assam.(not indexed).

Dev Sarma, M., Deka, D. K. and Borkakoty, M. R., 2000. Occurrence of hydatidosis and porcine cysticercosis in Guwahati City. J. Vet. Parasitol., 14: 173-174.

Dev Sarma, M., Deka, D. K., Rahman, H., Sarma, S. and Borkakoty, M. R., 2002. Purification and partial characterization of cystic stages of *Echinococcus granulosus* and *Taenia hydatigena*. J. Vet. Parasitol., 16: 63-64.

Devasena, B., Ravikumar, K. and D'Souza, P. E., 1998. Occurrence of *Coenuras gaigeri* in a three month old kid. Curr. Res., Univ Agri. Sci., Bangalore. 27: 198.

Devendra Kumar, 1987. Immunodiagnostic and immunoprophylactic studies on *Cysticercus cellulosae* infection in pigs.Thesis submitted to G. B. Pant Univ. of Agri. and Tech., Pantnagar- 263145.

Devi, C. S. and Parija, S. C., 2003. Latex agglutination test (LAT) for antigen detection in the cystic fluid for the diagnosis of cystic echinococcosis. Diag. Microbiol. Infect. Dis., 45: 123-126.

Devi, C. S., Tarachand, P., Devi, S. L., Suryakumari, G. and Reddy, C. R. R. M., 1971. Chemical analysis of hydatid fluid in relation to the presence or absence of the live scolices. Indian Med. Sci., 25: 460-463.

Dewan Muthu Mohammad, M. S. and Rajendran, E. I., 1974. Incidence of *Coenurus* in sheep and goats. Cheiron 3: 168-169.

Dey, D., 1909. *Coenurus serialis* in a goat. J. Trop. Vet. Sci., 556-560.

Dey, P. C., Nayak, D. C., Mohanty, D. N., Nayak, S. and Pattanayak, G. M., 1988. A brief note on massive infection of *Coenurus gaigeri* cysts in a desi goat. Indian Vet. J., 65: 166.

Dhabolkar, D. B., Gafoor, M. A. and Narsapur, V. S., 1968. *Cysticercus ovis* in the heart of sheep. Indian Vet. J., 45: 169.

Dhakshayani, C. N., 1995. A comparative study on the percentage of infertile hydatid cysts in sheep, goats, buffaloes and cattle slaughtered at Madras, Seventh Natl. Cong. Vet. Parasitol., 19-21, August, 1995, Dept. Parasitol., Madras Vet. Coll., Madras -7, Abstr. SI 39, p.92.

Dhanalakshmi, H., 2003. Phenotypic and genotypic characterization and immunodiagnosis of *Taenia solium* bladder worm of pig. Ph.D. Thesis submitted to Univ. of Agri. Sci., Bangalore-560 065, India.

Dhanalakshmi, H., Jagannath, M. S. and D' Souza, P. E., 2005. Intermediate form of *Taenia solium* cysticercosis in pigs. Intas Polivet., 6: 282-283.

Dhanalakshmi, H., Jagannath, M. S., Isloor, S. and D' Souza, P. E., 2006. Protein-A ELISA in the serodiagnosis of porcine cysticercosis. J. Vet Parasitol., 20: 33-35.

Dhar, S., 1995. Studies on the biology and immunology of bovine hydatidosis. Ph. D. Thesis. Deemed Univ., Indian Vet. Res. Inst., Izatnagar. Abst., J. Vet. Parasitol., 9: 149-150.

Dhar, S. and Singh, B.P., 1995. Studies on the biology and immunology of bovine hydatidosis. J. Vet. Parasitol. 9: 149-150.

Dhar, S. and Singh, B. P., 1996a. Prevalence of hydatid disease in bovines in Uttar Pradesh. Global Meet on parasitic diseases, New Delhi, India 18-22 March, 1996. Abstr., J. Parasit. Dis., 20: 97.

Dhar, S. and Singh, B. P., 1996b. Taxonomic evaluation of *Echinococcus granulosus* metacestodes of bovines in India. Global Meet Parasit. Dis., New Delhi, India 18-22 March, 1996. Abstr., J. Parasit. Dis., 20: 110-111.

Dhar, S., Singh, B. P. and Raina, O. K., 1996. Hybridoma derived antibodies for the diagnosis of *Echinococcus granulosus* infection. J. Vet. Parasitol., 10: 153-157.

Dhoot, V. M. and Upadhye, S. V., 2002. Hydatidosis in a lion. Zoos' Print, 17: 964.

Dhote, S. W., Ingle, A. D., Bhandarkar, B. G., Joshi, M. V. and Bhagwat, S. S., 1992a. Intensity and distribution of hydatidosis in buffalo and cattle. Indian J. Vet. Pathol., 16: 42-44.

Dhote, S. W., Patil, S., Sadekar, R. D., Joshi, M. V. and Bhagwat, S. S., 1992b. Incidence of morbid condition in liver of slaughtered bullocks. Indian J. Anim. Sci., 62: 744-746.

D'Souza, P. E., 1998. Studies on porcine cysticercosis with special reference to serodiagnosis. J. Vet. Parasitol., 12: 64-65.

D'Souza, P. E. 2009. Advances in diagnosis, treatment and control of *Taenia solium* Cysticercosis in man and pigs. XIX Natl. Cong. Vet. Parasitol. & Natl. Sym. National Impact on Parasitic Diseases of Livestock Health and Production, 3-5 February, 2009, IAAVP & Dept. Vet. Parasitol., Coll. Vet. Sci., Guru Angad Dev Vet & Anim. Sci. Univ., Ludhiana-141 004. 1LZ: pp.134-143.(Review, not indexed).

D'Souza, P. E. and Dhanalakshmi, H., 2005. Cysticercosis in pig-Epidemiology, immunodiagnosis, economic importance under Indian condition and control. Intas Polivet, 6: 291-296.

D'Souza, P. E. and Hafeez Md., 1998a. Incidence of *Cysticercus cellulosae* in pigs in an organized abattoir in Andhra Pradesh and its economic implications. Seventh Natl. Cong. Vet. Parasitol., 19-21 August, 1995, Dept. Parasitol., Madras Vet. Coll., Madras -7, Abstr. Sl 9, p.76

D'Souza, P. E. and Hafeez Md., 1998b. Studies on *Cysticercus cellulosae* in pigs in an organized abattoir in Andhra Pradesh, India. J. Vet Parasitol., 12: 33-35.

D'Souza, P. E. and Hafeez Md., 1998c. Incidence of *Cysticercus cellulosae* in pigs in three districts of Karnataka. Mysore J. agric. Sci., 32: 67-70.

D'Souza, P. E. and Hafeez Md., 1998d. Incidence of *Cysticercus cellulosae* in pigs in Andhra Pradesh. Mysore J. agric. Sci.,32: 75-78.

D'Souza, P. E. and Hafeez Md., 1999a. Comparison of routine meat inspection, CIEP and ELISA to detect *Cysticercus cellulosae* in pigs. Indian Vet. J., 76: 285-288.

D'Souza, P. E. and Hafeez Md., 1999b. An enzyme-linked immunosorbant assay in the detection of *Cysticercus cellulosae* in pigs. J. Vet. Parasitol., 13: 125-127.

D'Souza, P. E. and Hafeez Md., 1999c. Detection of *Cysticercus cellulosae* in pigs by counter current immunoelectrophoresis. Int. J. Anim. Sci., 14: 215-219.

D'Souza, P. E. and Hafeez Md., 1999d. Detection of *Taenia solium* cysticercosis in pigs by ELISA with an excretory-secretory antigen. Vet. Res. Commun., 23: 293-298.

D'Souza, P. E., 1996. Studies on porcine cysticercosis with special reference to serodiagnosis. Ph. D. Thesis submitted during 1996 to Andhra Pradesh Agri. Univ., Hyderabad-500 003, India. (For abstract, vide J. Vet. Parasitol., 1998, 12: 64.).

D'Souza, P. E., 2011. An update on taenid cysticercosis and echinococcosis. XIV. Natl. Trng. Prog. Vet. Parasitol., Trends and perspectives in the biology, ecology and control of parasitic diseases. 28th February to 20th March, 2011. Centr. Adv. Stud., Dept. Parasitol., Vet. Coll., Univ. Agri. Sci., Hebbal, Bangalore-560 024. pp.115-129. (Review paper, not included in Index list).

Dwivedi, J. N. 1963. Studies on the pathology of pneumonia and associated pulmonary diseases of cattle and buffaloes. Agra Univ. J. Res., 12: 239-240.

Ebenezer Raja, E., 1971. *Coenurus* sp. from the peritoneal cavity of the gerbil, *Tatera indica* (Hardwicke) in Madras. Curr. Sci., 40: 521-523.

Ebenezer Raja, E., 1974. Parasitic infection communicable to man and animals. Ph. D. Thesis, Madras University, Madras, Abstract: Cheiron 3: 173-174.

Ebenezer Raja, E. and Anantaraman, M., 1984. Host-parasite relationship in *Taenia hydatigena* infection in dogs.

Edwards, J. T., 1927. Some diseases of cattle in India. 64-68. Supdt., Publications, Calcutta.

Endrejat, E., 1964. Helminths and helminthic infection in Assam. Indian Vet. J., 41: 538-542.

Farooqi, H. U., Mashkoor Ahmad and Shameem, M., 1986. Biocoenotically-linked varying patterns of livestock hydatidosis in the Indian region. Seventh Natl. Cong. Parasitol., Ravi Shankar Univ., Raipur, 26-28, December, 1986, Abstr. B-32, p.88.

Gabhane, G. K., Awandkar, S. P. and Rangnekar, M. N., 2009. Occurrence of *Multiceps multiceps* infection at Gadchiroli, Maharashtra. J. Bombay Vet. Coll., 19: 82-83.

Gahlod, B. M., Patil, S. N., Dakate, S. S., Marudwar, S. S., Kukure, N. V. and Kolte, S. W., 1999. Coenurus cyst at an unusual location in goat and its treatment. Indian J. Vet. Surg., 20: 61.

Gahlot, T. K. and Purohit, S., 2005. Surgical management of gid (coenurosis) in goat: Case report. Intas Polivet. 6: 297-298.

Geiger, S. H., 1907. *Coenurus serialis* found in two goats in India. J. Trop.Vet. Sci.(Calcutta), 2: 316.

Gaiger, S. H. 1910. J. Trop.Vet. Sci.(Calcutta), 5: 65-71.

Gaiger, S. H., 1915. J. Comp. Pathol. Ther., 28: 67. (*Coenurus* in sheep and camels) Cited by Rao, B. V. and Parihar, N. S., 1973. Indian Vet. J., 50: 318-319.

Gangulee, H. C., 1922. Echinococcus of the heart in a bullock. Vet Res.

Ganguly, N. K., Mahajan, R. C., Wangoo, A., Bose, S. M., Dilawari, J. B. and, 1986. Potential experimental model of unilocular hydatid disease. Indian J. Med. Res., 84: 210-212.

Ganorkar, A. G., Kolte, S. W. and Kurkure, N. V., 1997. Occurrence of hydatid cysts in a lion. Indian J. Vet. Pathol., 21: 64.

Gatne, M. L., 1987. Studies on immunodiagnosis of hydatidosis in bovines. P. G. Thesis, Bombay Vet. Coll., Bombay. (not indexed).

Gatne, M. L., 2001. Studies on antigenic profiles of hydatidosis in farm animals with particular emphasis on specific immunodiagnostics. Ph.D. Thesis submitted to Dr. Balasaheb Sawant Konkan Krishi Vidya Peeth, Dapoli, Maharashtra.

Gatne, M. L. and Gaurat, R. P., 2012. Immunodiagnosis of hydatidosis in food animals by ELISA. XXII Natl. Cong. Vet. Parasitol., March 15-17, 2012, Coll. Vet. Sci. and Anim. Husb., U. P. Pandit Deen Dayal Upadhyaya Pashu Chikitsa Vignan Vishwavidyalaya Evam Go Anusandhan Sansthan (DUVASU), Mathura-281001(UP.), India, Abst. No.III.07, p 36-37.

Gatne, M. L. and Narasapur, V. S., 1992. Hepatic hydatidosis in cattle. Pashudhan, 7(4). 1.

Gatne, M. L., Narasapur, V. S., Deshpande, V. S. and Niphadkar, S. M., 1990. Protein content and electrophoretic patterns of hydatid fluid. Indian Vet. J., 67: 169-170.

Gatne, M. L., Narasapur, V. S., Niphadkar, S. M. and Deshpande, V. S., 1989. Incdence of hydatidosis in cattle slaughtered at Bombay abattoirs. J. Maharashtra Agri. Univ., 14: 390.

Gatne, M. L., Narasapur, V. S., and Niphadkar, S. M., 1991. Evaluation of double diffusion and scolex precipitation tests for diagnosis of hydatidosis in cattle. J. Comp. Microbiol. Immunol. Infect. Dis., 12: 62-64.

Gatne, M. L., Palampalle, H. Y., Gaurat, R. P., Jayraw, A. K., and Pednekar, R. P., 2010. Prevalence of metacestodes in food producing animals slaughtered at Mumbai abbatoir. XX NCVP, Parasitology Today-Ecology to Molecular Biology, 18-20 February, 2010, Dept. Vet. Parasitol., CVS, CCS, Haryana Agri. Univ., Hissar-125004, pp.75-76.

Gaur, S. N. S., 1985. Immunological studies of *Cysticercus cellulosae*, a larval cestode in pigs and other domestic animals. Final report of research project, Nainital, India.

Gaur, S. N. S., 1976. Prevalence of *Cysticercus bovis* in cattle and buffaloes. Indian J. Anim. Res., 10: 47-48.

Gaur, S. N. S., Tewari, H. C. and Om Prakash, 1977. Occurrence of hydatid cyst in a suckling calf. Indian J. Anim. Res., 11: 109.

Gaur, S. N. S., Sethi, M. S., Tewari, H. C. and Om Prakash, 1980. Note on the incidence of *Cysticercus tenuicollis* in sheep and goats in certain parts of Uttar Pradesh. Indian J. Anim. Res., 14: 73-75.

Gaurat, R. P., 2002. Studies on the helminthic parasites of domestic pigs (*Sus scrofa domestica*) in Maharashtra. P. G. Thesis, Bombay Vet. Coll., Parel, Bombay. (not indexed).

Gaurat, R. P. and Gatne, M. L., 2005. Prevalence of helminth parasites in domestic pigs (*Sus surfa domestica*) in Mumbai: An abattoir survey. J. Bombay Vet. Coll., 13: 100-102.

Gaurat, R. P. and Gatne, M. L., 2009. Gross and histopathological lesions associated with helminthic infections in pigs. J. Bombay Vet. Coll., 17: 68-69.

Geetha Devi, L., Aparajita, C., Victoria Chana, K. and Patra, G., 2010. Echinococcosis infection in lungs of Nilgai (*Boselaphus tragocamelus*). Vet Scan, 5, Article 72.

Ghorui, S. K. and Sahai, B. N., 1989. Studies on the incidence of hydatid disease in ruminants. Indian J. Anim. Hlth., 28: 39-41.

Ghorui, S. K., Prasad, R. L., Pal, A. K. and Sahai, B. N., 1989. Studies on the biochemical constituents of hydatid fluid. J. Vet. Parasitol., 3: 145-147.

Ghorui, S. K., Prasad, R. L., Pal, A. K., and Sahai, B. N., 1990. Proteins and glycoprotein distribution in cyst wall and germinal layer of hydatid cyst. Indian Vet. J., 67: 802-804.

Ghorui, S. K., Pal, A. K., Prasad, R. L. and Sahai, B. N., 1992. Studies on some enzyme activities in germinal layers of bovine hydatid cyst of *Echinococcus granulosus*. Indian Vet. J., 69: 401-403.

Ghorui, S. K., Pal, A. K., Prasad, R. L. and Sahai, B. N., 1995. Lipids in the various fractions of bovine hydatid cyst of *Echinococcus granulosus*. Indian Vet. J., 72: 994-996.

Ghorui, S. K., Sanjay Kumar, Narnaware, S. D., Kashinath and Patil, N. V., 2012. Hydatidosis in camel: An important cause of production losses. XXII Natl. Cong. Vet. Parasitol., March 15-17, 2012, Coll. Vet. Sci. and Anim. Husb., U. P. Pandit Deen Dayal Upadhyaya Pashu Chikitsa Vigyan Vishwavidyalaya Evam Go Anusandhan Sansthan (DUVASU), Mathura-281001(UP.), India, Abstr. No. I.03, pp.2.

Ghosh, R. C., Dubey, S., Mandal, S. C. and Sharma, N., 2005. Occurrence of *Coenurus gaigeri* cyst in a goat. Indian Vet. J., 82: 90- 91.

Gill, H. S., 1967. Unusual brood capsules from hydaytid cysts in Indian buffalo. J. Helminthol., 41: 111-114.

Gill, H. S., 1968. A study on the parasitic zoonosis in India. Proc. 59th Indian Sci. Cong. Assc., Part III. Abstr., 37, p.549.

Gill, H. S., 1968. Secondary echinococcus in an Indian water buffalo (*Bos bubalis*). J. Parasitol., 54: 949-955.

Gill, H. S. and Rao, B. V., 1967a. On the biology and morphology of *Echinococcus granulosus* (Batsch, 1786) of buffalo-dog origin. Parasitol., 57: 695-704.

Gill, H. S. and Rao, B. V., 1967b. Incidence and fertility rate of hydatid cysts in Indian buffalo (*Bos bubalis*). Bull. Off. Int. Epizoot., 67: 989-997.

Gill, H. S. and Rao, B. V., 1967c. Unusual brood capsules from hydatid cysts in Indian buffalo. J. Helminthol., 41: 111-114.

Gill, H. S. and Rao, B. V., 1968. Some observations on the endogenous daughter cyst formation in hydatid cysts on buffaloes. Indian Vet. J., 45: 479-484.

Gill, H. S. and Rao, B. V., 1969. A study of hydatid cysts occurring in buffaloes in India. Ceylon Vet. J., 17: 31-33.

Gita Biswas, Sen, G. P., Thapa, D. and Lakhar, A., 1989. Hydatidosis in meat animals in Calcutta. Indian Vet. J., 66: 78-80.

Godara, R., Borah, M. K., Sharma, R. L. and Jangir, B. L., 2011. Caprine coenurosis with special reference to hepatic coenurosis. Comp. Clin. Pathol., 20: 277-280.

Gogoi, D., Lahon, D. K., Bhattacharya, M., Mukit., A. and Lakharu, J. C., 1992. A correlative study on the location of coenurosis in goat brain and symptoms exhibited. Indian J. Vet. Med., 16: 66-70.

Gogoi, A. R., 1974. Incidence of helminth in domesticated animals and poultry. Vetcol., 15: 33-36.

Gogoi, D. K., Bhattacharya, M. and Lakharu, J. C., 1991. Histopathological studies of coenurosis in goat. Indian J. Anim. Sci., 61: 283-285.

Gohain Borua, Pal, S. S., Doimari, S., Begum, S. and Barua, C. C., 2010. Prevalence of cystic in domestic animals. Indian Vet. J., 87: 932-933.

Gopala Krishna Rao, D., 1985. Hydatidosis of animals. Livestock Adviser, 10 (11): 40-41. (vide Rao, G. K., 1985).

Gopala Krishna Rao, D., Kamlapur, P. N. and Seshadri, S. J., 1983. Histopathological studies on renal hydatidosis of bovines. 5th Natl. Cong. Parasitol., June 25-27, 1983, Coll. Vet. Sci., A. P. Agri. Univ., Tirupati, A. P. and Indian Soc. Parasitol., Abstr. C-18, p. 62.

Goswami, A, Das, M and Laha, R., 2013. Characterization of immunogenic proteins of *Cysticercus tenuicollis* of goats. Vet. World, 6(5): 267-270

Goswami, R., Singh, S.M., Kataria, M. and Somvanshi, R. 2010. Clinicopathological studies on spontaneous *Cysticercus fasciolaris* infection in wild and laboratory rats. Indian J. comp. Microbiol. Immunol. infect. Dis., 31: 51-58.

Gudewar, J., Pan, D., Bera, A. K., Das, S. K., Konar, A., Rao, J. R., Tiwari, A. K. and Bhattacharya, D., 2009. Molecular characterization of *Echinococcus granulosus* of Indian animal isolates on the basis of nuclear and mitochondrial genotype. Mol. Biol. Rep. 36: 1381-1385.

Gupta, J. C., Nabrath, C. L. and Salgia, K. M., 1966. Hydatid diseases in man. J. Indian Med. Assc., 46: 649-651.

Gupta, P. P., 1979. Report of echinococcosis in a camel from India. Indian J. Parasitol., 3: 81.

Gupta, P. P. and Chowdhury, N., 1985. Cerebral coenuriasis in a buffalo. Indian Vet. J., 62: 613-614.

Gupta, P. P. and Singh, B., 1975. A note on unusual case of echinococcosis in a buffalo (*Bos bubalis*). Zbl. Vet. Med. B., 22: 793-795.

Gupta, V. K., Bist, B., Agrawal, R. D. and Gupta, P., 2011. Buffalo hydatidosis in Agra city of Uttar Pradesh. J. Vet. Parasitol., 25: 88-89.

Hafeez Md., 1997. Epidemiology and transmission of cystic echinococcosis: India. Arch. Int. Hydatid., 32: 54-55.

Hafeez Md., 2001. Helminth parasites of public health importance – Cestodes. J. Parasitic Dis., 25: 2-7. (A review, not cited in index).

Hafeez Md., Reddy, P. R., Hasina, S., Prasad, K. L. G., Nirmala Devi, K. and Thayeeb, M.D., 1994. Fertility rate of hydatidosis in cattle, buffaloes, sheep and pigs. Indian J. Anim. Sci., 64: 46-47.

Hafeez, Md., Subba Reddy,V., Ramesh, B., Aruna Devi, D. and Subhosh Chandra, M., 2004a. Prevalence of porcine cysticercosis in Andhra Pradesh, Indian J. Anim., Sci., 74 (12):

Hafeez Md., Subba Reddy, V., Ramesh, B., Aruna Devi, D. and Subhash Chandra, M., 2004b. Prevalence of porcine cysticercosis in South India. J. Parasit. Dis., 28: 118-120.

Hardev Singh and Rao, B. V., 1967. Some observations on the histochemistry of daughter cysts from hydatids of buffaloes. Indian Vet. J., 44: 933-938.

Harikrishnan, T. J., Ramathilgam, G. and Anandan, R., 1999. A note on the evaluation of antigenic response of *Cysticercus tenuicollis* for use in immunoprecipitation tests. J. Vet. Parasitol., 13: 63-64.

Hegde, K. S., Rahman, S. A., Rajasekhariah, G. R. and Jagannath, M. S., 1974. A study of the incidence of hydatid disease in animals and human beings in Bangalore City. Mysore J. agric. Sci., 8: 418-422.

Hegde, K. S., Rahman, S. A., Rajasekhariah, G. R. and Jagannath, M. S., 1975. Study of incidence of hydatid disease in animals and human beings in Bangalore City. J. Christian Med. Assc. India, 50: 296-299.

Hemant Kumar, H., Sharma, A. K., Das, L. L., Vinod Kumar and Shivendra Kumar, 2009. Abberent location of coenurus in goat. Indian J. Vet. Surg., 30: 68.

Gaur, S. N. S., 1985. Immunological studies of *Cysticercus cellulosae*, a larval cestode in pigs and other domestic animals. Final report of research project, Nainital, India.

Gill, H. S., 1972. A study on parasitic zoonosis in India. 59th Proc. Indian Sci. Cong. (Calcutta), Part III Abstr.37, p. 549.

Gill, H. S. and Rao, B. V., 1972. On the biology and morphology of *Echinococcus granulosus* (Batsch, 1786) of buffalo-dog origin. Parasitol., 57: 695-704.

Gupta, S., Jain, M. K., Katiyar, J. C. and Maitra, S. C., 1992. Substituted methyl benzimidazole carbamate: efficacy against experimental cysticercosis. Annl.Trop. Med. Parasitol., 86: 51-57.

Goswami, A., Das, M. and Laha, R., 2013. Characterization of immunogenic proteins of *Cysticercus tenuicollis* of goats. Vet. World, 6 (5): 267-270.

ICAR Progress Report, 2 000-2001. Studies on hydatidosis / cysticercosis in different animals of North Eastern Region. Dept. Parasitol., CVS AAU, Khanapara, Guwahati-29, Assam.

ICAR Progress Report, 2001- 2002. Studies on hydatidosis / cysticercosis in different animals of North Eastern Region. Dept. Parasitol., CVS AAU, Khanapara, Guwahati-29, Assam.

Ingole, R. S. Pathak, V. P. and Joshi, M. V., 2009. Coenurosis in a goat. Indian Vet. J., 86: 1296-1297.

Irshadullah, M. and Nizami, W. A., 1989. *In vivo* developmental pattern of *Echinococcus granulosus* protoscolices from buffalo liver and lung origin in experimental puppies. Proc. 9th Natl. Cong. Parasitol., Ujjain, Abstr. No.58.

Irshadullah, M. and Rani, M., 2010. Histochemical studies on the metacestode of *Echinococcus granulosus* (Cestoda: Taenidae) from Indian buffalo origin. J. Vet. Parasitol., 24: 33-37.

Irshadullah, M., Nizami, W. A. and Macpherson, C. N., 1989. Observations on the suitability and importance of the domestic intermediate hosts of *Echinococcus granulosus* in Uttar Pradesh, India. J. Helminthol., 63: 39-45.

Islam, S., Kalita, D., Bhuyan, D., Rahman, T. and Saleque, A., 2006. Ocular coenurosis in goat. J. Vet. Parasitol., 20: 53-56.

Jagannath, M. S. and D' Souza, P. E., 2000. Hydatidosis and cysticercosis – two important animal parasitic infections of public health significance. Intas Polivet, 1: 72-77.

Jain, M. K., Gupta, S., Katyar, J. C. and Srivastrava, V. L. M., 1989a. Status of the drug metabolising system in rat liver parasitized with *Cysticercus fasciolaris*. Med. Sci. Res.,17: 633-635.

Jain, M. K., Gupta, S., Katiyar, J. C., Singh, J. and Mitra, S. C., 1988. Methyl 5(6) −(á-hydroxyphenyl methyl) Benzimidazole carbamate and *Cysticercus fasciolaris* in rats. Chemotherapeutic and electron microscopic studies. Proc. VIII Natl. Cong. Parasitol., Calcutta Sch.Trop. Med., Calcutta-700 073, 10-12 February, 1988, Abstr., p.72.

Jain, M. K., Batra, S., Gupta, S., Katyyar, J. C., Srivastrava, V. L. M., 1989b. Impact of *Cysticercus fasciolaris* infection on reactive oxygen intermediates metabolising enzymes in rat liver. Med. Sci. Res.,17: 1051-1053.

Jain, M. K., Gupta, S., Katiyar, J. C., Mitra, S. C., Singh, J. and Bhakuni, D. S., 1989. Methyl 5(6)-(α-hydroxyphenyl methyl) Benzimidazole carbamate and cysticercosis: Chemotherapeutic and electron microscopic studies. Indian J. Exp. Biol., 27: 454-459.

Jain, P. C., 1972. Observations on the pathology of ovine liver infected with helminthic parasites. Orissa Vet. J. 7: 61-66.

Jain, P. C. and Shah, H. L., 1982. A note on the occurrence of *Coenurus gaigeri* from the gluteal muscles of sheep. Livestock Adviser, 7: 31-32.

Jana, D. and Jana, M., 2012. Occurrence of *Coenurus cerebralis* cyst in the eye of a black Bengal young kid. Indian J. Field Vet., 7(4): 68.

Janardhan Pillai, K., Narayana Rao, P. L. and Surya Rao, K., 1986. (Vide below, Pillai, K. J., Rao, P. L. N. and Rao, K. S., 1986)

Jeyathilakan, N., Karunakaran, P., Mathivanan, R. and Karunithi, K., 2000. Prevalence of parasites in small ruminants under farm condition. Bhubaneswar. XI Natl. Cong. Vet. Parasitol., Bhubaneswar, 4-6 February, 2000, p.58.

Jeyathilakan, N., Abdul Basith, S., Lalitha John., Daniel Joy Chandran, N. and Dhinakar Raj., 2010. Analysis of cross antigenicity among the metaceastode of sheep *Coenurus cerebralis* and *Cysticercus taeniformis*. XX NCVP Parasitology Today-Ecology to Molecular Biology, 18-20 February, 2010, Dept. Vet. Parasitol., CVS, CCS Haryana Agri. Univ., Hisar-125004, p.31.

Jeyathilakan, N., Abdul Basith, S., Lalitha John, Daniel Joy Chandran, N. and Dhinakar Raj, G., 2011. Development and evaluation of flow through technique for diagnosis of cystic echinococcosis in sheep. Vet. Parasitol.,180: 250-255.

Jithendran, K. P., 1996. Occurrence of hydatidosis and various liver infections in sheep and goats in Kangara Valley: An abattoir study. J. Vet. Parasitol., 10: 63-67.

Jithendran, K. P. and Somvanshi, R., 1998. Experimental infection of mice with *Taenia taeniaformis* eggs from cats – course of infection and pathological studies. J. Exp. Biol. 36: 523-525.

Jithendran, K. P. and Somvanshi, R., 1999. Studies on occurrence of *Cysticercus fasciolaris* in laboratory mice and rats. J. Vet. Parasitol., 13: 61-62.

Johri, L. N., 1959. On the occurrence of the larval form, hydatid cysts of *Echinococcus granulosus* (Batsch, 1786) from the sheep of Delhi State. Proc. Indian Acad. Sci., Section B, 29, pp.58-61.

Joshi, B. P. and Gupta, G. C., 1970. A case of pressure syndrome due to *Cysticercus cellulosae* in the brain of a dog. Indian Vet. J., 47: 366-367.

Joshi, S. C. and Sharma, R. K., 1988. A note on the parasites of rats (*Rattus rattus*) in M. P. Proc. VIII Natl. Cong. Parasitol., Calcutta,10-12 February, 1988, Abstr., p 37.

Juyal, P. D., Ruprah, N. S. and Chhabra, M. B., 1987. Epidemiology of *Sarcocystis* (microcyst) infection in slaughter goats (*Capra hircus*) and its transmission to dogs at Hissar (Haryana), India. Berlin GDR: Proc. 13th WAAVP Conf., 1989: 27.

Juyal, P. D., Singh, M. and Gupta, P. P., 1993. A case report on *Coenurus cerebralis* – A metacestode of *Taenia multiceps* in a goat (*Capra hircus*) in Punjab. Indian Vet. Med. J., 17: 165-166.

Juyal, P. D., Singh, N. K. and Kaur, P., 2005. Hydatidosis in India: a review from veterinary perspective. J. Parasit. Dis., 29: 97-102. (Review, not indexed).

Juyal, P. D., Sharma, R., Singh, N. K. and Singh, G., 2008a. Potential trends in epidemiology, diagnostics and control strategies against zoonotic cysticercosis (*Taenia solium*). In: Compendium of recent advances in diagnosis of parasitic diseases of livestock and poultry. Winter school, IVRI Izatnagar, December 2-22, 2008. (Review, not indexed).

Juyal, P. D., Sharma, R., Singh, N. K. and Singh, G., 2008b. Epidemiology and control strategies against cysticercosis (due to *Taenia solium*) with special reference to swine and human in Asia. J. Vet. Anim. Sci., 1: 1-10. (Review, not indexed).

Juyal, P. D., Aulakh, G. S., Sharma, R., Juyal, P. D. and Singh, J., 2009. Concurrent infection of *Taenia taeniaeformis* and *Isospora felis* in a stray kitten: a case report. Veterinarni Medicina, 54 (2): 81-83.

Kakru, D. K., Sofi, B. A. and Assadullah, S., 2008. Novel route of infection in experimental model of hydatid disease. Indian J. Pathol. Microbiol., 51: 373-375.

Kalai, K., Nehete, R.S., Ganguly, S., Ganguli, M., Dhanalaksmi, S. and Mukhopadhyay, S. K., 2012. Investigation of parasitic and bacterial diseases in pigs with analysis of hematological and serum biochemical profile. J. Parasit. Dis., 36: 129-134.

Kalita, D., 1997. Surgical correction of gid in goats. *Indian Vet. J.*,74: 682–684.

Kanwar, J. R. and Kanwar, R., 1994. Purification and partial immunochemical characterization of a low molecular mass, diagnostic *Echinococcus granulosus* immunogen for sheep hydatidosis. FEMS Immunol. Med. Microbiol.,9 (2): 101-107.

Kanwar, J. R., Kaushik, S.P., Sawhney, I.M., Kamboj, M.S., Mehta, S.K. and Vinayak, V. K., 1992. Specific antibodies in serum of patients with hydatidosis recognised by immuoblotting. J. Med. Microbiol., 36(1): 46-51.

Katiyar, R. D and Sinha, A. K., 1981. Hydatid disease in livestock in Sikkim. Livestock Adviser, 6: (11) 57-58.

Katoch, R. and Jithendran, K. P., 1999. Occurrence of *Cysticercus pisiformis* in rabbit. Indian Vet. J., 76: 747.

Katoch, R. and Singh, B. P., 1994a. Immuno-deficiencies of *Echinococcus granulosus* experimentally infected pups. Indian Vet. J., 71: 1179-1181.

Katoch, R. and Singh, B. P., 1994b. *In vitro* scolicidal activity of some anthelmintic against protoscoleces of cattle hydatid cysts. Indian Vet. J., 72: 457-1181.

Kaur, M., Joshi, K., Ganguly, N. K., Mahajan, R. C. and Malla, N., 1995. Evaluation of efficacy of albendazole against larvae of *Taenia solium* in experimentally infected pigs and kinetics of the immune response. Int. J. Parasitol., 25: 1443-1450.

Kaur, M., Mahajan. R. C., Malla, N., 1999. Diagnostic accuracy of rapid enzyme linked immunosorbent assay for the diagnosis of human hydatidosis. Indian J. Med. Res., 110: 18-21.

Khan, N. A., 1996. Prevalence of echinococcosis in buffaloes and applicability of gel diffusion and immunoelectrophoresis test in diagnosis of disease. M. V. Sc. Thesis submitted to Rajasthan Agri. Univ., Bikanir.

Khan, N. A. and Purohit, S. K., 2006. Prevalence of echinococcosis in buffaloes. Vet Scan (ISSN 0973-6980), 1: 1-2.

Kolte, G. N. Vegad, J. L. and Awadhiya, 1981. Note on unusual case of cysticercosis (*Cysticerus bovis*) in a Gir cow. Indian J. Anim. Sci. 51: 370-372.

Kolte, S. W., Ganorkar, A. G. and Kurkure, N. V., 1998. *Cysticercus tenuicollis* in spotted deer (*Cervus porcinus*). Indian Vet. J.,75: 834.

Kolte, S. W., Sanweer Khatoon, Hedaoo, V. R., Prajapati, B. S., Kurkure, N. V., Bhandarkar, A. G. and Maske, D. K., 2011. Studies on the occurrence of *Cysticerus fasciolaris* (Stobilocercus) infections in laboratory rats (*Rattus norvegicus*) in Nagpur. XII Natl Cong. Vet. Parasitol., 5-7th Jan, 2011, IAAAP and Dept. Vet. Parasitol. & Bombay Vet. Coll., Mumbai, S2- 015, p.87.

Konapur, S. G., 1996. Diagnosis of Hydatidosis in cattle and buffaloes by Avidin Biotin Enzyme Linked Immunoabsorbant Assay. M. V. Sc. Thesis (Vet. Med.), Univ. Agri. Sci., Bangalore.

Konapur, S. G., Panduranga, G. L., Rajasekhar, M., Jagannath, M. S. and Sastry, K. N. V., 1996. Avidin-biotin enzyme-linked immunoabsorbant assay in the diagnosis of hydatidosis in cattle and buffaloes. 8th Natl. Cong.Vet. Parasitol., Hisar, December 9-11, 1996.

Konapur, S. G., Panduranga, G. L., Rajasekhar, M., Jagannath, M. S. and Sastry, K. N. V., 1999a. Diagnosis of hydatidosis in cattle and buffaloes by avidin-biotin enzyme-linked immunoabsorbant assay. J. Vet. Parasitol., 13: 19-22.

Konapur, S. G., Panduranga, G. L., Sastry, K. N. V., Jagannath, M. S. and Rajasekhar, M., 1999b. Evaluation of agar-gel immunodiffusion and counter current immuno electrophorosis in the diagnosis of hydatidosis in cattle and buffaloes. Indian J. Anim. Sci., 69: 662-663.

Kosalaraman, V. R. and Ranganathan, M., 1980. A survey of disease condition of lungs of buffaloes. Cheiron, 9: 281-284.

Koshy, T. J. 1984. Taenid infection in dogs. Ph. D. Thesis submitted to Tamilnadu Agri. Univ., Coimbatore, India.

Krishna Murthy, D., 1949. *Cysticercus cellulosae,* their incidence in canines. Indian Vet. J., 25: 367-370.

Krishnan, K. R., Ranganathan, M. and Ramamurthi, R., 1984. Occurrence of viable *Cysticercus bovis* in cattle and buffaloes. Cheiron, 13: 109-111.

Krishnan, K. R. and Ranganathan, M., 1972. A survey of incidence of *Cysticercus bovis* in bovines slaughtered at the Madras Corporation slaughter house. Indian Vet. J., 49: 1182-1184.

Kulkarni, D. and Satyanarayana Shetty, M., 1983a. Observations on the anthelmintic efficacy of Niclosan against *Multiceps multiceps and M. gaigeri* in dogs. 5th Natl.

Cong. Parasitol., June 25-27, 1983, Coll. Vet. Sci., A. P. Agri. Univ., Tirupati, A. P. and Indian Soc. Parasitol., Abstr. C-14, pp.60-61.

Kukarni, D. and Satyanarayana Shetty, M., 1983b. Anthelmintic efficacy of Dicestal against *Multiceps multiceps* in dogs. 5th Natl. Cong. Parasitol., June 25-27, 1983, Coll. Vet. Sci., A. P. Agri. Univ., Tirupati, A. P. and Indian Soc. Parasitol., Abstr. C-15, p.61.

Kulkarni, V.G., Deshpande, B. B. and Gafoor, M. A., 1984. A note on parasitic infection of liver of cattle and buffaloes. Indian J. Vet. Pathol., 8: 66-68.

Kulkarni, D., Gangadhar Rao, Y. V. B. and Rao, B. V., 1974. A note on certain aberration in the scolex of a *Coenurus* of a goat. Indian Vet. J., 51: 739.

Kulkarni, D., Muralidharam, S. R. G., Mahendar, M. and Rao, V.M., 1983. Incidence of hydatidosis in cattle, buffaloes, sheep and goats in Andhra Pradesh with particular reference to different agro-climatic conditions. 5th Natl. Cong. Parasitol., June 25-27, 1983, Coll. Vet. Sci., A. P. Agri. Univ., Tirupati, A. P. and Indian Soc. Parasitol., Abstr. D-16, p.86.

Kulkarni, D., Muralidharam, S. R. G., Mahendar, M. and Rao, V. M., 1986. Incidence and prevalence of hydatidosis in cattle, buffaloes, sheep and goats in Andhra Pradesh with particular reference to different agro-climatic conditions. Livestock Adviser, 11(3): 37-40.

Kulkarni, D., Satyanarayana Chetty, M., Mahendar, M., Dhananjaya Reddy, B. and Ramakrishna, K., 1985. Observations on anthelmintic efficacy of Niclosamide against *Multiceps multiceps* and *Multiceps gaigeri* in dogs (an experimental study). Livestock Adviser, 10 (6): 21-23.

Kumar, A., Rana, R., Vihan, V. S. and Arora, N., 2002. Cerebrospinal fluid analysis and serum biochemistry in coenurosis in goats for clinical appraisal. In: Proceedings of the V Natl. Semr. on "Strength challenges and opportunities in small ruminants diseases in new millennium". Organized by ISSGPU at Jaipur on December 30–31, 2002. p.78

Kumar, D., 1987. Immunodiagnostic and immunoprophylactic studies on *Cysticercus cellulosae* infection in pigs. Ph. D. Thesis submitted to G.B. Pant Univ. of Agri. & Tech., Pantnagar, 263145. Abstr., J. Vet. Parasitol., 1: 84-85.

Kumar, D. and Gaur, S. N. S., 1987a. Serodiagnosis of porcine cysts by enzyme-linked immunosorbant assay (ELISA) using fractionated antigen. Vet. Parasitol., 24: 195-202.

Kumar, D. and Gaur, S. N. S., 1987b. Counter-immunoelectrophoresis in the diagnosis of *Taenia solium* cysticercosis in pigs using fractionated antigens. J. Vet. Parasitol., 1: 43-46.

Kumar, D. and Gaur, S. N. S., 1988. Immunization of pigs against cysticercosis of *Taenia solium*. Second Natl. Parasitol. Cong., 3-5 March, 1988, Dept. Parasitol. Vet. Coll., Bangalore, India. Stream II. Abstr. p.23.

Kumar, D. and Gaur, S. N. S., 1989. Comparative evaluation of various immunodiagnostic tests for the diagnosis of *Taenia solium* cysticercosis in pigs using fractionated antigens. J. Helminthol., 63: 13-17.

Kumar, D. and Gaur, S. N. S., 1994. *Taenia solium* cysticercosis in pigs. Helm. Abst., 63: 365-383. (Review, not cited in Index).

Kumar, D., Gaur, S. N. S. and Pathak, K. M. L., 1987. Fractionation and characterization of the cysticercus of *Taenia solium*. Res. Vet. Sci., 43: 395-397.

Kumar, D., Gaur, S. N. S. and Pathak, K. M. L., 1987. Immunization of pigs against cysticercus of *Taenia solium* using fractionized first and second peaks of *Cysticercus cellulosae* scolex antigen. Indain J. Anim. Sci., 57: 932-935.

Kumar, D., Gaur, S. N. S. and Varshney, K. C., 1988. Host tissue response in *Taenia solium* cysticercosis in pigs. Second Natl. Cong. Parasitol., 3-5 March, 1988, Dept. Parasitol., Vet. Coll., Bangalore, India. Stream II. Abstr., p.15.

Kumar, D., Gaur, S. N. S. and Varshney, K. C., 1991. Host tissue response against *Taenia solium* cysticercosis in pigs. Indian J. Anim. Sci. 61: 270-273.

Kumar, G. S., and Parihar, N. S., 1987. Pulmonary hydatidosis in Indin water buffaloes. Indian Vet. Parasitol., 2: 160-161.

Kumar, H., Mallick, A. I. and Hajra, S., 2003. Prevalence of *Echinococcus granulosus* (hydatid disease) infection in domestic animals. J. Vet. Pub. Hlth., 1: 81-82.

Kumari, N., Prasad, L. N. and Singh, B.K., 2002. A note on pulmonary hydatidosis in goats. Indian J. Vet. Pathol., 24: 130.

Kuppuswamy, P. B. and Gupta, B. N., 1964. Gid in bovines. Indian Vet. J., 41: 54-58.

Kurkure, N.V., Kolte, S. W., Ganorkar, A. G., Bhandarkar, A. G. and Chopde, S. S. 2000. *Cysticercus tenucollis* in Osmanabadi goat. Intas Polivet, 1: 241.

Lal, S. S., Seema Rani and Madhu Bhatnagar., 1983. Histochemical (glycogen and lipid) changes in liver of black rat due to *Taenia taeniaeformis* infections. Proc. Fifth Natl. Cong. Parasitol., June 25-27, 1983, A. P. Agri. Univ. Coll. Vet. Sci., Tirupathi, A. P., Abstr., B-16.

Lekharu, J. C., Lahon, D. K. and Gogoi, D., 1968. Vetcol, 25: 31-32.(Incidence in goats).

Lekharu, J. C., Deka K. N., Lahon, D. K., Pathak, S. C., 1989. Surgical technique of craniotomy in gid in goats. *Indian Vet. Med. J.*,13(3): 161–166.

Lodha, K. R., Raisinghani, P. M. and Vyas, V. K., 1982. Notes on echinococcosis in Indian camel (*Camelus dromedaries)* Indian J. Anim. Sci., 52: 613-616.

Londhe, M. S., Tripathi, K. K., Agrawal, R. D., 2012. Cardiac *Cysticercus tenuicollis* in Barbari goat - A case report. Indian J. Vet. Pathol., 36: 92-93.

Madan Joshi, Bhandarkar, A. G., Raote, Y. V., Deshmukh, D. T. and Bhagwat, S. S., 1989. Hydatid cysts in myocardium of two cross-bred bullock. Livestock Adviser, 14: (5) 36-38.

Madhu, D. N., Tamil Mahan, Sudhakar, N. R., Mayura, P. S., Banerjee, P. S., Shivani Sahu, S and Pawde, A.M., 2013. *Coenurus gaigeri* cyst in the thigh of a goat and its successful management. J. Parasit. Dis., DOI 10.1007/s12639-013-0242-4. (Accepted: 5 January 2013)

Maisha Mathur, G. D., Sharma, G. D. and Sandeep Khare, 2001. Histopathological observations in hydatidosis in kidney of sheep. Indian Vet. Med. J., 25: 177-178.

Maity, B. and Bandopadhyaya, 1991. Prevalence of Coenurus of *Multiceps gaigeri* in goat and its predilection and distribution. Indian J. comp. Microbiol. Immunol. infect. Dis., 12 (1&?): 71 72.

Maity, A., Bhattacharya, D, Batabyal, S., Chattopadyay, S., Bera, A. K. and Karmakar, P. K., 2007. *Echinococcus granulosus* of buffalo in India: Partial characterization of excretory-secretory and germinal membrane antigens.Vet. Res. Comm., 31: 457-460.

Malsawmtluangi, C. and Tandon, V., 2009a. Occurrence of the metacestode of *Taenia* sp. and the nematode, *Capillaria hepatica* in liver of wild rodents from bamboo growing areas of Mizoram. In: Currents Trends in Parasitology. Veena Tandon, Arun K.Yadav and Bishnupada Roy (Eds.) Proc. 20th Natl. Cong. Parasitol., Shillong, India (November 3-5, 2008), Panima Publishing Corporation, New Delhi. pp.127-136.

Malsawmtluangi, C. and Tandon, V., 2009b Helminth parasites spectrum in rodent hosts from bamboo growng areas of Mizoram, North-east India. J. Parasit. Dis., 33: 28-35.

Malla, N., Kaur, M., Ganguli, N. K. and Mahajan, R. C., 1996. Pathobiology of cysticercosis: an experimental study. Abstract: Global Meet Parasit. Dis., New Delhi, 18-22 March, 1996, J. Parasit. Dis., 20: 71.

Malla, N., Kaur, M., Joshi, K., Ganguli, N. K. and Mahajan, R. C., 1995. Experimental cysticercosis: Evaluation of the efficacy of albendazole and kinetics of immune response in pigs. 12th Natl. Cong. Parasitol., Panaji, Goa, 23-25 January, 1995, Abstr., J. Parasit. Dis., 19: 94.

Mandakhalikar, K. D., Jangir, D. and Nasakar, V. S., 2009. Prevalence of cysticercosis in pigs and economic losses to condemnation of pork. J. Bombay Vet. Coll., 17: 82-83.

Mandal, S. and Mandal, M. D., 2012. Human cystic echinococcosis: epidemiologic, zoonotic, clinical, diagnostic and therapeutic aspects. Asian Pacific J. Trop. Med., 6: 253-260 (Review).

Mandal, S. C., 1977. Fatal hydatid disease with involvement of cerebrum in buffaloes. Zentralbl. Veterinarmed., 224: 678-679.

Mandal, S. C., Tiwari, S. K., Shakya, S. and Bhonsle, D., 2004. Outbreak of coenurosis (gid or sturdy) in local goats. Indian Vet. Med. J., 28: 183.

Manjunatha, D. R., Mahesh, V. and Ranganath, L., 2010. An unusual case of *Coenurus gaigeri* cyst in the eye of goat. Indian J.Vet. Surg., 31: 70.

Mapelstone, P. A. and Bahaduri, N.V., 1940. The helminth parasites of dogs in Calcutta and their bearing on human parasitology. Indian J. Med. Res., 28: 595-604.

Mathur, K. N. and Khanna, V. K. 1977. Incidence of hydatid disease in sheep and goats in the city of Jaipur. Sci. and Cult., 43: 371-372.

Mathur, P. B. and Dutt, S. C., 1969. A fatal infection of *Coenurus gaigeri* (Hall, 1916) in a sheep. Indian Vet. J., 46: 259-260.

Mathur, V. C. Aiyasami, S. S. Latha B. R. and Lalitha John., 2003. Western blot analyses of *Cysticercus tenuicollis* antigens. Indian J. Anim. Sci., 73: 837-839.

Mir, M. S., Darzi, M. M., Kamal, S. A., Nashiruddulla and Saleem Iqubal, 2006. Pathology of *Taenia pisiformis* infestation in Angora rabbits. J. Vet. Parasitol., 20: 129-132.

Mishra, K. C., 1977. *Coenurus cerebralis* in goat. *Orissa Vet. J.*, 11: 495–501.

Misra, S. C. and Rupra, N. S., 1968. Incidence of helminthes in goats at Hissar. Punjab Agri. Univ. J. Res., 5: 279-285.

Moghaddar, N., Gaur, S. N. S. and Tavalay., B., 2001. Studies on *Taenia saginata* in cattle. Twelfth Natl. Cong. Vet. Parasitol., 26-27August, 2001, Dept. Parasitol., Coll. Vet Sci., Acharya Ranga Agri. Univ., Tirupati-517 502, India, Abstr. S2-L: 3, p.23.

Mohinder Singh, Nigam, J. M., Kishtwaria, R. S. and Rao, V. N., 1999. Surgical management of cyst in goral (*Nemarhaedus goral*). Indian Vet. J. 76: 175-176

Mohammed, C. R., 1974. Cheiron, 3: 168-169.

Mohan, V. R., Tharmalingam, J., Muliyal, A. Oommen, P., Dorny, P., Vercruysse, J. and Vendam, R., 2012. Prevalence of porcine cysticercosis in Vellore, South India. Trans. R. Soc. Trop. Med., 107: 62-64.

Mohd Irfan Naik, Bashir Ahmad Fomda, Rajesh Kumar Tenguria, Javid Ahmad Bhat., 2013. Analysis of specific IgG, IgM, IgE and IgG subclass antibodies for serological diagnosis of human cystic *Echinococcosis*. Internatl. J. Microbiol. Adv. Immunol., (IJMAI)1(1): 130.

Montosh Banerjee, Gupta, P. P. and Nem Singh., 1985. Incidence and type of pulmonary affection in sheep and goats of Ludhiana (Punjab). Indian J. Anim. Hlth., 24: 21-23.

Mudaliyar, S. V. and Alwar, V. S., 1947. Check-list of parasites (Class: Trematoda and Cestoda) in the Department of Parasitology, Madras Veterinary College Laboratory. Indian Vet. J., 23: 423-434.

Munde, D. K., 1999. Prevalence of zoonotic bladder worms (Metacestodes) in food animals and their economic implications. M.V.Sc. Thesis submitted to Maharashtra Anim. Fish. Sci. Univ., Nagpur.

Muralidhar, A. and Shastri, K. N. V., 1988a. Prevalence of hydatid in cattle and sheep. Second Natl. Cong. Vet. Parasitol., 3-5 March, 1988, Bangalore, Souvenir, Abstr., Stream. I, p.5.

Muralidhar, A. and Shastri, K. N. V., 1988b. Prevalence of hydatid in sheep and goats. Proc. 2nd Natl. Seminar on sheep and goats disease.19-22 September, 1988, CIRG, Makhdoom, Mathura, India, p. 202.

Muralidhar, A and Shastri, K. N. V., 1988c. Prevalence of hydatid in cattle and sheep. J. Bombay Vet. Coll., 6: 75.

Muraleedharan, K., Seshadri, S. J., Srinivasa Gowda, R. N. and Appaji, 2010. Cancer of eye in a Jaguar (*Panthera onca*). Zoos' Print, 25: 33.

Nageswar Rao, K., 1970. A survey of the incidence of *Cysticercus cellulosae* among pigs in certain parts of Andhra Pradesh. Thesis submitted for the award of M. Sc. (Vet) Degree to Andhra Pradesh Agri. Univ., Hyderabad.

Nama, H. S. and Parihar A., 1976.Quantitative and qualitative analysis of helminth fauna in *Rattus rattus rufescens*. J. Helminthol., 50: 99 102.

Narasapur, V. S., 1996. Anoplocephaline cestodes of domestic animals. In: Recent Trends in Diagnosis and Control of Parasitic Diseases. Natl. Trng. Prog. Teachers / Scientists.25th-30th March, 1996. Centr. Adv. Stud., Dept. Parasitol., Vet. Coll., Univ. Agri. Sci., Bangalore-560 024. pp.11-21. (Review, not indexed).

Nath, S., Pal, S., Sanyal, P. K., and Mandal, S. C., 2009a. Prevalence of *Taenia hydatigena* cysticercosis in goat in Durg, Chhattisgarh. XIX Natl. Cong. Vet. Parasitol. & Natl. Sym. National Impact on Parasit. Dis, Livestock Hlth. Prod., February 3-5, 2009, IAAVP& Dept. Vet. Parasitol., Coll. Vet. Sci., Guru Angad Dev Vet & Anim. Sci. Univ., Ludhiana-141 004, Abstr., OE26, p.38.

Nath, S., Pal, S., Sanyal, P. K. and Mandal, S. C., 2009b. Prevalence of caprine *Taenia hydatigena* cysticercosis in Durg. Indian J. Field Vet., 5: 64-66.

Nath, S., Pal, S., Sanyal, P. K., and Mandal, S. C., 2010a. Determination of immune reactive proteins in the fluids and membranes of *Cysticercus taeniaeformis*. XX NCVP Parasitology Today-Ecology to Molecular Biology, 18-20 February, 2010, Dept. Vet. Parasitol., CVS, CCS Haryana Agri. Univ., Hissar-125004, p.34.

Nath, S., Pal, S., Sanyal, P. K., and Mandal, S. C., 2010b. Western blot analysis of membrane analysis of *Cysticercus taeniaeformis* of goat origin. XX NCVP Parasitology Today-Ecology to Molecular Biology, 18-20 February, 2010, Dept. Vet. Parasitol., CVS, CCS, Haryana Agri. Univ., Hissar-125004, p.35.

Nath, S., Pal, S., Sanyal, P. K., Mandal, S. C., Dutta, G. K., Maiti, S. K. and Tiwari, S. K., 2009c. Chemical and biochemical contents of cystic fluid of *Cysticercus tenuicollis* of goat origin. XIX Natl. Cong. Vet. Parasitol. & Natl. Sym. Natl. Impact on Parasit. Dis, Livestock Hlth. Prod., February 3-5, 2009, IAAVP& Dept. Vet. Parasitol., Coll. Vet. Sci., Guru Angad Dev Vet. & Anim. Sci. Univ., Ludhiana-141 004, Abstr., OE25, p.38.

Nath, S., Pal, S., Sanyal, P. K., Mandal, S. C., 2010. Chemical and biochemical characterization of *Taenia hydatigena* cysticerci in goats.Vet. World, 3: 312-314.

Nath, S., Pal, S., Sanyal, P. K., and Mandal, S. C., 2011a. Determination of immune reactive proteins in cystic membranes of *Cysticercus taeniformis* of goat origin. Indian J. Anim. Sci., 81: 822-833.

Nath, S., Pal, S., Sanyal.P. K., Ghosh, R. C. and Mandal, S. C., 2011b. Histopathology of hepatitis cysticercosa in goat. J. Vet. Parasitol., 25: 187-188.

Naveen Kumar, Aithal, H. P., Gupta, S. C. and Mogha, I.V., 2003. Occurrence of unusually large *Coenurus* cyst in subcutaneous tissues of sheep. J. Vet. Parasitol.,17: 73-74.

Nayak, B. C., Rao, A. T. and Das, B. C., 1973. Myocardial cysticercosis in bovines – A report of two cases. Indian Vet. J., 50: 749-750.

Nayar, S. R., 1974. Hypertrophy of the liver due to hydatid cyst- a probable cause for recurrent tympany in a cross-bred bull. Indian Vet. J., 51: 161-163.

Neena Singla, Singla, L. D. and Gupta, K., 2013. Pathological alterations in natural cases of *Capillaria hepatica* infection alone and in concurrence with *Cysticercus fasciolaris* in *Bandicota bengalensis*. J. Parasit. Dis., 37: 16-20.

Neena Singla, Singla, L. D. and Parshad, V.R., 2010. Prevalence of adult metacestodal infections of cestode parasites in *Rattus rattus* collected from poultry farms around Ludhiana (Punjab). In: XX Nat. Cong. Vet. Parasitol., February18-20, 2010, Dept. Vet. Parasitol., Coll. Vet. Sci., Chaudhary Charan Singh Haryana Agri. Univ., Hisar-125 004, Abstr., p 16.

Neena Singla, Singla, L. D., Gupta, K. and Sood, N. K., 2012. Azithromycin, clindamycin, histopathological changes pathological alterations in natural cases of *Capillaria hepatica* infection alone and concurrent with *Cysticercus fasciolaris* in *Bandicota bengalensis*. XXII Natl. Cong. Vet. Parasitol., March 15-17, 2012, Coll. Vet. Sci. and Anim. Husb., U. P. Pandit Deen Dayal Upadhyaya Pashu Chikitsa Vigyan Vishwavidyalaya Evam Go Anusandhan Sansthan (DUVASU), Mathura-281001(UP.), India, Abstr. No.VI.58, p.112.

Nichal, A. M., Maske, D. K. and Radhakrishna, S., 2003. Prevalence of caprine cysticercosis in Nagpur. J. Vet. Parasitol., 17: 171-172.

Nimbalkar, R. K., Shinde, S. S., Kamtikar, V. N and Muley, S. P., 2011. Study of *Taenia hydatigena* in the slaughtered sheep (*Ovis bharal*) and goats (*Capra hircus*) in Maharashtra, India. Global Veterinaria, 6(4): 374-377.

Niphadkar, S. M. and Rao, S. R., 1966. Observations on helminthic fauna of rats in Bombay. Indian Vet. J., 43: 844-845.

Nishant, Sinha, B. K., Sinha, S. R. P. and Sucheta Sinha, 2005. Histopathological studies on hydatidosis in buffalo livers at abattoirs in Patna. Intas Polivet, 6: 301-302.

Pal, A. K. and Sinha, P. K., 1970. Multilocular hydatidosis in a bullock. Indian Vet. J., 47: 910-912.

Pal, S., 1996. Indian J. comp. Microbiol. Immunol. infect. Dis., 16: 58-60.

Paliwal, O. P. and Singh, K. P., 1971a. An outbreak of acute coenurosis in sheep – A pathogenic study. Indian Vet. J., 48: 783-785.

Paliwal, O. P., Singh, S. P. and Singh, K. P., 1971b. Occurrence of *Multiceps* cyst over the kidney and brain of a goat. Indian Vet. J., 48: 1278-1279.

Pan, D., Bera, A. K., Bandyopadhay, S., Das, S.K., Rana, T., Das, S. K., Bandyopadhay, B., Manna, D.and Bhattacharya, D., 2011. Molecular characterization of antigen

B2 submit in two genotypes of *Echinococcus granulosus* from Indian bubaline isolates, its stage specific expression and serological evaluation. Mol. Biol. Rep., 38: 2067-2073.

Pan, D., Bera, A. K., Bandyopadhyay, S, Das, S. K., Bandyopadhyay, S., Bhattacharyya, S., Manna, B, De, S. and Bhattacharya, D., 2009a. Relative expression of antigen B coding gene of bubaline isolates of *Echinococcus granulosus* in fertile and sterile cysts. J. Helminthol. 28: 1-4.

Pan, D., Bera, A. K., Das, S.K., Bandyopadhyay, S., Manna, B. and Bhattacharya, D., 2009b. Polymorphism and natural selection of antigen B1 of *Echinococcus granulosus* isolated from different host assemblages in India. Mol. Biol. Rep., 37: 1477-1482.

Pan, D., Bera, A. K., De, S., Bandypadhyay, S., Das, S.K., Manna, B., Sreevatsava, V. and Bhattacharya, D., 2010. Relative expression of 14-3-3 gene in different morphotypes of cysts of *Echinococcus granulosus* isolated from the Indian buffalo. J. Helminthol. February 15, 1-4. [Epub ahead of print].

Pan, D., Bhattacharya, D., Bera, A. K., Gudewar, J., De, S. and Das S. K., 2008. Stressor-induced changes to the protoscolesces of *Echinococcus granulosus* of Indian buffalo origin. J. Helminthol., 82: 309-311.

Panda, M. R., Ghosh, S. and Varma, T. K., 2000. Potential diagnostic test for developmental and natural ovine *Taenia hydatigena*. Acta Vet. Hungeria, 48: 173-182.

Panda, M. R., Ghosh, S. and Varma, T. K., 2005. Immunodiagnosis of cysticercosis by ELISA in lamb infected with *Cysticercus tenuicollis*. Indian J. Anim. Sci., 69: 747-749.

Panda, M. R., Sardar, K.K., Panda, D.N. and Das, A. K., 2005. Study of metacestodes infection in ruminants in and around Bhubaneswar. Intas Polivet, 6: 286-290.

Pandey, N. N., Nauriyal, D. C. and Rathor, S. S., 1978. Pulmonary echinococcosis in buffaloes – (*Bubalus bubalis*) – a clinical entity. Indian Vet. J., 55: 823-826.

Pandey, V. S., 1966. Observation on the morphology, biology and pathogenicity of *Echinococcus granulosus* (Batsch, 1786). M. V. Sc., Thesis submitted to Bihar Vet. Coll., Patna, Bihar, India. (not indexed).

Pandey, V. S., 1971a. Biochemical observations on in hydatid fluid – a preliminary report. Indian Vet. J., 48: 899-901.

Pandey, V. S., 1971b. Observations on echinococcosis in Bihar. Indian J. Anim. Sci., 41: 596-599.

Pandey, V. S., 1971c. Pathology of hydatid infection in the goat and *Echinococcus granulosus* in the dog. Ann. Med. Vet., 115: 519-527. (vide Helm. Abstr. 41: 3336.)

Pandey, V. S., 1972. Observations on the morphology and biology of *Echinococcus granulosus* (Batsch, 1786) of goat-dog origin. J. Helminthol., 46: 219-233.

Parihar, N. S., 1998. Pathology of *Coenurus cerebralis* in ovine subclinical infection. *Indian J. Anim. Sci.*, 56: 539-543.

Panisup, A. S., Gaba, S. C. and Bhatnagar, V. K., 1979. Incidence of coenuriosis in sheep in Hissar. Haryana Vet. J., 48: 53-55.

Parija, S. C., 1991. Recent trends in the diagnosis of hydatid diseases. Southeast Asian J. Trop. Med. Pub. Hlth., 22(Suppl.): 371-376. (Review, not indexed).

Parija, S. C., 1998. A review of some simple immuno assays in the sero diagnosis of cystic hydatid disease. Acta Trop. 70, 17-24. (Review, not indexed).

Parija S. C. and Sheela Devi, C., 1999. Current concepts in the diagnosis of cystic echinococcosis in humans and livestock and intestinal echinococcosis in canine hosts. J. Vet. Parasitol.,13: 93-102. (Review, not indexed).

Parija, S. C. and Swarna, S. R., 2005. Echinococcosis in India. In: Asian Monograph, The Federation of Asian Parasitologists, Japan. 2: 283-324.

Pathak, K. M. L., 1992. Ocurrence of *Cysticercus tenuicollis* in the spleen and gall bladder of goats (India). Agri. Sci. Dig., 2: 128-130.

Pathak, K. M. L., 1994. New approaches in the diagnosis of porcine cysticercosis (*Taenia solium*). Sixth Natl. Cong. Vet. Parasitol, 22-24 October, 1994, Dept. Parasitol., Coll. Vet. Sci., JNKVV, Jabalpur-482 001, M. P., India, Abstr. S-3-23, p.40.

Pathak, K. M. L., 1996. Recent advances in the diagnosis of *Taenia solium* cysticercosis. Global Meet Parasit. Dis., New Delhi, India 18-22 March, 1996. Abstr., J. Parasit. Dis., 20: 85.

Pathak, K. M. L., Allan, J. C., Ersfeld, K., and Craig, P., 1994. A western blot and ELISA assay for the diagnosis of *Taenia solium* in pig. Vet. Parasitol., 53: 209-217.

Pathak, K. M. L. and Chhabra, M. D., 2012. *Taenia solium* cysticercosis in India: A vetero-medical update. Indian J. Anim. Sci., 82 (9): (Review, not indexed).

Pathak, K. M L. and Gaur, S. N. S., 1981a. Clinical signs associated with experimental *Cysticercus tenuicollis* infection in kids. Haryana Vet., 20: 22-23.

Pathak, K. M L. and Gaur, S. N. S., 1981b. Serum proteins in goats experimentally infected with *Cysticercus tenuicollis* of *Taenia hydatigena*. Haryana Vet., 20: 109-113.

Pathak, K. M L. and Gaur, S. N. S., 1981c Serum levels of GOT, GPT and OCT enzymes in goats infected with *Cysticercus tenuicollis*.Vet. Parasitol., 8: 95-97.

Pathak, K. M. L. and Gaur, S. N. S., 1981d. Chemical composition of *Cysticercus tenuicollis*. Agri. Sci. Dig., India. 1(1): 53-54.

Pathak, K. M. L. and Gaur, S. N. S., 1981e.Haematological changes associated with *Cysticercus tenuicollis* infection in worm-free kids. Indian J. Anim. Sci.,51 (9): 850-854.

Pathak, K. M L. and Gaur, S. N. S., 1982a. The incidence of adult and larval stages of *Taenia hydatigena* in Uttar Pradesh (India).Vet. Parasitol., 10: 91-95.

Pathak, K. M. L. and Gaur, S. N. S., 1982b. Occurrence of *Cysticercus tenuicollis* in the spleen and gall bladder of goats (India). Agri. Sci. Dig., 2: 128-130.

Pathak, K. M. L. and Gaur, S. N. S., 1983a. Diagnostic importance of serum enzymes in pigs infected with *Cysticercus cellulosae*. 5th Natl. Cong. Parasitol., June 25-27, 1983, Coll. Vet. Sci., A. P. Agri. Univ., Tirupati, A. P. and Indian Soc. Parasitol., Abstr. B-23, p.41.

Pathak, K. M. L. and Gaur, S. N. S., 1983b. Haematological, biochemical and pathological changes associated with infection in dogs. Indian J. Anim.Sci., 53: 635-639.

Pathak, K. M. L. and Gaur, S. N. S., 1985a. Biochemical composition of *Cysticercus* of *Taenia solium* 6th Natl. Cong. Parasitol., India, Pantnagar, Abstr. p 24.

Pathak, K. M. L. and Gaur, S. N. S., 1985b. Fractionation and characterization of cysticercus antigen of *Taenia solium*. Indian J. Comp. Microbiol.Immunology infect. Dis., 6: 45-47.

Pathak, K. M. L. and Gaur, S. N. S., 1985c. Changes in serum enzyme activities in pigs naturally infected with the metacestodes of *Taenia solium*. Vet. Res. Comm., 9: 143-146.

Pathak, K. M. L. and Gaur, S. N. S., 1986a. Evaluation of some serodiagnostic tests in porcine cysticercosis. Indian J. Anim. Sci., 56: 1059-1061.

Pathak, K. M. L. and Gaur, S. N. S., 1986b. Observation on the epidemiology of adult and larval stages of 7th Natl. Cong. Parasitol., Ravi Shankar Univ. Raipur, 26-28, December. 1986, Abstr. B-14, p 70.

Pathak, K. M. L. and Gaur, S. N. S., 1988. Surveillance of human taeniasis and porcine cysticercosis in certain parts of Uttar Pradesh. Second Natl. Cong. Vet. Parasitol., 3-5 March., 1988, Bangalore, Souvenir, Abstr., Stream I, pp. 5.

Pathak, K. M. L. and Gaur, S. N. S. 1989. Prevalence and economic implications of *Taenia solium* taeniasis and cysticercosis in Uttar Pradesh State of India. Acta Leiden., 57: 197-200.

Pathak, K. M. L. and Gaur, S. N. S., 1990. Immunization of pigs with culture antigen of *Taenia solium*. Vet. Parasitol., 34: 353-356.

Pathak, K. M. L., Gaur, S. N. S. and Garg, S. K., 1984a. Counter-current immunoelectrophoresis, a new method for the rapid serodiagnosis of porcine cysticercosis. J. Helminthol., 58: 321-324.

Pathak, K. M. L., Gaur, S. N. S. and Garg, S. K., 1984b. Diagnosis of porcine cysticercosis by immunoelectrophoresis. Indian J. Parasitol., 8: 79-80.

Pathak, K. M. L., Gaur, S. N. S. and Gupta, R. S., 1983. Antigenic components of *Taenia hydatigena* cyst fluid and their relevance to diagnose to ovine cysticercosis by immunoelectrophoresis. Trop. Vet. Anim. Sci. Res., 1: 247-248.

Pathak, K. M. L., Gaur, S. N. S. and Kumar, D., 1984c. The epidemiology of strobilar and cystic phase of *Taenia solium* in certain parts of Uttar Pradesh (India). Indian J. Vet. Med., 4: 17-18.

Pathak, K. M. L., Gaur, S. N. S. and Sharma, S. N., 1982. The pathology of *Cysticercus tenuicollis* infection in goats. Vet. Parasitol., 11: 131-139.

Pathak, V. P., Raygude, D. R., Joshi, M. V., Deore U. B. and Ingole, R.S., 2004. Incidence of hydatid cysts in domestic ruminants – A postmortem study. Indian Vet. Med. J., 28: 173-175.

Patil S. B., Gatne, M. L., Palampalle, H. Y. and Gudewar, J. G. 2010. Prevalence of metacestodes in food producing animals slaughtered at Mumbai abbatoir. XX NCVP Parasitology Today-Ecology to Molecular Biology, 18-20 February, 2010, Dept. Vet. Parasitol., CVS, CCS, Haryana Agri. Univ., Hissar-125004, Abstr. p.78.

Patro, D. N., Suhani, B., Sahoo, P. K., Nanda, S. K., Pradhan, R. K., Nayak, B. C. and Tripathy, S. B., 1997. Incidence of generalized *Coenurus gaigeri* infection in a goat farm. Indian Vet. J., 74: 68-69.

Pattanayak, G. M. and Gupta P. P., 1981. Parasitic infections in pigs- note. J. Res, Punjab Agri. Univ., 18: 457-461.

Pease, H.T., 1902. Agric. Ledger, (Calcutta), Special Vet.Series No. 5: 9-15.

Pednekar, R. P., 2008. Molecular characterization of hydatidosis in food animals of Maharashtra State. M. V. Sc. Thesis submitted to Bombay Vet. Coll., Parel, Bombay. (Not indexed).

Pednekar, R. P., Gaurat, R. P. and Gatne, M. L., 2009. Prevalence, fertility rate and protein content of hydatid cysts in food animals slaughtered at Deonar abattoir. J. Bombay Vet. Coll., 17: 55-59.

Pednekar, R. P., Gatne, M. L., Thompson, R. C. A. and Traub, R. J., 2009. Molecular and morphological characterization of *Echinococcus* from food producing animals in India. Vet. Parasitol., 165: 58-65.

Pethkar, D. K. and Hiregaudar, L.S., 1972. Helminthic infection of cattle and buffaloes in Gujarat State. Gujvet, 6: 30-31.

Pillai, K. J., Rao, P. L. N. and Rao, K. S., 1986. A study on the prevalence of hydatidosis in sheep and goats at Tirupati Muncipality slaughter house. Indian J. Pub. Hlth., 30: 160-165.

Pillai, M.V., 1928. Echinococcosis hepatitis in a bullock. Indian Vet. J., 5: 279-280.

Placid E. D'Souza (D'Souza, P E., vide supra).

Plain, B., 1991. Studies on the incidence of Cysticercosis in pigs and cattle with special reference to taeniasis in human beings in Assam and Meghalaya. M. V. Sc. Thesis submitted to Assam Agri. Univ., Khanapara, Guwahati-22 (Assam).

Plain, B., Nath, N. C. and Gogoi, A. R., 1992. Protein pattern of cystic fluid of *Cysticercus cellulosae*. J. Vet. Parasitol., 6: 5-8.

Prabhakaran, P., Soman, Iyer, R. P. and Abraham, J., 1980. Common disease conditions among cattle slaughtered in Trichur Municipal slaughter house: a preliminary study. Kerala J. Vet Sci., 11: 159-163.

Prakash, A., Saikumar, G., Rout, M., Nagarajan, K. and Ram Kumar, 2007. Neurocysticercosis in free roaming pigs – a slaughterhouse survey. Trop. Anim. Hlth. Prod., 39: 391-394.

Pramanik, A. K., Bhattacharya, H. M. and Sengupta, D. N., 1985. Occurrence of *Cysticercus cellulosae* in slaughtered pigs in Calcutta and its public health significance. Indian J. Anim. Hlth. 24: 143-146.

Prasad, B. N., 1981. Prevalence of hydatidosis among pigs of Bihar and its public health importance. Haryana Vet., 20: 24-28.

Prasad, B. N. and Mandal, L. N., 1978. Hydatidosis in goats in India. Philippine J. Vet Med., 17: 191-196

Prasad, B. N. and Mandal, L. N., 1979. Incidence of hydatidosis in buffaloes in Bihar. Kerala J. Vet Sci., 10: 220-225.

Prasad, B. N. and Mandal, L. N., 1982. Hydatidosis in sheep in Bihar. Indian J. Pub. Hlth., 10: 220-225.

Prasad, B. N. and Prasad, L. N., 1980. Notes on the pathology of hydatidosis in sheep and goats. Indian Vet. Med. J., 4: 83-84.

Prasad, K. N., Chawla, S., Jain, D., Pande, C. M., Pal., L., Pradhan, S. and Gupta, R. K., 2002. Human and porcine *Taenia solium* infection in rural Northern India. Trans. Roy. Soc. Trop. Med. Hyg., 96: 515-516.

Prasad, K.N., Chawla, S., Prasad, A., Tripathi, M., Husain, N. and Gupta, R. K., 2006. Clinical signs for identification of neurocysticercosis in swine naturally infected with *Taenia solium*. Parasitol. International, 55: 51-54.

Prasad, M. C. and Srivastava, S. N., 1967. Gid in a bullock. Indian Vet. J., 44: 981-982.

Prasanna, K. M., Jagannath, M. S. and D'Souza, P. E., 2001a. Detection of *Taenia solium* cysticercosis in naturally infected pigs by latex agglutination test. Twelfth Natl. Cong. Vet. Parasitol., August 26-27, 2001, Dept. Parasitol., Coll. Vet. Sci., Acharya Ranga Agri. Univ. Tirupati-517 502, India. Abstr. S-3: 22, p.110.

Prasanna, K. M., Jagannath, M. S. and D'Souza, P. E., 2001b. Diagnosis of *Taenia solium* cysticercosis in pigs by latex agglutination test. J. Vet. Parasitol., 15: 47-49.

Prasanna, K. M., Jagannath, M. S. and D'Souza, P. E., 2001c. Evaluation of dot-ELISA in the detection of *Taenia solium* cysticercosis in pigs. Indian J. Anim. Sci., 71: 235-236.

Prasanna, K., Gangadhara Rao, Y. V. B., Subbarao and Rao, B. V., 1992. Sharing of antigen between hydatid and *Cysticercus tenuicollis*. Indian J. Anim. Sci., 62: 942-943.

Pratibha Jumde, Anita Katikyayi, Praniti Pantavne, Dudhe Nitin, Sanveer Khatun and Maske, D. K., 2010. Cysticercosis in goats: An abattoir study. XX NCVP Parasitology Today-Ecology to Molecular Biology, 18-20 February, 2010, Vet. Parasitol., CVS, CCS Haryana Agri. Univ., Hissar-125004, p.81.

Preet, S. and Prakash, S., 2000a. Experimental cysticercosis and intrahepatic venous systems in *Rattus norvegicus*. Parasitol. Res., 86: 509-513.

Preet, S. and Prakash, S., 2000b. Cysticercosis in rat infected with *Cysticercus fasciolaris* (Rud.): The possible role of plasma and cysts anions in diagnosis. J. Parasit. Dis., 24: 135-139.

Pujatti, D., 1950. *Cysticercus fasciolaris*, Rudolphi (1808) in Muridi del Sud-India (Cestoda). J. Doriana [supplement to Annali del. Museo Civico di Storia Naturale "G-Doria"],Vol.1 (8): 4.

Puttalakshmamma, G. C., Dilip Kumar, D. and Shiva Prakash, B. V., 2011. *Coenurus gaigeri* in subcutaneous tissue of sheep: A report. XII Natl. Cong. Vet. Parasitol., 5-7th January, 2011, IAa AAP and Dept. Vet. Parasitol., & Bombay Vet. Coll., Mumbai, S1-015, p.30.

Puvarajan, B., Anna, T., Umarani, R. and Muruganandan, B., 2013. Occurrence of coenurosis in sheep in Tamilnadu. Indian Vet. J., 90: 83-85.

Pythal, C., 1974. Studies on cestodes of ruminants. Thesis (M.Sc.) submitted to Univ. Kerala, Trivandrum, Kerala.

Qazi, M., Shah, K. A., Anjum, A., Qazi, N. and Muyeen, A. C., 2011. Hydatid cyst of muscle in goat - A rare site. Vet Scan, 6 (2): Article 97.

Radhakrishna Reddy, K. and Gouher Ali Khan, K., 1970. Hydatidosis in a great Indian squirrel (*Ratafa indica maxima*). Indian Vet. J., 47: 1024.

Rahimuddin, M., 1935. Gid in sheep. Indian Vet. J., 11: 280.

Rahimuddin, M., 1941. Coenurosis. Indian Vet. J., 17: 229-303.

Rai, M. K. and Anzari, M. Z., 1988. Prevalence of helminth parasites of yak (*Bos poephagus*) in Sikkim and histopathological studies of liver, lung and caecum infected with helminth parasites. 2nd Natl. Cong. Vet. Parasitol., Bangalore, India, Stream I, Abstr. p.6.

Rai, M. K., 1988. Studies on helminth parasites of yak (*Bos poephagus*) in Sikkim. M.V.Sc. Thesis submitted to Birsa Agri. Univ., Ranchi, Abstr., J. Vet. Parasitol., 2: 77-78.

Raina, O. K. and Singh, B. P., 1995. Antigenic characterization of buffalo hydatid cystic fluid. SIII 8, Seventh Natl. Cong. Vet. Parasitol., 19-21 August, 1995. Dept. Parasitol., Madras Vet. Coll., Madras -7, p.84.

Raina, O. K. and Singh, B. P., 1995. Antigenic characterization of buffalo hydatid cystic fluid. J. Vet. Parasitol., 11: 185-187.

Rajasekar, M., Jagannath, M. S. and Sastry, K. N.V., 1999. Diagnosis of hydatidosis in cattle and buffaloes by Avidin-Biotin Enzyme-linked immunosorbent Assay. J. Vet. Parasitol., 13: 19-22.

Rajendra Singh Avapal, Sharma, J. K. and Juyal, P. D., 2003. Occurrence of *Cysticercus cellulosae* in pigs in and around Ludhiana city. J. Vet. Parasitol., 17: 69-70.

Rajinish Sharma (vide Sharma, R., *infra*).

Rajkhowa, C. and Bandopaydhyay, S., 2001. Prevalence of larval cestode of zoonotic importance in Meghalaya. Twelfth Natl. Cong. Vet. Parasitol., 26-27 August, 2001, Dept. Parasitol., Coll. Vet. Sci., Acharya Ranga Agri. Univ., Tirupati-517 502, India, Abstr.S-2: 39, p.67.

Rama, O. K. and Singh, B. P., 1997. Antigenic characteristic of buffalo hydatid cystic fluid. J. Vet. Parasitol., 11: 185-187.

Rama, U. V.S., Sehgalk, S., Bhatia, R. and Bharadwaj, N., 1986. Hydatidosis in animals in and around Delhi. J. Com. Dis., 18: 116-119.

Ramamurthy, R., 1964. Dissertation, Univ. Madras, Madras.

Raman, M. and Lalitha John, 2003. Prevalence of hydatidosis in sheep and goats in Chennai, India. Indian J. Anim. Res., 37: 57-58.

Raman, M. and Rajvelu, G., 2000. Studies on metacestodes of sheep and goats in Chennai. Natl. Cong. Vet. Parasitol., Bhubaneswar, 4-6 February, 2000, Abstr., XI, p.78.

Ramanujachari, G. and Alwar, V. S., 1954. A check-list of parasites (Classes Trematoda, Cestoda and Nematoda) in the Department of Parasitology, Madras Veterinary College (additions since 1947). Indian Vet. J., 31: 46-56

Ramanujachari, G. and Alwar, V. S., 1955. A check-list of parasites in the Department of Parasitology, Madras Veterinary College. Indian Vet. J., 32: 47.

Rana, U. V. S., Sehgal, S., Bhatia, R. and Bhardwaj, M., 1986. Hydatidosis in animals in and around Delhi. J. Com. Dis., 18: 116-119.

Ranga Rao, G. S. and Sharma, R. L., 1995. Cyclophyllid cestodiasis in camels (*Camel dromedaries*) in India. Indian J. Anim. Sci., 65: 261-265.

Ranga Rao, G. S., Sharma, R. L.and Hemaprasanth, 1994. Parasitic infections of Indian yak, *Bos (poephagus) grunniens* - an overview. Vet. Parasitol., 53: 75-82.

Rao, A. N., 1989. M. V. Sc. Thesis. Pathology Division, Indian Vet. Res. Inst., Izatnagar, Bareilly (Uttar Pradesh), India.

Rao, A. T. and Acharyyo, L. N., 1984. Diagnosis and classification of common diseases of captive animals at Nandankanan Zoo. Indian J. Anim. Hlth., 23: 147-152.

Rao, A. T., Misra, S. C. and Acharyyo, L. N., 1972. Pulmonary hydatidosis in captive animals at Nandankanan Zoo. Indian Vet. J., 49: 842-843.

Rao, B. T. (vide Bhaskara Rao, T.).

Rao, B. V., 1958. Studies on helminth parasites of carnivorous mammals. M. Sc. Thesis, Faculty of Vet. Sci., Univ. Madras, Madras.

Rao, B. V., 1968. Experimental transmission of *Echinococcus* of buffalo origin to foxes (*Vulpes bengalensis*). Vet. Rec., 83: 56-57.

Rao, B. V., 1970. A study of hydatid cysts occurring in buffaloes in India.Ceylon Vet. J., 27: 31-35.

Rao, B. V., 1971. Experimental infection of *Cysticercus fasciolaris* in laboratory animals. Ann. Parasit. Hum. Comp., 46: 11-14.

Rao, B. V. and Anantaraman, M., 1966. Observations on the biology and development of some taeniid cestodes. Indian J. Helminth., 18: 161-171.

Rao, B. V. and Mittal, K. R., 1973. Studies on hydatidosis in Indian buffaloes (*Bubalus bubalis*) 1. Some serological observations in natural infections. Z. Tropenmed. Parasitol., 24: 476-480.

Rao, B. V. and Parihar, N. S., 1973. Cerebral coenuriasis in a buffalo (*Bubalus bubalis*). Indian Vet. J., 50: 318-319.

Rao, D. G. and Kohli, S., 1976. Incidence of hydatid in bovines and histopathological changes of pulmonary tissues in hydatidosis. Indian J. Anim. Sci., 44: 437-440.

Rao, D. G. and Mohiyuddin, S., 1976. Incidence of hydatid cyst in bovines and histopathological changes of pulmonary tissue in hydatidosis. Indian J. Anim. Sci., 44: 437-440.

Rao, J. R. and Singh, B. P., 1994. An enzyme-linked immunosorbent assay for detection of experimentally induced secondary echinococcosis in BALB/C mice. J. Vet. Parasitol., 8: 9-13.

Rao, M. A. N., 1935. A note on *Taenia solium* Linneus, 1758 in the Madras Presidency. Indian J. Vet. Sci., 5: 28-29.

Rao, S. R., Bhatavdekar, M. Y. and Detha, K. T., 1957. Morphology and development of *Coenurus gaigeri*, Hall in a ewe with particular reference to the taxonomy of the genus *Multiceps*. Bombay Vet. Coll. Mag., 6: 12-18.

Rashid, A., Dar, L. M., Darzi, H. M., Kamil, S. A. and Khan, H. M., 2011. Prevalence of *Cysticercus tenuicollis* infections in slaughtered sheep of Kashmir region. XII Natl. Cong. Vet. Parasitol., 5th to 7th January, 2011, IAAVP and Bombay Vet. Coll., Mumbai, S1-P15, p.55.

Rashid, M, Kotwal, S. K. and Malik, M. A., 2005. Incidence of echinococcosis in Bakerwali sheep and goats, treatment and control. Intas Polivet, 6: 284-285.

Ratnam, S., 1975. Thesis, Univ. Calcutta, Calcutta. (Not indexed).

Ratnam, S. and Khanna, P. N., 1988. Incidence of *Cysticerus bovis* in slaughtered animals and its public health importance. Indian J. Comp. Microbiol. Immunol. infect. Dis. 9: 65-67.

Ravi Raidurg and Thimma Reddy, P. M., 2003. Parasitic cyst of *Coenurus gaigeri* in the lower eye lid of a kid. Intas Polivet,10 (2): 302-303.

Ray, D. K., Negi, S. K. and Srivastava, P. S., 1977. Note on the occurrence of metacestodiasis in goats in Tarai of Uttar Pradesh. Pantnagar J. Res., 2: 242.

Ray, S.W., Banerjee, T. P. and Banerjee, K. N., 1972. Extensive *Cysticercus cellulosae* infection in a pig. Indian J. Anim. Hlth., 11: 237.

Reddy, C. R. R. M., Narasaiah, I. L., Parvathi, G. and Somasundara Rao, M., 1968. Epidemiology of hydatid disease in Kurnool. Indian J. Med. Res., 56: 1105-1120.

Reddy, C. R., Suvarnakumari, G. and Reddy, M. R., 1970. Fertility rate of hydatid cysts in animals from Kurnool. Indian J. Med. Sci., 24: 357-360.

Reddy, K. R. and Ali Khan, K.G., 1970.(vide Radhakrishna Reddy supra).

Reddy, P. R., Hafeez, Md., Kumar, E. G. T. V. and Hasina, S., 1993. Prevalence of hydatidosis in food animals in Andhra Pradesh. Indian J. Anim. Sci., 63: 631-632.

Reddy, Y. A. and Rao, J. R., 1998. Restriction endonuclease analysis of genomic DNA of bovine and bubaline isolates of *Echinococcus granulosus*. J. Vet. Parasitol., 12: 36-39.

Reddy, Y. A., Rao, J. R., Butchaiah, G. and Sharma, B., 1998. Random amplified polymorphic DNA for the specific detection of bubaline *Echinococcus granulosus* by hybridization assay. Vet. Parasitol., 79: 315-323.

Rout, M. and Saikumar, G., 2012. Porcine cysticercosis: An underestimated zoonotic disease. Indian J. Vet. Pathol., 36 (1): 94-96.

Roy, B. and Tandon, V., 1989. Metacestodiasis in North East India: A study on the prevalenvce of hydatidosis and cysticercosis in Mizoram, Nagaland and Assam. Indian J. Anim. Hlth., 28: 5-10.

Roy, B. and Tandon, V., 1991. Prevalence of hydatidosis in the hilly mountainous region of North- East India: A Study of domestic intermediate hosts in Meghalaya. *Proc. Semr. Zoonoses*, 9-10 November, 1989. Reg. Med. Res. Centr. (ICMR), Dibrugarh, Assam, pp.83-91.

Roy, B., Naveen Kumar, Aethal, H. P., Gupta, S. C. and Mogha, I. V. 2003. Occurrence of unusually large cyst in subcutaneous tissues of sheep. J. Vet. Parasitol., 17: 73-74.

Rupinderjit Singh Avapal, Sharma, J. K. and Juyal, P. D., 2003. Occurrence of *Cysticercus cellulosae* in pigs in and around Ludhiana City. J. Vet. Parasitol., 17: 69-70.

Sachdeva, Y. and Talwar, J. R., 1960. Hydatid cyst of lung. Dis.Chest, 38: 638.

Sadana, J. R. and Kalra, D. S., 1973. Two cases of generalized cysticercosis in pigs with special reference to the lesion in lungs. Haryana Vet., 12: 39-40.

Sadarnashipur, P. O. and Lalgosa, P., 1991. Prevalence of coenurus of *Multiceps gaigeri* in goat and its predilection site distribution. Indian J. Comp Microbiol. Immunol. infect. Dis., 12: 71-72.

Saha, B., Batabyal, S. and Chattopadhyay, S. P., 2011. Purification and immunobochemical characterization of sheep hydatid cyst fluid antigen. Biotechnol. Bioinf. Bioeng., 1: 265-268.

Sahasrabudhe, V. K., Shaw, H. L., Dubey, J. P. and Shrivastava, H. O. P., 1969. Helminth parasites of dogs in Madya Pradesh and their public significance. Indian J. Med. Res., 57: 56-59.

Saikia, J., Pathak, S. C. and Barman, A. K., 1987. Coenurosis in goats. Indian Vet. Med. J., 11: 135-141.

Sajjan, S. A., Paturkar, A. M., Katkar, S. M. and Bhadrige, V.V., 2011.Characterisation of porcine hydatid fluid (HF) antigens by sodium dodecyl sulphate-polyacrylamide gel electrophoresis (SDS-PAGE). XII Natl. Cong. Vet. Parasitol.,

5-7ᵗʰ January, 2011, IAAAP and Dept. Vet. Parasitol., Bombay Vet. Coll., Mumbai, S3-P5, p.117.

Saleque, A., 1978. On experimental taeniasis-cysticercosis due to *Taenia solium* and *Taenia saginata*. M. V. Sc. Thesis submitted to Assam Agri. Univ., Guwahati, Assam. (not indexed).

Saleque, A., Chowdhury, N., Iyer, P. K. R. and Baruah, G., 1987. Induced *Taenia solium* cysticercosis in rhesus monkeys (*Macaca mulata*): a clinico-pathological study. Ann. Trop. Med. Parasitol., 82: 103-105.

Samanta, S., 2000. Immunological studies on protoscoleces antigen of *Echinococcus granulosus*. Ph. D. Thesis, Deemed Univ., Indian Vet. Res. Inst., Izatnagar-243122, India, Abstr. J. Vet. Parasitol., 14: 93.

Samanta, S., Singh, B. P. and Raina O. K., 2009. Diagnosis of hydatidosis in buffaloes using protoscoleces antigens. Indian Vet. J., 86: 850-851.

Samanta, S., Singh, B. P., Gupta, S. C. and Ghorui, S. K., 2001. Characterisation of protoscoleces antigen of *Echinococcus granulosus* hydatid cysts from buffalo, sheep and goat by double immune diffusion (DID). Twelfth Natl. Cong. Vet. Parasitol., 26-27 August, 2001, Dept. Parasitol., Coll. Vet. Sci., Acharya Ranga Agri. Univ., Tirupati-517 502, India, Abstr.S-4: 32, p.116.

Samanta, S., Singh, B. P., Ghorui, S. K., Gupta, S. C. and Chandra, D., 2003. Characterisation of protoscoleces antigen of hydatid cysts of buffalo, sheep and goat by double diffusion. J. Vet. Parasitol., 17: 101-103.

Samanta, S., Singh, B. P., Ghorui, S. K., Raina O. K. Gupta, S. C. and Tewari, A. K., 2008. Protective efficacy of affinity purified protoscolices antigens of *Echinococcus granulosus* against secondary hydatidosis in mice. 20ᵗʰ Natl. Cong. Parasitol., 3-5 November, 2008, Dept. Parasitol. and Indian Soc. Parasitol., North Eastern Hill Univ., Shillong-793 022, Abstr. P-95, p.176.

Samanta, S., Singh, B. P., Saikumar, G. and Raina, O. K., 2008. Secondary hydatidosis in swiss albino mice following infection with protoscolices of *Echinococcus granulosus* of buffalo origin. Indian J. Anim. Sci., 78: 373-375.

Sambamurthy, B., 1956. Cardiac hydatidiasis in the bovines. Indian Vet. J., 32: 347-350.

Sami, M. A. and Khan, D. G., 1938. Hydatid disease in Punjab. Indian Med. Gaz., 73(2): 90-94.

Sangaran, A., 1994. Immunodiagnosis of hydatidosis in some food animals and human beings. M. V. Sc. Thesis submitted to Tamilnadu Vet. Anim. Sci. Univ., Chennai, India.

Sangaran, A. and Lalitha John, 2009a. Prevalence of hydatidosis in sheep and goats in and around Chennai. Tamilnadu. J. Vet. Anim. Sci., 5: 208-210.

Sangaran, A. and Lalitha John, 2009b. Incidence of cystic echinococcosis in cattle. Indian J. Field Vet., 5: 65.

Sangaran, A. and Lalitha John, 2010. Incidence of cystic echinococcosis in buffaloes slaughtered at Chennai. J. Vet. Parasitol., 24: 93-94.

Sangaran, A. and Lalitha John, 2012. Antigen based detection of cystic echinococcosis in goats using enzyme linked immunosorbent assay (ELISA) and dot-enzyme immunoassay (Dot-EIA). XXII Natl. Cong. Vet. Parasitol., 15-17 March, 2012, Coll. of Vet. Sci. and Anim. Husb., U. P. Pandit Deen Dayal Upadhyaya Pashu Chikitsa Vigyan Vishwavidyalaya Evam Go Anusandhan Sansthan (DUVASU), Mathura-281001(UP.), India, Abstr. No.III.10, p 38.

Sangaran, A., Arunkumar, S. and Lalitha John, 2011. Incidence of hydatidosis in buffaloes slaughtered at Chennai. Tamil Nadu J. Vet. Anim. Sci., 7: 105-106.

Sangaran, A., Balasundaram, S., Jayakumar, R and Anandan, R., 1995. Incidence of hydatidosis in sheep and buffaloes at Madras. Seventh Natl. Cong. Vet. Parasitol., 19-21 August, 1995, Dept. Parasitol., Madras Vet. Coll., Madras -7, Abstr.SI 22, p.84.

Sanjiv Kumar, Verma, S. K.and Pawanjit Singh, 2010. Multisystem involvement of hydatid cyst. Vet. Practitioner, 11 (1): 7-8

Sanyal, P. K., 2011. Determination of immune reactive protein in cystic membrane of *Cysticercus tenuicollis* of goat origin. Indian J. Anim. Sci., 81:

Sanyal, P. K. and Sinha, P. K., 1983a. Caprine metacestodiasis: Incidence in West Bengal. Haryana Vet., 22: 38-40.

Sanyal, P. K. and Sinha, P. K., 1983b. A note on the prevalence of hydatidosis in cattle and buffaloes in West Bengal. Haryana Vet., 22: 47-50.

Sanyal, P. K., Nath, S., Pal, S. and Mandal, S. C., 2011. Determination of immune reactive proteins in cystic membrane of *Cysticercus tenuicollis* of goat origin. Indian J. Anim.Sci., 81:

Sarkar, A. and Mitra, D., 2008. Surgical management of a subcutanaeous coenurus cyst in a kid. Indian Vet. J. 85: 1009- 1010.

Sarkar, S., Babu, S., Dasgupta, C. K. and Dasgupta, R., 2001. *In vitro* efficacy of some germicides on protoscolices of buffalo pulmonary hydatid cysts. Twelfth Natl. Cong. Vet. Parasitol., 26-27 August, 2001, Dept. Parasitol., Coll. Vet. Sci., Acharya Ranga Agri. Univ., Tirupati-517 502, India, Abstr. S-5: 33, p.143.

Sarma, M. D. *et al.,* (Dev Sarma, M. *et al., 2000 & 2002, vide supra,)

Sasmal, N. K., Sarkar, A. and Laha, R., 2008. Transmission dynamics of pig cysticercosis and taeniasis in highly endemic tribal communities. Environ. Ecol., 26 (1): 76-80.(vide internet, google).

Satendra Kumar, 2009.Incidence and histopathological study of porcine cysticercosis in Patna and its surrounding area. M. V. Sc. Thesis submitted to Bihar Vet. Coll., Patna, Bihar, India. (Not indexed).

Sathasivam, S., Khan, P. N., Senthil Kumar, K. and Perrumal, K. P. M., 2009. Hydatidosis in a Jaugar (*Panthero onca*). Zoos' Print, 24: 16.

Sathianesan, V., 1996. Helminthiasis in poultry. In: Recent Trends in Diagnosis and Control of Parasitic Diseases. Natl. Trng. Prog. for Teachers/ Scientists. 25th-30th March, 1996. Centre of Advanced Studies, Dept. Parasitol., Vet. Coll. Univ. Agri. Sci., Bangalore-560 024. pp. 28-40. (Review, not indexed).

Satyaprakash, Ghose, J. N. and Varmani, B. M. L., 1967. *Echinococcus granulosus* infection in man and animals in and around Delhi with a brief review of work done in India. Bull. Indian Soc. Mal. Com. Dis., 4: 174-182.

Saxena, A. K. and Bhargava, A. K., 1973. A case of gid-false twirling in a goat. Indian Vet. J. 50: 952.

Selvam, P., D'Souza, P. E. and Jagannath, M. S., 2002. Detection of circulating antigens of the bladderworm of *Taenia solium* in pigs by co-agglutination test. Indian Vet. J., 79: 1012-1016.

Selvam, P., D'Souza, P. E. and Jagannath, M. S., 2004. Serodiagnosis of *Taenia solium* cysticercosis in pigs by indirect haemagglutination test. Vet. archiv., 74: 453-458.

Shahnawaz, M, Shahardar, R.A. and Wani, Z. A., 2011. Seasonal prevalence of ovine platyhelminth parasite infections in Ganderbal area of Kashmir Valley. XII Natl. Cong. Vet. Parasitol., 5th to 7th January, 2011, IAAVP and Bombay Vet. Coll., Mumbai, S1-P5, p.49.

Sharma, B., Baissya, M. and Dutta, B., 2001. Surgical removal of parasitic cyst following craniotomy in goat. Intas Polivet, 2: 46-47.

Sharma, G. C., 1977. Studies on incidence of human taenisis and animal cysticercosis in Greater Guwahati area of Assam. M. V. P. H. Thesis, Univ. Calcutta, West Bengal.

Sharma Deorani, V. P., 1967a. Proc. 54th Indian Sci. Cong. (Hyderabad), Part III, Abstr, p. 523.

Sharma Deorani, V. P., 1967b. Histopathological studies on hepatitis cysticercosa in sheep and a deer. Indian Vet. J., 44: 939-942.

Sharma, D. K. and Chauhan, P. P. S., 2006. Coenurus status in Afro-Asian region: A review. Sml. Rum. Res., 64: 197-202.(Review, not indexed; *Coenurus cerebralis* has been observed in sheep, goat, cattle, buffalo, camel and yak including some wild animals and dog is the DH of *Taenia multiceps*).

Sharma, D. K., Sanil, N. K., Agnihotri, M. K. and Nem Singh, 1995. Subcutanaeous coenurus in a Barbari goat. Indian Vet. J., 72: 1203-1205.

Sharma, D. K., Singh, N. and Tiwari, H. A., 1998. Prevalence and pathology of *Coenurus* in organized goat farms. J. Vet. Parasitol., 12: 30-32.

Sharma, G. C., 1977. Studies on incidence of human taeniasis and animal cysticercosis in Greater Guwahati area of Assam. M. V. P. H. Thesis, Univ. of Calcutta, West Bengal. (Not Indexed).

Sharma, H. K., Vohra, S., Yadav, A., Katoch R., Sood, S. and Kumar, J., 2008. Prevalence of *Cysticercus tenuicollis* in sheep of Jammu region. Vet. Practitioner, 9: 67-68.

Sharma, H. N., 1965. *Coenurus cerebralis* in goats. Indian Vet. J., 52: 137-139.

Sharma, H. N. and Tyagi, R. P. S., 1965. Diagnosis and surgical treatment of coenurosis in goat (*Capra hircus*). Indian Vet. J., 52: 482-488.

Sharma, H. N. and Tyagi, R. P. S., 1974. Technique of cerebral angioplasty in goats. Indian J. Anim. Hlth., 13: 177-178.

Sharma, H. N. and Tyagi, R. P. S., 1975. Diagnosis and surgical treatment of coenuriasis in goat (*Capra hircus*). Indian Vet. J., 52: 482-488.

Sharma, H. N., Bhargava, A. K. and Tyagi, R. P. S., 1974. Cerebral angioplasty for locating the space occupying lesions in goats. Indian Vet J., 51: 718-721.

Sharma, R. 2001. Prevalence and Immunodiagnosis of *Cysticerus cellulosae* in Swine and Its Public Health Significance. M. V. Sc. Thesis submitted to Punjab Agri. Univ., Ludhiana.

Sharma, R., Sharma, D. K. and P. D. Juyal., 2004a. Seroprevalence of swine cysticercosis in Ludhiana, Punjab. J. Vet. Parasitol., 18: 75-76.

Sharma, R., Sharma, D. K., Juyal, P. D. and Sharma, J. K., 2004b. Epidemiology of *Taenia solium* cystcercosis in pig of northern Punjab, India. J. Parasit. Dis., 28: 124-126.

Sharma, R., Sharma, D. K., Juyal, P. D., Asha Rani and Sharma, J. K., 2005a. Seroprevalence of swine cystcercosis in Ludhiana. J. Vet. Parasitol., 19: 143-145.

Sharma, R., Sharma, D. K., Juyal, P. D., Asha Rani and Sharma, J. K., 2005b. Comparative evaluation of counter immunoelectrophoresis and routine meat inspection for the diagnosis of swine cysticercosis. Indian J. Anim. Sci., 75: 792-793.

Sharma, S., 2010. A report of cystic echinococcosis in a murrah buffalo. Indian Vet. J., 87: 291-292.

Sharma, T. D. and Chitkara, N. L., 1963. Hydatid disease in Amristar- A study of potential human risk. Indian J. Med. Res., 51: 1015-1018.

Shastri, U. V., 1966. Helminth Parasites of Domestic Pig (*Sus scrofa domestica*) in Bombay (Maharashtra State). M.V.Sc. Thesis submitted to Univ. of Bombay, Bombay.

Shastri, U. V., Gafoor, M. A. and Gaffar, M. A., 1985. A note on a massive natural infection on *Coenurus gaigeri* cysts in a goat. Indian Vet. J., 62: 615-616.

Shinde, M. V., 1992. Immunodiagnosis of *Cysticercus cellulosae* in pigs together with observations on its incidence. Bombay Vet. Coll., Parel, Bombay. (not indexed).

Shinde, M. V., Gatne, M. L. and Narsapur, V. S., 1993a. Evaluation of IHA test for serodiagnosis of porcine cysticercosis. Proc. 5th Natl. Conf. Vet. Parasitol., Udgir, Maharashtra, India, April 21-23, 1993. Abstr.

Shinde, M. V., Gatne, M. L. and Narsapur, V. S., 1993b. Antigenic characterization of *Cysticercus cellulosae*. Proc. 5th Natl. Conf. Vet. Parasitol., Udgir, Maharashtra, India, April 21-23, 1993. Abstr.

Shinde, M. V., Narsapur, V. S. and Gatne, M. L., 1991. Evaluation of indirect haemagglutination in *Taenia solium* cysticercosis of pigs. The Blue Cross Book, 3: 4-7.

Shivaprakash, B. V. and Thimma Reddy, P. M., 2009. An outbreak of multiple subcutaneous coenurus cysts in goats. J. Vet. Parasitol., 23: 199-200.

Shivasharanappa, N., Gururaj, K., Sharma, D. K., Reddy G. B. Manjunatha, V., 2011. Acute hepatic and pneumonic cysticercosis in Barbari Goats. Indian J. Vet. Pathol., 35: 80-81.

Shukla, D. E. and Victor, D. A., 1974. Incidence of sarcosporidial infection in bovines carcasses in Madras slaughter houses. Gujrat Vet., 7: 84-87.

Shukla, N. Hussain, N., Agrawal, G. G. and Hussain, M., 2008a. Utility of *Cysticercus fasciolaris* antigen in DOT ELISA for the diagnosis of neurocysticercosis. Indian J. Med. Sci., 62: 222-227.

Shukla, N., Hussain, N., Jyotsna, Gupta, S., Hussain, M. and Victor, D. A., 2008b. Comparison between scolex and membrane antigen of *Cysticercus fasciolaris* and *Cysticercus cellulosae* larvae in neurocysticercosis. J. Microbiol. Immunobiol. Infect. Dis., 41: 519-524.

Singh, B. B. and Rao, B.V., 1966. Some biological studies on *Taenia taeniaeformis*. Indian J. Helminth., 18: 151-160.

Singh, B. B. and Rao, B.V., 1967a. On the development of *Cysticercus fasciolaris* in albino rats liver and its reaction on the host tissue. Ceylon Vet. J., 15: 121-129.

Singh, B. B. and Rao, B.V., 1967b. Chemotherapeutic trials with numural and yomesan in experimental *Taenia taeniformis* infection in cats. Indian Vet. J., 44: 208-212.

Singh, B. B. and Rao, B.V., 1968. The effect of cortisone in the development of *Cysticercus fasciolaris* infection in laboratory animals. Indian J. Anim. Hlth., 7: 265-270.

Singh, B. B., Sharma, R., Sharma, J. K. and Juyal. P. D., 2010. Parasitic zoonoses in India: an overview. Rev sci. tech. Off. Int. Epiz., 29: 629-637. (Review, not indexed; refers to cysticercosis and hydatidosis).

Singh, B. K., 1970. Larval cestodes from domestic pigs (*Sus scrofa*) in Bihar, India. Ceylon Vet, J., 18: 86-87.

Singh, B. P., 1993. Transmission studies of *Echinococcus* from buffalo-dog origin to calves and piglets. J. Vet. Parasitol., 7: 1-14.

Singh, B. P., 1994. Isolation, purification and characterization of hydatid antigen. J. Vet. Parasitol., 8: 107.

Singh, B. P. and Dhar, D. N., 1987. *In vitro* development of worms in buffalo-dog cycle. J. Vet. Parasitol., 1: 59-61.

Singh, B. P. and Dhar, D. N., 1988a. Indirect fluorescent antibody test for the detection of antibodies of *Echinococcus granulosus* in experimentally infected pups. Vet. Parasitol., 28: 185-190.

Singh, B. P. and Dhar, D. N., 1988b. *Echinococcus granulosus* in animals in northern India. Vet. Parasitol., 28: 261-266.

Singh, B. P. and Dhar, D. N., 1988c. Effect of gamma irradiation on the protoscolices of *Echinococcus granulosus* of sheep origin. Int. J. Parasitol., 18: 557-560.

Singh, B. P. and Pal, S., 1994. Immunodiagnosis of hydatid disease in goats using western blot. Sixth Natl. Cong. Vet. Parasitol, 22-24 October, 1994, Dept. Parasitol., Coll. Vet. Sci., JNKVV, Jabalpur-482 001, M. P., India, Abstr., S-3: 24, p. 40.

Singh, B. P., Joshi, P. and Singh, L. N., 1992a. Electrophoretic and immunological differentiation of Indian buffalo and sheep hydatidosis. Indian J. Parasitol., 16: 105-109.

Singh, B. P., Sharma Deorani, V. P. and Srivastava, V. K., 1987. Hydatidosis in lambs and kids with *Echinococcus granulosus* of buffalo-dog origin. Indian J. comp. Microbiol. Immunol. infect. Dis., 8: 142-144.

Singh, B. P., Srivastava, V. K. and Chattopadhyay, 1992b. Histopathological studies on porcine hydatidosis. J. Vet. Parasitol., 6: 41-45.

Singh, B. P., Srivastava, V. K. and Sharma Deorani, V. P., 1988a. Pig hydatidosis in Uttar Pradesh. Vet. Rec., 123: 299-300.

Singh, B. P., Ramkumar, Mukherjee, S. C. and Dhar, D. N., 1988b. Host reaction and pathology of hydatidosis in animals experimentally infected with *Echinococcus granulosus*. J. Vet. Parasitol., 2: 101-104.

Singh, D., Gulyani, R. and Bhasin, V., 1992. *Cysticercus pisiformis* in commercial rabbits. Indian Vet. Med. J., 16: 71.

Singh, G., Juyal, P. D., Rani, A. and Sharma, R., 2005. *Taenia solium* cysticercosis in the Indian subcontinent: problems, priorities and prospects for future study. In: Taeniasis/cysticercosis and Echinococosis. Internatl. Symp. with Focus on Asia and Pacific, 5-8 th July, 2005, Asahikawa, Japan, pp. 30. (Review, not indexed).

Singh, G. S., Prabhakar, S., Ito, A., Cho, S.Y. and Qiu, D., 2002. *Taenia solium* taeniasis and cysticercosis in Asia. In: *Taenia solium* cysticercosis: from basic to clinical science. CABI, Oxford, pp.111-128.

Singh, K. P. and Singh, C. M., 1970. Parasitic infections of bovine heart in Uttar Pradesh. Indian Vet. J., 47: 1023.

Singh, K. P. and Singh, S. P., 1972. Occurrence of *Multiceps* cyst in the lymph node of goat. Indian Vet. J., 49: 1156-1157.

Singh, P. N. and Kuppuswamy, P. B., 1969. Pathology of hydatidosis in caprine line. Indian J. Anim. Hlth., 8: 191-193.

Singh, M., *et al.*, 1999. (Vide Mohinder Singh).

Singh, S. and Banerjee, D. P., 1997. Role of wildlife in parasitic disease of man and animals. Zoos'Print,12 (4): 14-16.(review, not indexed).

Singh, S. P., Paliwal, O. P. and Singh, K. P., 1968. Occurrence of larval *Multiceps gaigeri* over the heart of goat. Indian J. Anim. Hlth., 10: 241-242.

Singh, S. P., Sharma Deorani, V. P. and Srivastava, V. K., 1986. Exceptionally high hydatid infection in liver of an Indian buffalo and its epidemiological significance. Seventh Nat. Cong. Parasitol., Ravi Shankar Univ., Raipur, 26-28 December, 1986, Abstr. B-6, p. 62.

Singh, S. P., Sharma Deorani, V. P. and Srivastava, V. K., 1988c. Prevalence of hydatid in buffaloes in India and report of a severe liver infection. J. Helminthol., 62: 124-126.

Singh, S. P., Sharma Deorani, V. P., Srivastava, V. K. and Ramaswami, K., 1987. Secondary hydatidosis with unusual brood capsule in buffalo. Indian J. Parasitol., 11: 23-24.

Singla, L. D., Aulakh, G. S., Sharma, R., Juyal, P. D. and Singh, J., 2009. Concurrent infection of *Taenia taeniformis* and *Isospora felis* in a stray kitten: a case report. Veterinari Medicina, 54: 81-83.

Singla, L. D., Neena Singla, Parshad, V. R., Sandhu, B. and Singh, J., 2003. Occurrence of pathomorphological observations on *Cysticercus fasciolaris* in lesser bandiccot rats in India. In: Rats, Mice and Man. Singleton, G. R., Hinds, L.A., Kreb, C.J., Spratt, D. M. (Eds). Rodent Biology and Management, Australian Centr. Internatl. Agri. Res. (ACIAR), Canberra, pp.57-59.

Singla, L. D., Neena Singla, Parshad, V. R., Juyal, P. D. and Sood, N. K., 2008. Rodents as reservoirs of parasites in India. Integr. Zool., 3: 21-26.

Sinha, B. K., 1966. Helminths and protozoan parasites of domestic pigs in Bihar. Bihar Anim. Husb. Bull., 9: 33-40.

Sinha, B. K., 1970. Larval cestodes from domestic pigs (*Sus scrofa domestica*). Ceylon Vet. J., 18: 86-87.

Sinha, B. P., Varma, R. R. and Ray, S. K., 1977. Multiple hydatids in the heart of an Indian buffalo (*Bubalis bubalis*). Trop. Anim. Hlth. Prod., 9: 18.

Sivakumar, V., Pradeep, M., Vijaya, N. and Valsala, K.V., 2003. *Cysticercus fasciolaris* in laboratory rat. J. Vet. Parasitol., 17: 75-76.

Somvanshi, R., Ranga Rao, G. S. C., Laha, R and Bhattacharya, D., 1994. Pathoanatomical changes associated with spontaneous *Cysticercus fasciolaris* infected wild rats. Indian J. comp. Microbiol. Immunol. infect. Dis., 15: 58-60.

Soundararajan, C., Sivakumar, T. and Palanidorai, R., 2005. Occurrence of *Cysticercus tenucollis* in Boer Crossbred goats. Indian Vet. J., 82: 674.

Souri, B. N., 1963. Dissertation, Univ. Madras.

Southwell, T., 1930. The Fauna of British India including Ceylon and Burma. Cestode. Vol II.,Taylor and Francis London.

Sreedevi, C., Hafeez, Md., Subramanyam, K. V., Ananda Kumar, P and Chengalva Rayalu, V., 2011a. Development and evaluation of flow through assay for detection of antibodies against porcine cysticercosis. XII Natl. Cong. Vet. Parasitol., 5-7th January, 2011, IAAVP and Dept. Vet. Parasitol., Bombay Vet. Coll., Mumbai, S5-O2, pp.161-162.

Sreedevi, C., Hafeez, Md., Subramanyam, K. V., Anandakumar, P. and Rayalu, V. C., 2011b. Development and evaluation of flow through assay for detection of antibodies against porcine cysticercosis. Trop. Biomed. 28(1): 160–170.

Sreedevi, C., Hafeez, Md., Anandakumar, P., Rayalu, V. C. and Subramanyam, K. V., 2011c. Validation of meat inspection results for *Taenia solium* cysticercosis in pigs by polymerase chain reaction. Compendium-cum-Souvenir, XII Natl. Cong. Vet. Parasitol., 5-7th January, 2011, IAAAP and Dept. Vet. Parasitol., & Bombay Vet. Coll., Mumbai, S4-O1, p.139.

Sreekrishnan, R., Kumar, D. and Das, S. S., 2008. *Cyclospora* infection in animals from India. J. Parasit. Dis., 32: 128-130

Sreekumaran, C., Kirubakaran, A., Venkitaraman, R., Selvan, P., Anilkumar, R. and Iyue, M., 2010. Spontaneous primary thoracic extrapulmonary hydatid cyst in a boiler rabbit. Helminthologia, 47: 193-195.

Sreemannarayana, O. and Christopher, K. J., 1977. A case of combined infection of cysticercosis and sarcosporidiosis in bullock. Indian J. Anim. Hlth., 17: 188.

Sreenivasa Murthy, G. S., D' Souza, P. E. and Jagannath, M. S., 1999. Enzyme linked immuno electro transfer blot in the diagnosis of *Taenia solium* cysticercosis in pigs. J. Parasit. Dis., 23: 85-88.

Sreenivasa Murthy, G.S., D'Souza, P. E and Jagannath, M. S., 2001. Diagnosis of *Taenia solium* cysticercosis in pigs and man by western blotting. Twelfth Nat. Cong. Vet. Parasitol., August 26-27, 2001, Dept. Parasitol., Coll. Vet Sci., Acharya Ranga Agri. Univ., Tirupati-517 502, India, Abstr. S-4: 38, p.120.

Sridhar, R., Murali Manohar, B., Shakir, S. A. and Sundararaj, A., 1996. *Cysticercus tenuicollis* Indian Vet. J., 73: 352-353.

Srivastava, H. O. P. and Shah, H. L., 1968. Studies on helminth parasites of desi pigs Madhya Pradesh. Indian Vet. J., 45: 698-699.

Srivastava, V. K. and Singh, B. P., 1994. Efficacies of various drugs against *Echinococcus granulosus*. Indian Vet. J., 71: 1175-1178.

Sudhan, N. A. and Shahardar, R. A. 1999. Occurrence of coenurus cyst in subcutaneous tissue of a goat. J. Vet. Parasitol., 13: 75-76.

Sundaram, R. H. and Natarajan, R., 1960. A study on the incidence of hydatid disease in cattle in the city of Madras. Indian Vet. J., 37: 19-24.

Tamuli, S. M., Chakraborty and Jamir, K. L., 2005. Cysticercosis in heart muscle of a heifer. J. Vet. Parasitol., 19: 55-56.

Tewari, A. N. and Iyer, P. K. R., 1960. Localised lesions in the omentum of goats due to *Taenia* sp. Indian Vet. J., 37: 627-630.

Thandaveswar, M. G., Subba Rao, H. S. and Setty, D. R. L., 1978. *Cysticercus cellulosae* in a dog. Curr. Res., 7: 24.

Thapar, G. S., 1954. Systematic survey of helminth parasites of domesticated animals in India. Indian J. Vet. Sci., 26: 221-271.

Thimmaiah, M. and Jagannath, M. S., 1982. Occurrence of *Coenurus cerebralis* cyst in the placenta of a cow. Indian Vet. J., 59: 824-825.

Tiwari, S. K., Mandal, S. C., Shakya, S., Ghosh, R. C. and Bhonsle, D., 2004. Successful surgical management of gid in a local goat- A case report. Indian Vet. Med. J., 28: 193.

Tripathi, K. P. and Ray, D. K., 1976. A note on the occurrence of *Cysticercus fasciolaris* in golden hamsters (*Mesocricetus auratus*). Indian Vet. J., 53: 369-330.

Tyagi, A. P. and Mishra, S. D., 1978.Occurrence of *Taenia taeniaeformis* cyst in wild rats, *Rattus rattus*. Indian J. Parasitol., 2: 169.

Upadhayay, A. N., Ahluwalia, S. S. and Asthana, V.S., 1977. A survey of parasites of pigs at Central Dairy Farm, Aligarh, India. Indian Vet. J., 54: 495-496.

Upadhaya, T. N., Datta, B. M. and Rahman, T., 1983. Studies on the pathology of caprine pneumonia and hydatidosis. Indian Vet. J., 60: 787-790.

Utpal Das and ArupKr. Das, 1998. Cystic hydatidosis of food animals of Calcutta. Indian Vet. J., 75: 387-388.

Varma, T. K., 1970. M. V. Sc., Thesis, Agra Univ., Agra, U. P.

Varma, T. K., 1978. Hydatid disease in food animals and man. Livestock Advisor, 3: 17-19.

Varma, T. K., 1982. A note on the incidence of helminthic parasites of domestic pigs. Livestock Adviser, 7: 56-57.

Varma, T. K., 1990. Prevalence of *Echinococcus granulosus* infection in domestic animals of district Gurgaon (Haryana), India. Philippine J. Vet. Sci., 27: 65-69.

Varma, T. K., 1991. Cysticercosis. Livestock Adviser, 16: 9-10.

Varma, T. K., 1993. Some observations on the prevalence of helminth parasites of domestic pigs in Bareilly, Uttar Pradesh. Livestock Adviser, 18(11):

Varma, T. K. and Ahluwalia, S. S., 1981. Potential danger of cysticercosis and taeniasis in man and animals. Livestock Adviser, 6: 54-56.

Varma, T. K. and Ahluwalia, S. S., 1985. Development of *Taenia solium* in cats and dogs. Livestock Adviser, 10 (12): 52-53.

Varma, T. K. and Ahluwalia, S. S., 1986a. Some observations on the prevalence and variations in the morphology and biology of *Cysticercus tenuicollis* of sheep, goat, pig and buffalo origin. Indian J. Anim. Sci., 56: 1135-1140.

Varma, T. K. and Ahluwalia, S. S., 1987. Some observations on the development of *Taenia hydatigena* in pups and laboratory animals. Indian J. Anim. Sci., 57: 804-811.

Varma, T. K. and Ahluwalia, S. S., 1989b. Incidence of *Cysticercus cellulosae* in slaughtered pigs and human taeniasis in western and central Uttar Pradesh. Indian Vet. J., 66: 673-674.

Varma, T. K. and Ahluwalia, S. S., 1990a. A note on the development of *Taenia solium* Linnaeus 1758 in cats and dogs. J. Vet. Parasitol., 4: 65-66.

Varma, T. K. and Ahluwalia, S. S., 1990b. Prevalence of *Taenia solium* cysticercosis in slaughtered pigs in central and western Uttar Pradesh, India. Rev. di Parassitol., 7: 263-268.

Varma, T. K. and Ahluwalia, S. S., 1990c. Prevalence of *Echinococcus granulosus* infection in domestic animals of western and central Uttar Pradesh. J. Vet. Parasitol., 4: 67-69.

Varma, T. K. and Ahluwalia, S. S., 1991. Immunization of lambs and kids against *Taenia hydatigena* Palias, 1766 using an extract of *T. hydatigena* eggs. Indian Vet. J., 68: 919-922.

Varma, T. K. and Ahluwalia, S. S., 1992. Development of *Taenia solium* Linnaeus, 1758 in golden hamsters. Indian J. Anim. Sci., 62: 48-49.

Varma, T. K. and Malaviya, H. C., 1986. Unusual occurrence of secondary echinococcosis in goat. Indian J. Parasitol., 10: 235-236.

Varma T. K. and Malaviya, H. C., 1989a. Studies of development of *Muticeps gaigeri* Hall, 1916 in pups and laboratory animals. J. Vet. Parasitol., 3: 41-43.

Varma T. K. and Malaviya, H. C., 1989b. Prevalence of coenuriosis in sheep, goats and pigs in Bareilly, Uttar Pradesh. J. Vet. Parasitol., 3: 69-71.

Varma T. K. and Malaviya, H. C., 1989c. The incidence of hydatid cysts in slaughtered animals in Bareilly, Uttar Pradesh, India. Rev. Parasitol., 49: 45-49.

Varma, T. K. and Malaviya, H. C., 1989d. Comparative study of serodiagnosis for hydatidosis in buffaloes. Rev. Parasitol., 50: 51-56.

Varma, T. K. and Malviya, H.C., 1992. Studies on the fertility rate of hydatid cysts from domestic animals and prevalence of *Echinococcus granulosus* and stray dogs Indian J. Parasitol., 16: 55-57.

Varma T. K. and Malaviya, H. C., 1994a. Unusual occurrence of *Coenurus cerebralis* in brain of a buffalo. Indian Vet. J., 71: 844- 845.

Varma, T. K. and Malviya, H. C. 1994b. Hydatid cyst from a swamp deer (*Cervus duvauceli duvauceli*). Abst. S-2: 17, Sixth Natl. Cong. Vet. Parasitol, 22-24 October, 1994, Dept. Parasitol., Coll. Vet. Sci., JNKVV, Jabalpur-482 001, M. P., India, p.16.

Varma T. K. and Rao, B.V., 1973. Certain epidemiological studies on larval and strobilar phases of the common dog tapewowm, *Taenia hydatigena*. Indian J. Anim. Sci., 43: 534-539.

Varma T. K. and Rao, B.V., 1974. Some observations on the effect of gamma irradiation on the cystic and stobilar phases of *Taenia hydatigena*. Indian Vet. J., 51: 47-53.

Varma T. K. and Sharma Deorani, V.P., 1980. Effect of gamma radiation of coenuri of *Multiceps gaigeri*. Indian J. Parasitol., 10: 77-78.

Varma T. K., Arora, B. M. and Malaviya, H. C., 1989. Prevalence of coenurosis in sheep, goats and pigs in Uttar Pradesh. J. Vet. Parasitol., 3: 69-71.

Varma T. K., Arora, B. M. and Malaviya, H. C., 1994. A note on *Coenurus gaigeri* (Hall, 1916) recovered from thigh muscles of a sambar (*Cervus unicolor*). Indian Vet. J., 71: 618.

Varma T. K., Arora, B. M. and Malaviya, H. C., 1995. On the occurrence of hydatid cyst in giant squirrel (*Rafuta indica*). Indian Vet. J., 72: 1305-1306.

Varma, T. K., Kulshreshtha, S. B. and Rao, B. V., 1974a. Primary studies on precipitation in active immunized rabbits, naturally infected sheep and goats against *Cysticercus tenuicollis*. Ceylon Vet J., 22: 10-13.

Varma, T. K., Kulshreshtha, S. B. and Rao, B. V., 1974b. Immunological studies on *Cysticercus tenuicollis* in sheep and goats with bentonite flocculation test. Rev. Parassitol., 35: 49-56.

Varma, T. K., Kulshreshtha, S. B. and Rao, B. V., 1974c. Indirect haemagglutination test in the diagnosis of cysticercosis caused by *Cysticercus tenuicollis* infection in sheep and goats. Rev. di Parassitol., 35: 103-111.

Varma, T. K., Malviya, H. C. and Arora, B. M., 1994. Hydatid cyst from a swamp deer (*Cervus duvauceli duvauceli*). J. Vet. Parasitol., 8: 99-100.

Varma, T.K., Kulshreshtha, S. B. and Rao, B. V.and Kumar, S., 1973. The conglutinating complement absorption test for the serodiagnosis of *Cysticercus tenuicollis* infection in sheep and goats. J. Helminthol., 47: 191-197.

Varma, T.K., Malaviya, H. C. and Ahluwalia, S. S., 1985. On comparative efficacy of Scolaban, Niclosan and Pantelmin against experimental *Taenia hydatigena* infection in pups. Indian Vet. J., 62: 231-234.

Varma, T.K., Malaviya, H. C. and Ahluwalia, S. S., 1986. Serodiagnosis of infection with metacestodes of *Taenia solium* in pigs and taeniasis in man and dogs by indirect haemagglutination assay. Indian J. Anim. Sci., 56: 621-627.

Varma, T.K., Panda, M. R. and Arora, B. M., 1998. *Cysticercus tenucollis* from mysentry of a black buck (*Antelope cervicapra*). Indian Vet. J., 75: 337-378.

Varma, T.K., Prasad, A., and Arora, B. M., 1998. Morphology, transmission and experimental studies on *Echinococcus granulosus* from a giant squirrel. Indian J. Vet. Res., 7: 18-24.

Venkatachala Ayyer, V. V., 1941. A case of multiple hydatid cysts in the internal organs of a cow. Indian Vet. J., 18: 27-28.

Verma, B. B., Mishra, S. S., Das, U. L., Singh, D. P. and Sinha, B. P., 1973. A clinical case of gid in a she goat with particular reference to clinico-pathological studies. Indian Vet. J., 50: 1052-1055.

Vijayasmitha, R., 1991. Studies on hydatidosis in food animals and diagnostic utility of leucocyte migration test. M.V.Sc. Thesis. Submitted to Univ. Agri. Sci., Bangalore.

Vijayasmitha, R., Jagannath, M. S., Rahman, S. A. and Honnappa, T.G., 1993. The study of leucocyte migration test in the diagnosis of hydatidosis in food animals. Indian J. Anim. Sci., 63: 596-599.

Vinodh Kumar, O. R., Swain, N., Rajapandi, S., Rajendiran, A. S., Senthilvel, K. and Harikrishnan, T. J., 2011. Indian J. Vet. Med., 31: 57.(Coenurus cyst).

Wangoo, A., Ganguly, N. K. and Mahajan, R. C., 1987. *In vivo* efficacy of mebendazole in containment of larval cyst mass in early stages of hydatid disease due to *Echinococcus granulosus*. Trans. R. Soc. Trop. Med. Hyg., 81: 965-966.

Wangoo, A., Ganguly, N. K. and Mahajan, R. C., 1988. Phagocytic function of monocytes in experimental hydatid disease in mice with *E. granulosus*. Proc. VIII Natl. Cong. Parasitol., Calcutta Sch. Trop. Med., Calcutta-700 073, 10-12 February, 1988, Abstr., pp.90-91.

Wangoo, A., Ganguly, N. K, and Mahajan, R. C., 1989. Phagocytic function of monocytes in murine model of *Echinococcus granulosus* of human origin. Indian J. Med. Res. 89: 40-42.

Wilson, F. D., 1962. Cytophlegia due to *Cysticercus cellulosae* in the brain of a dog–a case report. Indian Vet. J., 39: 393-396.

Yadav, R. P. and Ahluwalia, S. S., 1973. Some morbidities among food animals slaughtered in western Uttar Pradesh. Indian Vet. J., 50: 645-647.

2.9 FAMILY: DIPHYLLOBOTHRIIDAE

Rajavelu, G. and Ebenezer Raja, E., 1995.Incidence of *Spirometra mansonoides* in cats in Madras and its biology. Seventh Natl. Cong. Vet. Parasitol., 19-21, August, 1995. Dept. Parasitol., Madras Vet. Coll., Madras -7, Abstr. Sl 37, p. 91.

Ramachandra Raju, P., 1974. On the *Sparganum* (Pleurocercoid) of *Spirometra* sp. (Psuedophyllidae, Diphyllobothriidae) from *Tropidonotus piscator* Wall. Curr. Sci., 43: 193.

Saleque, A., Juyal, P. D. and Bhatia, B.B., 1990. *Spirometra* sp. in a domestic cat in India. Vet. Parasitol., 35: 273-276.

Uppal, S. K., 1974. *Diphyllobothrium* sp. pleurocercoid larvae (Cestode: Psuedophyllidae, Diphyllobothriidae) from fresh water fish, *Heterpneustus fossilis* (Bloch). Proc. Indian Sci. Cong., Part III, 61: 122-123.

2.10 GENERAL REFERENCE

Chowdhury, N. 2001.Helminths of domesticated animals in Indian sub continent. In: Perspectives on Helminthology. (Chowdhury, N. and Tada. I. (Eds.) Oxford & IBH Publishing Co. Pvt. Ltd.New Delhi. (Review, not Indexed, vide Table 4, pp.101-102.)

Mudaliar, S. V. and Alwar, V. S., 1947. A check-list of parasites (Class-Trematoda and Cestoda) in the Department of Parasitology, Madras Veterinary College laboratory. Indian Vet. J., 23: 423-431.

I.B PHYLUM: NEMATHELMINTHES

3.CLASS: NEMATODA

3.1 FAMILY: HETERAKIDAE

David Jacob, P., George Varghese, C., Georgekutty, P. T. and Peter, C. T, 1970. A preliminary study on the role of grasshoppers (*Oedaleus abruptus* and

Spathosternum parasiniferum) in the transmission of *Ascaridia galli* (Schrank, 1788) Freeborn, 1922 in poultry. Kerala J. Vet. Sci. 1: 65-70.

3.2 FAMILY: SUBULURIDAE

Arora, G. S. and Rai, P., 1972a. Proc. 59th Indian Sci. Cong. (Calcutta), Part III, Abstr., 29, p.545.

Arora, G. S. and Rai, P., 1972b. Experimental development of *Subulura brumpti* in laboratory raised chicks. Proc. 59th Indian Sci. Cong. (Calcutta), Part III, Abstr., 30, pp.545-546.

Arora, G. S. and Rai, P., 1972c. Studies on fowl tapeworm, *Subulura brumpti*. I. Experimental development in laboratory-raised chicks. Orissa Vet. J., 7: 125-138.

Arora, G. S. and Rai, P., 1972d. Studies on fowl tapeworm, *Subulura brumpti*. II. Compartive efficacy of three anthelmintics in experimental infection. Orissa Vet. J., 7: 139-144.

Karunamoorthy, G., Chellappa, D. J. and Anandan, R., 1994. The life history of *Subulura brumpti* in the beetle, *Alphitobius diaperinus*. Indian Vet. J., 71: 12-15.

Mathur, S. C., 1967. Observations on experimental infections with some of the tapeworms and nematode parasites in laboratory-raised chicks on normal and deficient rations. M. V. Sc. Thesis, Agra University, Agra, U. P., India. (Not indexed, but provided details for IH, vide infra, Mathur, S. C. and Pande, B. P., 1967).

Mathur, S. C. and Pande, B. P., 1967. On adults of the subulurid infective larva from tenebrionid beetle with remarks on the validity of *Subulura minetti,* Bhalerao. Curr. Sci., 36: 422-424.

Srivastava, J. S. 1966. Observations on the cysticercoids and larval nematodes from some of the common insects in deep litter and studies on helminth parasites of fish-eating-birds – Kingfishers. M. V. Sc. Thesis, Agra University, Agra, U. P., India. (Not indexed, but provided details for IH, vide infra, Srivastava, J. S. and Pande, B. P., 1967).

Srivastava, J. S. and Pande, B. P., 1967. The tenebrionid beetle, *Gonocephalum depressum*, F., from poultry units, as the vector of some of the larval helminths. Indian J. Ent., 29: 205-212.

3.3 FAMILY: TRICHINELLIDAE

Deodar, N. S., Narasapur, V. S. and Ajinkya, M., 1968. A survey of *Trichinella* and sarcosporidium in pigs in Bombay. Bombay Vet. Coll. Mag., 15: 38-40.

Kalapesi, R. M. and Rao, S. R., 1954. *Trichinella spiralis* infection in a cat that died in the Zoological Garden, Bombay. Indian Med. Gaz., 89: 578-580.

Maplestone, P. A. and Bahaduri, N. V., 1942. A record of *Trichinella spiralis* (Owen, 1835) in India. Indian Med. Gaz., 77: 193-195.

Niphadkar, S.M., 1964. Further observations on helminthic fauna of rats in Bombay. P. G. Thesis, Bombay Veterinary College, Bombay.

Niphadkar, S. M., 1973. *Trichinella spiralis* (Owen, 1835) in *Bandicota bengalensis* (Gray) in Bombay. Curr. Sci., 42: 135-136.

Niphadkar, S. M., 1977a. Epidemiology and rediscovery of *Trichinella spiralis* (Owen, 1835) in Bombay. Proc. 2nd Natl. Seminar on Zoonosis in India and their control, Abstr. 8.10., p.50.

Niphadkar, S. M., 1977b. Epidemiological transmission of *Trichinella spiralis* isolated from bandicoots in Bombay. Proc. 2nd Natl. Seminar on Zoonosis in India and their control, Abstr 8.11., p.50.

Niphadkar, S. M., Pradahan, M. H. and Deshpande, V. S., 1979. Rediscovery of *Trichinella spiralis* (Owen, 1835) in domestic pigs in India. Curr. Sci., 48: 372-373.

Pethe, R. S., 1992. Observations on haematology, immune response and treatment of *Trichinella spiralis* in experimental animals. P. G. Thesis. Bombay Vet. Coll., Parel, Bombay.

Parmeter, S.N., Schad, G.A., Chowdhury, A.B., 1968. Another record of *Trichinella spiralis* in Calcutta. Wiad. Parazytol. 14, 239.

Pradhan, M.H., 1978. Studies on the life cycle of *Trichinella spiralis* (Owen, 1835) Railliet, 1895, in rats. P. G. Thesis, Bombay Vet. Coll., Parel, Bombay.

Pradhan, M. H. and Niphadkar, S. M., 1978. Preliminary studies on the life cycle of *Trichinella spiralis* isolated from bandicoots in Bombay. 1st Asian Cong. Parasitol., 23-26 February, 1978,Bombay, Abstr., IV, p.89.

Ramamurthi, R. and Ranganathan M., 1968. A survey of incidence of trichinosis in pigs in Madras city. Indian Vet J.,45: 740-742.

Ranade, D. R. and Bhalchandra, D. V., 1976. A note on the natural infection in rat flea, *Xenopsylla cheopsis* (Roths) with *Trichinella spiralis* (Owen, 1835). J. Comm. Dis., 8: 77-80.

Schad, G. A. and Chowdhury, A. B., 1967. *Trichinella spiralis* in India. I. Its life history in India, rediscovery in Calcutta and the ecology of its maintenance in nature. Trans. R. Soc. Trop. Med. Hyg., 61: 244-248. (Helm. Abstr., 1967, 36, Abstr., 2945)

Schad, G. A., Nundy, S., Chowdhury, A. B. and Bandyopadhyay, A. K., 1967. *Trichinella spiralis* in India. II. Characteristic of strain isolated from a civet cat in Calcutta. Trans. Roy. Soc. Top. Med. Hyg., 61: 249-258. (Helm. Abstr., 1967, 36, Abstr., 2922 and internet).

Shaikenov, B. S.and Boev, S.N., 1983. Distribution of *Trichinella* species in the old world. Wiad. Parazytol. 29, 595–608.

Sharma, K. D. 2011. Ten die from trichinosis after eating wild boar meat. Sources: Sharma, K. D., Deputy Chief Medical Hospital, Paudi Dist. Hospital, Uttarkhand.The Pioneer, News Paper, 17th Oct., 2011.http: //Outbreaknews.

Singh, B.B., Sharma, R., Sharma, J. K. and Juyal. P.D., 2010. Parasitic zoonoses in India: an overview. Rev. sci. tech. Off. Int. Epiz. 29: 629-637. (Review, not indexed).

Temjenmongla and Yadav, A. K., 2005. Nematocidal activity of *Gynura angulosa* DC against *Trichinella spiralis* infection in mice. Abst. 17ᵗʰ Natl. Cong. Parasitol, 24-26 October, 2005, Reg. Med. Res. Centre, N. E. Region (ICMR), Dibrugarh-786 001, Assam, pp.12-13.

Yadav, A. K. and Temjenmongla., 2008. Herbicidal remedy of Trichinellosis: A pilot study on medicinal plants on Swiss Albino mice. Natl. Cong. Parasitol., November 3-5, 2008, Dept. Parasitol. and Indian Soc. Parasitol., North Eastern Hill Univ., Shillong-793 022, Abstr.No. PP-112., p.176.

3.4 FAMILY: STEPHANURIDAE

Sinha, P. K., 1967. Earthworm, *Eutypheus waltoni* and *Pheretima* sp. as transport hosts of swine kidney worm, *Stephanurus dentatus* in Bihar, India (Preliminary observations). Ceylon Vet. J. 15: 130-132.

3.5 FAMILY: ANCYLOSTOMATIDAE

Banrjee, D. and Om prakash, 1972. Experimental infection of cyclops with the larvae of *Ancylostoma caninum*. Indian J. Med. Res., 60: 218-220.

3.6 FAMILY: PROTOSTRONGYLIDAE

Bhalerao, G. D., 1945. Interesting mode of life cycle of the lungworm, *Varestrongylus pnuemonicus* Bhalerao, 1932. Curr. Sci., 106-107.

Bhalerao, G. D.and Kapoor, B. N., 1944. Some observations on the life history of *Varestrongylus pnuemonicus* (Bhalerao, 1932). Proc. 31ˢᵗ Indian Sci. Cong. (Delhi), Part III, Sect. VIII, Abstr. 35. p.112.

Sharma Deorani, V. P., 1965. Studies on the embryonation of eggs of *Varestrongylus pnuemonicus* (Protostronglidae, Nematoda) and guinea pig infection with larvae from terrestrial snails. Indian J. Helminth., 17: 85-88.

3.7 FAMILY: METASTRONGYIDAE

Bhattacharya, H. M., Sinha, P. K. and Sarkar, P. B., 1971. Studies on the incidence of *Metastrongylus salmi* in West Bengal with observations on its life cycle. Indian Vet. J., 48: 993-996.

Thomas, P. C., 1971. Studies on the common nematodes in pigs (*Sus scorfa domestica*). Thesis for Master's Degree. University of Kerala, Trivandrum, Kerala. (refers to IH of *Metastrongylus salmi* also).

Thomas, P. C. and Peter, C.T., 1975. Studies on the common nematodes in pigs (*Sus scorfa domestica*). Indian Vet. J., 52: 668-669.

Subramaniyam, T., D'Souza, B. A. and Victor, D. A., 1967. Bronchopnuemonia in baby pigs due to *Metastrongylus apri*. Indian Vet. J., 44: 121-127.

3.8 FAMILY: SPIROCERCIDAE

Gupta, V. P., 1981. *Cyathospirura* from pups infected with encapsulated larvae from paratenic host. Indian J. Anim. Sci., 51: 526-534.

Gupta, V.P. and Pande, B. P., 1981. *Cynathospirura chaubaudi*, n. sp. from pups infected with re-encapsulated larvae from a paratenic host. Indian J. Anim. Sci., 51: 526-534.

3.9 FAMILY: SPIRURIDAE

Ajitkumar, K. G., Reghu Ravindran, Ajith Jacob, George, Joju Johns, Aparna, M., Mithin, U. C. and Devada, K., 2010. Detection of *Spirocerca lupi* in paratenic host. XX NCVP Parasitology Today-Ecology to Molecular Biology, 18-20 Feb, 2010, Dept. Vet. Parasitol., CVS, CCS Haryana Agri. Univ., Hisar-125004, pp.83.

Anantaraman, M. and Jayalakshmi, N., 1963. On the life history of *Spirocerca lupi* (Rudolphi, 1809), a nematode of dogs in India. Proc. Indian Acad. Sci., 58: 137-147.

Anantaraman, M. and Sen, K., 1966. Experimental spirocercosis in dogs with larvae from a paratenic host, *Calotes versicolar*, the common garden lizard in Madras. J. Parasitol., 52: 911-912.

Carter, H. J., 1861. On a bi-sexual nematoid worm which infests the common house fly (*Musca domestica*) in Bombay. Ann. Mag. Nat. Hist. ser. (3), 7: 29-33.

Bhatia, B. B., Chauhan, P. P. S., Agarwal, R. D. and Ahluwalia, S. S., 1979. On helminthic infection in Indian duck and their significance. Vet. Res. Bull., 2: 129-135.

Chhabra, R. C. and Singh, K. S., 1972a. Diagnosis, treatment and control of spirocercosis in dog. Indian J. Anim. Sci., 42: 203-207.

Chhabra, R. C. and Singh, K. S., 1972b. Development and pathogenecity of *Spirocerca lupi* in experimentally infected kids and lambs. Indian J. Anim. Sci., 42: 272.

Chhabra, R. C. and Singh, K. S., 1972c. On *Spirocerca lupi* infection in some paratenic hosts infected experimentally. Indian J. Anim. Sci., 42: 297-304.

Chhabra, R. C. and Singh, K. S., 1972d. Development of resistance in *Spirocerca* infection. Indian J. Anim. Sci., 42: 355-357.

Chhabra, R. C. and Singh, K. S., 1972e. Study on the life history of *Spirocerca lupi*: Histotropic juveniles in dogs. Indian J. Anim. Sci., 42: 628-636.

Chhabra, R. C. and Singh, K. S., 1972f. On the adult *Spirocerca lupi* in experimentally infected dogs. Indian J. Anim. Sci., 42: 636-641.

Chhabra, R. C. and Singh, K. S., 1973a. A study on the life history of *Spirocerca lupi*: Intermediate hosts and their biology. Indian J. Anim. Sci., 43: 49-54.

Chhabra, R. C. and Singh, K. S., 1973b. The life history of *Spirocerca lupi*: Development and biology of infective juvenile. Indian J. Anim. Sci., 43: 178-184.

Chowdhury, N., 1968. On some of the encysted nematode larvae from dung beetles and their development in experimental animals-rabbit and guinea pig. M. V. Sc. Thesis, Agra University, Agra, U. P., India.(Not indexed but for details of IH, vide infra, Chowdhury, N. and Pande, B. P., 1969).

Chowdhury, N. and Pande, B. P., 1969. The development of the infective larva of the canine oesophageal tumour worm, *Spirocerca lupi* in rabbits and its histopathology. Z. Parasitkde. 32: 1-10.

Gupta, V. P., 1971. Studies on certain aspects in the life-history, histopathology and taxonomy of some of the spiruroid nematode parasitic in domestic animals including poultry. Ph. D.Thesis, Agra University, Agra, U. P., India. (Not indexed).

Gupta, V. P. and Pande, B. P., 1970. Natural intermediaries and paratenic hosts of *Simondsia paradoxa* – an important stomach worm of pig. Curr. Sci., 39: 536.

Gupta, V. P. and Pande, B. P., 1981. Mammalian phase in life-cycle of *Simondsia paradoxa* Cobbold, 1864 (Ascaropinae, Thelazidae, Spiruridae) and its histopathology in experimental infected piglets. Indian J. Anim. Sci., 51: 855-868.

Jayalakshmi Jagannathan, 1980. *Musca domestica* L. as a vector of the nematode *Habronema muscae* Carter, 1861(Spiruroidea) in Madras. Curr. Sci.,49: 877.

Joju John S., Mithin, U. C., Aparna, M., Harshal, P.T., Mathew Abraham, Ajith Kumar, K. G., Reghu Ravindran and Devada, K., 2010. Detection of larval stages of *Spirocerca lupi* in garden lizards (*Calotes versicolor*). Fishery Vet. Sci., Proc.22nd Kerala Sci. Cong., 28-31 January, 2010, KFRI, Peechi, Paper No. 02-21, p.141

Joseph, S. A., 1979. Indian barn owl (*Tyto alba javanica*) as a new paratenic host for *Spirocerca lupi* (Rudolphi, 1809) Raillet and Henry, 1911. Cheiron, 8: 168-170.

Katiyar, J. C. and Pande, B. P. 1965. On *Spirocerca lupi* (Rud., 1809) Chitwood, 1933, in *Equus asinus*. Agra Univ. J. Res. (Sci.) 14: 151-156.

Muralidhar, P. and Rao, P. N. 1991. A note on a new vector of horse nematode, *Habronema* sp. from a dipteran host (Empididae) from Hyderabad, A. P. Curr. Nematol., 2: 85-86.

Narsi Reddy,Y and Narayana Rao, P., 1982. *Musca crassirostris* Stein an invertebrate vector of a mammalian eye worm in South India. In: Vector and Vector-borne Diseases. Proc. All India Sym., Trivandrum, Kerala, India.

Pande, B. P., Rai, P. and Bhatia, B. B., 1961. Nematode affecting the aorta in Indian caprine and equine hosts. J. Parasitol., 47: 951.

Pujatti, D., 1953. Opisti intermedi *Spirocerca lupi* (Rudolphi, 1809 nel Sud India (Nematoda). Atti Soc. Ital. Sci. Nat. 92: 30-32. Refer Helm. Abstr. 22: 331a, 1953.

Sen, K. and Anantaraman, M., 1971. Some observations on the development of *Spirocerca lupi* in its intermediate and definitive hosts. J. Helminth., 45: 123-131.

Srivastava, J. S. and Pande, B. P., 1967. The tenebrionid beetle, *Gonocephalum depressum*, F., from poultry units, as the vector of some of the larval helminths. Indian J. Ent., 29: 205-212.

3.10 FAMILY: ASCAROPIDAE

Chowdhury, N. and Pande, B. P., 1968a. On a third stage larva *Physocephalus sexalutus* from dung beetle and its development in rabbit with remarks on histopathology. Indian J. Vet. Sci., 38: 747-752.

Chowdhury, N. and Pande, B. P., 1968b. On the dung beetle acting as intermediate hosts of *Gongylonema pulchrum*. Indian J. Anim. Hlth., 7: 163-165.

Chowdhury, N. and Pande, B. P., 1969. Intermediate hosts of *Ascarops strongylina* in India and the development of the parasite in the rabbit and guinea pig. Indian J. Anim. Sci., 39: 139-149.

Gupta, V. P., 1969. Pond-snake and ground squirrel as paratenic hosts of *Ascarops strongylina*. Curr. Sci., 38: 548-549

Gupta, V. P., 1970a. *Musca domestica*, an intermediate host of two spiruroid nematode parasites of livestock. Orissa Vet., 5: 147-150.

Gupta, V. P., 1970b. Experimental development of the oesophageal worm of cattle in rabbit. Curr. Sci., 39: 237-238.

Sunderam, R. K., 1971. Study on spiruroids of fowl (*Gallus gallus domesticus*). Ph. D. Thesis. University of Kerala, Trivandrum, Kerala.

Thomas, P. C., 1971. Studies on the common nematodes in pigs (*Sus scorfa domestica*). Thesis for Master's Degree. University of Kerala, Trivandrum, Kerala (refers to *Physocephalus sexalutus* also).

Thomas, P. C. and Peter, C. T., 1975. Studies on the common nematodes in pigs (*Sus scorfa domestica*). Indian Vet. J., 52: 668-669. (refers to *Physocephalus sexalatus* also)

Varma, S., Malik, P. D. and Lal, S. S., 1976. White mice as resevoirs of swine stomach worm, *Ascarops stongylina*. Haryana Agri. Univ. J. Res., 51: 134-135.

Varma, S., Malik, P. D. and Lal, S. S., 1977. Some new intermediate hosts of *Ascarops stongylina*. J. Helminth., 51: 134-135.

3.11 FAMILY: ACUARIIDAE

Chandrasekharan, K., 1977. Studies on the biology, pathogenicity and treatment of important nematodes of domestic ducks. Thesis (Ph. D). University of Kerala, Trivandrum, Kerala.

Muraleedharan, K., 1999. Role of certain insects like ants, lice, grass-hoppers and dragon flies as intermediate hosts of helminth parasites. In: Vectors and Vector-borne Parasitic Diseases. IV Natl. Trng. Prog., 25-1-1999 to 8-2-1999. Centr. Adv. Stud., Dept Parasitol., Vet. Coll., Univ. Agri. Sci., Bangalore-560 024. pp.125-131 (Review, not indexed).

Muraleedharan, K., 2000. Biology, economic importance, and control of lice, ants, grass-hoppers and dragon flies. In: Acarines and Insects of Veterinary and Medical Importance. V. Natl. Trng. Prog., Centr. of Adv. Stud., Dept Parasitol., Vet. Coll., Univ. Agri. Sci., Bangalore-560 024. pp.18-25. (Review, not indexed).

Ramaswamy, K. and Sunderam, R. K., 1979. *Porcellio laevis* (Isopoda) as a new intermediate host for *Tetrameres mohtedai*, Bhalerao and Rao, 1994. Indian Vet. J., 56: 363-366.(Mixed experimental infection with *Acuria spiralis*)

Ramaswamy, K. and Sunderam, R. K., 1985. Histopathological changes in the proventriculus of fowl given experimental monospecific infection with *Acuaria spiralis*.Vet. Parasitol., 17: 309-317.

Sunderam, R. K., 1971. Study on spiruroids of fowl (*Gallus gallus domesticus*). Ph. D. Thesis. University of Kerala, Trivandrum, Kerala.

Sunderam, R. K., 1977. Life cycle of some spiruroid parasite of fowl. Abstr. 1st Natl. Cong. Parasitol., Baroda, 24-26 Feb. 1977. (Helm. Abstr., 46: 4531, 1977).

3.12 FAMILY: RICTULARIIDAE

Gupta, V. P. and Pande, B. P., 1970. *Hemidactylus flaviviridis*, a paratenic host of *Rictularia cahirensis*. Curr. Sci., 39: 535-536.

Srivastava, H. D., 1940. An unrecorded spiruroid nematode worm *Rictularia cahirensis* Jagerskiold, 1904 from the intestine of an Indian civet cat. Indian J. Vet. Sci., 10: 133-114

3.13 FAMILY: TETRAMERIDAE

Chandrasekharan, K., 1967. Studies on the common nematodes encountered in domestic ducks. Thesis (M. Sc.), Univ. Kerala, Trivandrum, Kerala.

Chandrasekharan, K., 1977. Studies on the biology, pathogenecity and treatment of important nematodes of domestic ducks. Thesis (Ph. D), Univ. Kerala, Trivandrum, Kerala.

Chandrasekharan, K. and Peter, C. T., 1969. Studies on the common nematodes of domestic ducks (*Anas platyrhynchos domesticus*). Indian Vet. J., 46: 454.

Mukherje, G. S. and Sinha, P. K., 1965. On some intermediate hosts of *Tetrameres mohtedai*. Bengal Vet., 13: 11.

Muraleedharan, K., 1999. Role of certain insects like ants, lice, grass-hoppers and dragon flies as intermediate hosts of helminth parasites. In: Vectors and Vector-borne Parasitic Diseases. IV Natl. Trng. Prog., 25-1-1999 to 8-2-1999. Centr. Adv. Stud., Dept. Parasitol., Vet. Coll., Univ. Agri. Sci., Bangalore-560 024. pp.125-131. (Review, not indexed)

Muraleedharan, K., 2000. Biology, economic importance, and control of lice, ants, grass-hoppers and dragon flies. In: Acarines and Insects of Veterinary and Medical Importance. V. Natl. Trng. Prog., Centr. Adv. Stud., Dept. Parasitol., Vet. Coll., Univ. Agri. Sci., Bangalore-560 024. pp.18-25. (Review, not indexed)

Radhakrishna Reddy, K and Rao, B. V., 1984. Experimental trials of an irradiated vaccine against *Tetrameres mohtedai*, Bhalerao and Rao, 1994 infection in chicks and certain host responses. Indian Vet. J., 61: 1018-1023.

Radhakrishna Reddy, K and Rao, B. V., 1985. The biological development and pathogenecity of *Tetrameres mohtedai*, Bhalerao and Rao, 1994 in fowls. Indian Vet. J., 62: 127-132.

Ramaswamy, K., 1971.Studies on the pathogenecity of *Tetrameres* mohtedai and *Acuaria spiralis* in fowl Thesis. Univ. Kerala, Trivandrum, Kerala.

Ramaswamy, K. and Sunderam, R. K., 1979. *Porcellio laevis* (Isopoda) as a new intermediate host for *Tetrameres mohtedai*, Bhalerao and Rao, 1994. Indian Vet. J., 56: 363-366.

Sunderam, R. K., 1971. Study on spiruroids of fowl (*Gallus gallus domesticus*). Ph. D. Thesis. Univ. Kerala, Trivandrum, Kerala.

Sunderam, R. K. 1977. Life cycle of some spiruroid parasite of fowl. Abstr. 1ˢᵗ Natl. Cong. Parasitol., Baroda, 24-26 Feb. 1977. (Helm. Abstr., 46: 4531, 1977).

Sunderam, R. K., Radhakrishnan, C. V., Sivasubramaniyam, M. S., Padmanabhan, R. and Peter, C. T., 1963. Studies on the life cycle of *Tetrameres mohtedai*, Bhalerao and Rao, 1994. Indian Vet. J., 40: 7-15.

3.14 FAMILY: PHYSALOPTERIDAE

Gupta, V. P. and Pande, B. P., 1970. Partial life cycle of a physalopterid nematode parasitic in the stomach of carnivores. Curr. Sci., 39: 399-400.

3.15 FAMILY: THELAZIIDAE

Gupta, V. P., 1970a. *Musca domestica*, an intermediate host of two spiruroid nematode parasites of livestock. Orissa Vet., 5: 147-150.

Narsi Reddy, Y. and Narayana Rao., P, 1982. *Musca crassirostris* Stein an invertebrate vector of a mammalian eye worm in South India. In: Vectors and Vector-borne Diseases. Proc. All India Sym., Trivandrum, Kerala, India. Feb., 26-28, 1992, p. 123-124.

Rao, S. R., 1949. A short note on *Musca vicina* as possible vector of *Thelazia rhodesii*. Indian Vet. J., 25: 399-401.

Sunderam, R. K., 1971. Study on spiruroids of fowl (*Gallus gallus domesticus*), Ph. D. Thesis. Univ. Kerala, Trivandrum, Kerala.

3.16 FAMILY: GNATHOSTOMATIDAE

Baruah, N., Islam, S. and Gogoi, A. R., 2010. *Gnathostoma* and gnathostomiasis in Northern India: 1. Prevalence of *Gnathostoma spinigerum* Owen 1836, larvae in fishes, experimental infection in first and second intermediate host with a description of an advance third stage larva recovered from *Ophiocephalus punctatus*. XX NCVP, Parasitology Today-Ecology to Molecular Biology, 18-20 February, 2010, Dept. Vet. Parasitol., CVS, CCS Haryana Agri. Univ., Hisar-125004. pp.17-18.

Baruah, N., Islam, S. and Gogoi, A. R., 2011. Gnathostomiasis in Assam: 1. On the prevalence of *Gnanothostoma spinigerum* larvae in fishes, experimental infection in first and second intermediate host with a description of an advanced third stage larva recovered from *Ophiocephalus punctatus*. J. Vet. Parasitol., 25: 11-17.

Baruah, N., Talukdar, S. K. and Gogoi, A. R., 1988. Survivability and infectivity of first stage larvae of *Gnathostoma spinigerum*. Second Natl. Parasitol, March 3-5, 1988, Dept. Parasitol. Vet. Coll. Bangalore, India. Abstr., Stream.,1, p.1.

Bhatia, B. B., 1991. Current status of food-borne parasitic zoonoses in India. In. Cross, J. H. ed. Emerging problem in food-borne parasitic zoonoses impact on agriculture and public health. Proc. 33ʳᵈ SEAMED – TROPMED regional Seminar – Southeast Asian Trop. Med. Publ. Hlth., 22 (Suppl): 36-41.(Review not index; Ref to *Gnanothostoma spinigerum*).

Chellappa, D. J. and Anantaraman, M., 1970. The development of *Gnathostoma spinigerum*, Owen, 1836, in the first intermediate host, *Mesocyclops leuckarti*, Claus, 1857 in India with remarks on its zoonotic importance. Curr. Sci., 39: 199-201.

Chellappa, D. J. and Anantaraman, M., 1978. The endoparasites of domestic cats in Madras and their zoonotic importance. Asian Cong. Parasitol., 23-26 February., 1978, Bombay, Abstr., Sect., III, 46, pp.165-166.

Gogoi, A. R. and Baruah, N., 1988. *Gnathostoma spinigerum*, a pathogenic zoonotic nematode and the role of fishes in its life cycle in Assam. Second Natl. Parasitol., March 3-5, 1988, Dept. Parasitol., Vet Coll., Bangalore, India. Abstr., Stream 1, p.2.

Rai, P., 1976. Fish borne zoonosis with special reference to helminth diseases. UP Vet. J. 4: 52-54

3.17 FAMILY: FILARIIDAE

Gogoi, A. R., 2002. Filarids of animals in India. J. Vet. Parasitol.,16: 131-138.(Review, not indexed).

Lucy Sabu and Subramanian, H., 2006. Prevalence and vector potentialities of mosquitoes collected from Thrissur. Proc. XVII Natl. Cong. Vet. Parasitol., November, 15-17, 2006. Dept. Parasitol., Rajiv Gandhi Coll. Vet. Sci. & Anim. Sci, Kurumbapet, Puducherry- 605 009. Abstr., S.II.20, p.63.

Lucy Sabu and Subramanian, H., 2007. Prevalence and vector potentialities of mosquitoes from Thrissur, Kerala. J. Vet. Parasitol., 21: 165-167.

Malaviya, H. C., 1966. Studies on the life history of *Artionema labiatopapillosa* (Nematoda, Filariidae) – preliminary report. Sci. Cult., 32: 99-100.

Megat Abd Rani, P. A., Irwin, P. J. and Rebecca, J.T., 2011. *Hippobosca longipennis*- a potential intermediate host of a species of *Acanthocheilonema* in dogs. Parasites and Vectors 4: 143-150. (Paper presented at the 6th Sym.on Canine vector-borne diseases, Nice, France 10th-13th April, 2011).

Megat Abd Rani, P. A., Irwin, P. J., Mukulesh Gatne, Coleman, G. T., Linda, M. M. and Rebecca, J.T., 2010a. Canine vector-borne diseases in India: a review of the literature and identification of existing knowledge gaps. Parasites and Vectors, 3: 28. (Review, not cited in index).

Megat Abd Rani, P. A., Irwin, P. J., Mukulesh Gatne, Coleman, G. T., Rebecca, J. T. and Linda, M. M., 2010b. A survey of canine filarial diseases of veterinary and public health significance in India. Parasites and Vectors 3: 30. (Review on vectors, not cited in index)

Sahai, B. N. and Singh, S.P., 1971. A preliminary note on the development of *Parafilaria bovicola* in *Musca vitripenis*. Indian J. Anim. Hlth., 10: 243-245.

Shastri, U. V., 1983. Incidence of infection of nymphs of *Hylomma anatolicum anatolicum* with filarioid larvae in Marathwada (Maharashtra) and skins of cattle as a possible source of tick infection. Ticks and Tick-borne Diseases, Seminar, 21-22 March, 1983. Dept. Parasitol., Madras Vet. Coll., Madras, Abstr. Session III, p.17.

Shastri, U. V., 1984. Natural infection of *Hyalomma anatolicum anatolicum* tick with larvae of hitherto undescribed filarioid nematode in Parbhani (Maharashtra). Cheiron 13: 158-160.

Shastri, U. V., 2001. On the identity of a filarid involving ixodid ticks, *Hyalomma anatolicum anatolicum*, as a vector. J. Vet. Parasitol., 15: 165-166.

Shastri, U. V. and Gafoor, M. A., 1979. Ticks as possible vectors of filarioid nematode of cattle in India. Second Natl. Cong. Parasitol., Abstr. II, 43, p.25.

Shastri, U. V. and Gatoor, M. A., 1981. Hard ticks, *Hylomma anatolicum anatolicum* and *Boophilus microplus* as possible vectors of filarioid nematode of cattle in India. Indian Vet. J., 58: 774-776.

3.18 FAMILY: SETARIIDAE

Agrawal, M. C., 1977. Sudies on prevalence, morphology, transmission, pathology and chemotherapy of *Stephanofilaria zaheeri* Singh 1958 the parasite of ear-sore in buffaloes. M. V. Sc. Thesis, Jawarharal Nehru, Agri. Univ., Jabalpur. Thesis Abstracts. 3: 128-130

Ansari, J. A., 1964. Studies on *Setaria cervi* (Nematoda): Filroidea). Part II. Its peritoneal transplant and periodicity of the microfilariae in white rats. Z. Parasitenkde., 24: 105-111.

Choudhury Phukan, S., 2003. Studies on humpsore in cattle in Assam. Ph. D. (Veterinary Parasitology) Thesis, 2002, Assam Agri. Univ., Guwahati-781 022, India, Abstr., J. Vet. Parasitol., 17: 82.

Choudhury Phukan, S., Das, M and Borkakoty, M. R., 2005. Humpsore in cattle in Assam. J. Vet. Parasitol., 19: 19-22.

Dutt, S. C., 1970. Preliminary studies on life history of *Stephanofilaria zaheeri* Singh 1958. Indian J. Helminth., 22: 139-143.

Dutta, P. K., 1972. Studies on epidemiology, histopathology and chemotherapy of stephanofilariasis in cattle. M. V. Sc. Thesis, Assam Agri. Univ., Khanapara, Guwahati, Assam.

Khan, F. A. and Singh, H. S., 2003. Encapsulation of *Setaria cervi* larvae in *Aedes vittatus*. J. Vet. Parasitol., 19: 109-111.

Malaviya, H. C., 1972. Stephanofilarial infection in cattle and buffaloes in Andaman islands. Indian J. Helminth., 24: 68-71.

Nabaneel Baruah, Saidul Islam and Ambu Ram Gogoi, 2010, *Gnathostoma* and Gnathostomiasis in North East India: On the prevalence of *Gnathostoma spinigerum* Owen, 1836 larvae in fishes, experimental infection in first and second intermediate host with a description of an advanced third stage larva recovered from *Ophiocephalus punctatus*. XX NCVP, CCS HAU, February 18-20, 2010. pp.17-18.

Patnaik, B., 1966. Stephanofilariasis in animals. Indian Vet. J., 43: 761-762.

Patnaik, B., 1973. Studies on stephanofilariasis in Orissa. III. Life cycle of *Stephanofilaria assamensis* Pande, 1936. Z. Tropenmed. Parasitol., 24: 457-466.

Patnaik, B. and Roy, S. P., 1966a. Experimental infection of laboratory bred *Musca conducens* Walker, 1859, with microfilariae of *Stephanofilaria assamensis* Pande, 1936. Proc. 53ʳᵈ Indian Sci. Cong.(Chandigarh), Part III, pp.352-353.

Patnaik, B. and Roy, S. P., 1966b. On the life cycle of filariid, *Stephanofilaria assamensis* Pande, 1936, in the arthropod vector, *Musca conduscens* Walker, 1859. Indian J. Anim. Hlth. 5: 91-101.

Patnaik, B. and Roy, S. P., 1967. On the life cycle of the bovine filariid, *Stephanofilaria assamensis* Pande, 1936. Proc. 54ᵗʰ Indian Sci. Cong.(Hyderabad), Part III, p.516.

Patnaik, B. and Roy, S. P., 1969. Studies on Stephanofilariasis in Orissa. IV. Rearing of *Musca conducens* Walker, 1859, the arthropod vector of *Stephanofilaria assamensis* in the laboratory. Indian J. Anim. Sci., 39: 455-462.

Patnaik, B. and Vinaykumar., 1972. A research note on the occurrence of juveniles of *Stephanofilaria*, the cauasative parasite of ear sores in buffaloes, in *Musca autumnalis*, de Geer (1767). Indian J. Anim. Sci., 42: 351-352.

Phukan, S. C. and Das M., 2010. In search of vector of *Stephanofilaria assamensis* Pande, 1936. XX NCVP Parasitology Today-Ecology to Molecular Biology, 18-20 February, 2010, Dept. Vet. Parasitol., CVS, CCS Haryana Agri. Univ., Hisar-125004, p.17.

Rai, R. B., Neeraj Srivastava, Jai Sunder, Kundu, A. and Jeykumar, S., 2010. Stephanofilariasis in bovines – Prevalence and control and eradication in Andaman and Nicobar Islands, India. Indian J. Anim. Sci., 80: (Review, not indexed).

Sahai, B. N. and Srivastava,V. K., 1966. On *Stephanofilaria* and Stephanofilariasis (life history, pathogenecity and chemotherapy). Bihar Anim. Husb. Bull., 8: (2) 44-46.

Sharma Deorani and Tiwari, H. C., 1968. Contribution to the parasitic part of life cycle of *Stephanofilaria assamensis* Pande, 1936, a skin parasite of cattle in India. Arch. Vet., 5: 11-18.

Srivastava, H. D. and Dutt, S. C., 1963. Studies on the life history of *Stephanofilaria assamensis* Pande, the causative parasite of 'hump sore' of Indian cattle. Indian J. Vet. Sci., 33: 173-177.

Srivastava, H. D., Dutt, S. C. and Malviya, H. C., 1967. The preinfective larval stages of *Stephanofilaria assamensis*, the causative parasite of hump sore of Indian cattle. Indian J. Helm., 19: 81-86.

Varma, A. K., Sahai, B. N., Singh, S. P., Lakra, P. and Srivastava,V. K., 1971. On *Setaria digitata* (Von Linston, 1906), its specific characters, incidence and development in *Aedes vittatus* (new vector) and *Armigeres obturbans* in India with a note on its ectopic occurrence in kidney of a cow. Z. Parasitenk., 36: 62-72.

Wajihulla, 2001. Comparative account of developing larvae of *Setaria cervi* and *Diplotriaena tricuspia*. J. Vet. Parasitol., 15: 117-120.

3.19 FAMILY: DRACUNCULIDAE

Anon, 1861. Cited by Chowdhury, N., Helminths of domesticated animals in Indian subcontinent. In: Chowdhury, N. and Tada. I., 2001. (ed.). Perspective on Helminthology. Oxford & IBH Publishing Co. Pvt. Ltd., New Delhi. p.531.

Mehta, S and Gupta, A. N., 1982. Dracunculiasis *vis-a vis* the carrier host. Abst. All India Sym.on vector and vector-borne diseases, 26-28th February, 1982, Trivandrum, Kerala.

Moorthy, V. N. and Sweet, W. C., 1936a. A note on the experimental infection of dogs with Dracontiasis. Indian Med. Gaz., 71: 437-442.

Moorthy, V. N. and Sweet, W. C., 1936b. Guinea worm infection of cyclops in nature. Indian Med. Gaz., 71: 568-570.

Moorthy, V. N. and Sweet, W. C., 1938. Further notes on the developmental infection of dogs with dracontiasis. Amer. J. Hyg., 27: 301-310.

Moorthy, V. N., 1938. Observations on the development of *Dracunculus medinensis* larvae in cyclops. Amer. J. Hyg., 27: 437-458.

Saroj Bapna, 1985. Relative susceptibility of cyclops from Rajasthan State to guinea worm (*Dracuncus medinensis*) larvae. Bull. Wld. Hlth. Org., 63: 881-886.

I.C. PHYLUM: ACANTHOCEPHALA

4.1 FAMILY: POLYMORPHIDAE

Foetoder, D. N., Raina, M. K. and Dhar, R. L., 1977. Life cycle studies and experimental transmission of *Polymorpus magnus* Skrjabin, 1913 from *Gammarus pulex* to *Anas platyrhynchos domesticus* in Kashmir, India. Abstr. 1st Natl. Cong. Parasitol., Baroda, 24-26, February, 1977. (Helm. 46: Abst., 4004).

4.2 FAMILY: OLIGOACANTHORHYNCHIDAE

Padmavati, P., 1967. Likely paratenic avian hosts for the thorny- headed worm, *Oncicola* in and around Madras. Indian Vet. J., 44: 727-728.

4.3. GENERAL REFERENCE

Chowdhury, N. 2001. Helminths of domesticated animals in Indian sub continent. In: Perspectives on Helminthology. Chowdhury, N. and Tada. I. (Eds.) Oxford & IBH Publishing Co. Pvt. Ltd.New Delhi.(Review, not Indexed, refer, Table 4, pp.101-106.)

SECTION II–ARTHROPOD PARASITES

II. PHYLUM: ARTHROPODA

Note: Role of intermediate hosts of various insects and acarines of this phylum have been indexed under different families of class Cestoda and Nematoda.Two reviews have been noted below to provide further information. (Eds.)

Anon, 1999. Vectors and Vector-borne Parasitic Diseases. IV Natl. Trng. Prog., 25-1-1999 to 8-2-1999. Centr. Adv. Stud., Dept. Parasitol., Vet. Coll., Univ. Agri. Sci., Bangalore-560 024. pp.125-131. (Reviews, not indexed).

Anon, 2000. Acarines and Insects of Veterinary and Medical Importance. V. Natl. Trng. Prog., Centr. Adv. Stud., Dept. Parasitol., Vet. Coll., Univ. Agri. Sci., Bangalore-560 024. pp.18-25. (Reviews, not indexed).

5. CLASS: PENTASTOMIDA

5.1 FAMILY: POROCEPHALIDAE

Linguatula serrata

Alwar, V. S., 1958. Parasites of pigs (*Sus scrofa domestica*) in Madras. Indian Vet J., 35: 112-116.

Banerjee, P. S., Yadav, C. L., Stuti Vatsya, Kumar, R. R. and Rajat Garg, 2005. Incidence of *Linguatula serrata* infection in small ruminants in Udham Singh Nagar, Uttaranchal. Ab-8, Proc. XVI Natl. Cong. Vet. Parasitol., December 6-8, 2005, Dept. Parasitol., Coll. Vet. Sci. & Ani. Sci., Indira Gandhi Agri. Univ., Anjora, Durg, 491 001, p.114.

Bindu Lakshman, Reghu Ravindran and Subramanian, H., 2006. *Linguatula serrata* among ruminants of Wayanad, Kerala. Proc. XVII Natl. Cong. Vet. Parasitol., November 15-17, 2006. Dept. Parasitol. Rajiv Gandhi Coll. Vet. Sci. & Ani. Sci, Kurumbapet, Puducherry- 605009. Ab S II. 14, p. 60.

Chauhan, P. P. S. and Bhatia, B. B., 1978. Occurrence of *Linguatula serrata* in a new host *Funambulus palmarum*. Indian J. Parasitol., 2: 121.

Ebenezer Raja, E., 1974. Parasitic infestation in rodents especially those communicable to man and animals. Cheiron, 3: 173-174.

Gaiger, S. H., 1909. J.Trop. Vet. Sci., 4: 528-531. (Dealt as *Linguatula taenioides*).

Gaiger, S. H., 1911. J. Trop. Vet. Sci., 6: 292. (Dealt as *Linguatula taenioides*).

Gaiger, S. H., 1911. Vet. News 8: 634-636. (Dealt as *Linguatula taenioides*).

Gill, B. S., 1972. A study on parasitic zoonosis in India. Proc. 59th Indian Sci. Cong. (Calcutta), Part III, Abstr., 37, p.549.

Gill, H. S., Rao, B. V. and Chhabra, R. C., 1968. A note on the occurrence of *Linguatula serrata* (Frohlich, 1789) in domestic animals. Trans. Roy. Soc. Trop. Med. Hyg., 62: 506-508.

Lal Krishnan, Kalicharan and Paliwal, O. P., 1973. A pathological study on the larval forms of *Linguatula serrata* (Frohlich, 1789) in goats. Indian Vet J., 50: 317-318.

Leese, A. S., 1911. *Linguatula* larvae and peritonitis. J. Trop. Vet. Sci., 4: 267.

Muraleedharan, K. and Syed Zaki, 1975. Occurrence of *Linguatula serrata* in cattle. Curr. Sci., 44: 430.

Ramanujachari, G. and Alwar, V. S., 1954. The role of domestic pigs in the life cycle of *Linguatula serrata* (Frohlich, 1789) (tongue worms). Indian Vet J., 30: 515-516.

Radhakrishna Reddy, K., Ramakrishna, K., Eswaraiah, Raghavan, R. S., and Gowhar Ali Khan, 1971. Occurrence of the nymphal stages of *Linguatula serrata* in Kashmiri goat. Indian Vet. J., 48: 1280-1282.

Ravindran, R., Bindu Lakshman, Ravishankar, C. and Subramanian H., 2006. Prevalence of *Linguatula serrata* in domestic ruminants in South India. Southeast Asian J. Trop. Med. Pub. Hlth., 39: 808-812.

Singh, S. P. Paliwal, O.P, and Singh, K. P., 1973. *Linguatula serrata* (Frohlich, 1789) in Indian goats. Indian J. Anim. Hlth., 12: 181-182.

Sangwan, A, K., Choudhari, S. S. and Bisla, R. S., 2007. Visceral pentastomosis in a goat. J. Vet Parasitol., 21: 83-84.

Sivakumar, P., Sankar, M., Nambi, P. A. Praveena, P. and Singh, N., 2005. The occurrence of the nymphal stages of *Linguatula serrata* in water buffaloes (*Bubalis bubalis*): Nymph micrometry and lymph node pathology. J. Vet. Med., Series A, 52: 506- 509.

Stuti Vatsya, Banerjee, P. S., Yadav, C.L. and Kumar, R. R., 2011. Prevalence of *Linguatula serrata* infection in small ruminants in and around Pantnagar, Uttarakhand. Indian J. Anim. Sci. 81: 249-250.

SECTION III–PROTOZOAN PARASITES

III. A PHYLUM: SARCOMASTIGOPHORA

6.1 FAMILY: TRYPANOSOMATIDAE

Alder, S. and Ber, M., 1941. The transmission of *Leishmania tropica* by the bite of *Phlebotomus papatasii*. Indian J. Med. Res., 29: 803-809.

Balarama Menon, P., 1952. A preliminary report on the incidence of surra and *Tabanus* flies in Rajasthan. Proc. 42nd Indian Sci. Cong. (Baroda), Part III, Sect., VIII, p.350.

Baldrey, F. S. H., 1911. The evolution of *Trypanosoma evansi* through the fly: *Tabanus* and *Stomoxys*. J. Trop. Vet. Sci., 6: 271-182.

Basu, B. C., 1947. Studies on the arthropod transmission of surra. Proc. 34th Indian Sci. Cong. Asso. (Delhi), III. 172.

Basu, B. C., Balarama Menon, P. and Sen Gupta, C. M., 1952. Regional distribution of *Tabanus* flies in India and its relationship to the incidence of surra. Indian J. Vet. Sci., 22: 273-292.

Chaudhury, R. P., Pavan Kumar and Khan, M. H., 1966. Role of stable fly, *Stomoxys calcitrans* in the transmission of surra in India. Indian J. Vet. Sci., 36: 18-28.

Christophers, S. R, Shott, H. E. and Barraud, J. P., 1925. The development of the parasite of Indian kala-azar in the sandfly, *Phlebotomus argentipes* Annandale and Brunetti. Indian J. Med. Res., 12: 605-607.

Cross, H. E., 1923. Further notes on surra transmission experiments with *Tabanus albimedicus* and tick.Trans. Roy. Soc. Trop. Med. Hyg., 16: 469.

Cross, H. E. and Abdulla Khan, 1923. Dept. Agri., Punjab Vet. Bull.,12,1-11.

Cross, H. E. and Patel, P. G., 1921a.vide, Gill, B. S. 1977.

Cross, H. E. and Patel, P. G., 1921b. A note on the transmission of surra by ticks—Vet. Bull. No. 6. Dept. Agri., Punjab.

Cross, H. E. and Patel, P. G., 1921c. A note on the transmission of surra by *Tabanus nemocallosus*. Vet. Bull. No. 8, Dept. Agri., Punjab, Lahore.

Cross, H. E. and Patel, P. G., 1921d. A note on Argasidae found in the Punjab. Vet. Bull. No. 6, Dept. Agri., Punjab, Lahore.(not indexed).

Cross, H. E and Patel, P. G., 1922a. Notes on transmission of Surra.Vet. Bull.,No. 8, Dept. Agri. Punjab.

Cross, H. E and Patel, P. G., 1922b.Vet. J., 78: 469, vide Sen & Flether.

Cross, H. E and Patel, P. G., 1922c. Vet. Bull. 8, Dept. Agri., Punjab, Lahore.

Cross, H. E and Patel, P. G., 1922d. vide Gill, B. S., 1971.

Cross, H. E and Patel, P. G., 1923. vide Gill, B. S., 1971, vide infra.

Dinesh, D. S., Kar, S. K., Kishore, K., Palit, A., Verma, N., Gupta, A. K., Chauhan, D. S., Singh, D., Sharma, V.D. and Katoch,V.M., 2000. Screening sandflies for natural infection with *Leishmania donovani* using a non-radioactive probe based on the total DNA of the parasite. Ann. Trop. Med. Parasitol., 94: 447-451.

Flectcher, T. B., 1916. Report of the imperial pathological entomologist. Rep. Agri., Res. Inst., Pusa, 1915-16., p.78.

Fletcher,T. B. and Senior-White,1921. Surra and Biting flies; A review: Report Proc. IVth Ento. Meeting held at Pusa. pp.222-235.(Not indexed).

Gill, B. S., 1977., Trypanosomes and Trypanosomiasis of Indian Livestock. ICAR, New Delhi.

Kahan Singh, K., 1925. Dept. Agri., Punjab Vet. Bull.,16: 1-8.

Kahan Singh, K., 1926. Dept. Agri., Punjab Vet. Bull.,16: 1-20.

Kahan Singh, K., 1929.vide, Gill, B. S. 1977.

Kumar, V., Bimal, S., Kesari, S., Kumar, A. J., Bagchi, A. K., Akbar, M. A., Kishore, K., Bhattacharya S. K.and Das, P., 2005. Evaluation of a dot-immunoblot assay for detecting leishmanial antigen in naturally infected *Phlebotomus argentipes* (Diptera: Psychodidae). Ann. Trop. Med. Parasitol., 99: 1-6.

Leese, A. S., 1909. Experiments regarding the natural transmission of surra, carried out at Mohand in 1908. J. Trop. Vet. Sci., 4: 107-132.

Leese, A. S., 1912. Biting flies and surra. J. Trop.Vet. Sci., 7: 19-32.

Lingard, A., 1906.Through what agency is *Trypanosma evansi* carried over from one surra season to another ? J. Trop.Vet. Sci.,1: 92-112.

Megat Adb Rani, P. A., Irwin, P. J., Gatne, M. L., Colman, G. T., Mcinnes, L. M. and Tromb, R. J., 2010. Canine vector-borne diseases in India: a review of the literature and identification of existing knowledge gaps. Parasites and Vectors 3: 28. (Review, not cited in Index, sand flies, *Leishmania tropica*).

Napier, L. V., Smith, R. C. A. and Krishnan, K. V., 1933b. The transmission of Indian kala-azar to hamsters by the bites of sand fly, *Phlebotomus argentipes*. Indian J. Med. Res., 21: 299-304.

Rogers, L., 1901. The transmission of the *Trypanosoma evansi* by horse flies and other experiments pointing to the probable identity of surra of India and Nagana or tsetse fly disease of Africa. Proc. Roy. Soc. London, 68: 163-170.

Sarma, R. D., 1979. High incidence of trypanosomiasis in Jammu and Kasmir State and its analysis. Livestock Adviser, 4: (5) 33.

Sarwar, S. M., 1936. The control of bovine surra, with special reference to possible vectors. Anim. Res. Workers' Conf., New Delhi,17-19 February, 1936. (unpublished; not indexed).

Sen, S. K., 1938. Experiments on the transmission of surra by the tick, *Ornithodoros tholozani*, Laboulbene and Megnin. Proc. Indian Sci. Cong.,Calcutta, January 1938.

Sen, S. K., 1946. Experiments on the transmission of surra by the tick, *Ornithodoros tholozani*, Laboulbene and Megnin. Indian J.Vet. Sci.,16: 165-166.

Sen, S. K. and Fletcher, T. B., 1962. Veterinary Entomology and Acarology for India. Indian Council of Agricultural Research, New Delhi.The Caxton Press Pvt. Ltd., New Delhi, pp. 668.(Not indexed; reference only).

Shastri, U. V. and Deshpande, P. D., 1981. *Hyalomma anatolicum anatolicum* (Koch, 1844) as a possible vector for transmission for *Trypanosoma theileri*, Laveran, 1902 in cattle. Vet. Parasitol., 9: 151-155.

Shastri, U. V., Deshpande, P. D. and Deshpande, M. S., 1981. *Hyalomma anatolicum anatolicum* (Koch, 1844) as a possible vector for transmission for *Trypanosoma theileri*, Laveran, 1902 in cattle and buffaloes. All India Symposium: Perspective in Research on Protozoa, Dept. Zool., Aurangabad, Abst., A27, p.14.

Shortt, H. E., 1928. The life history of *Leishmania donovani* in its insect and mammalian host. Trans. 7ᵗʰ Cong. Far Eastern Asso. Trop. Med., 3: 12-18. Thacker's Press, Calcutta.

Shortt, H. E., Smith, R. O. A. and Swaminath, C. S., 1932. Miscellaneous experiments with *Phlebotomus argentipes* in relation to transmission of kala-azar. Indian Med. Res. Mem., No. 25: 90-102.

Shortt, H. E., Smith, R. O. A., Swaminath, C. S. and Krishnan, K. V., 1931. Transmission of Indian kala-azar by the bites of *Phlebotomus argentipes*. Indian J. Med. Res., 18: 1373-1375.

Sinton, J. A., 1925. Notes on some Indian species of the genus *Phlebotomus*. Part IX. The role of insects of the genus *Phlebotomus* as carriers of disease with special reference to India. Indian J. Med. Res. 12: 701-729.(Not indexed).

Smith, R.O. A., Halder, K. C. and Ahmed. I., 1940. Further investigations on the transmission of kala-azar, Part III. The transmissions of by the bite of the sand fly, *Phlebotomus argentipes*. Indian J. Med. Res., 28: 585-591.

Smith, R. O. A., Halder, K. C. and Ahmed. I., 1941. Further investigations on the transmission of kala-azar, Part VI. A second series of transmissions of *Leishmania donovani* by *Phlebotomus argentipes*. Indian J. Med. Res., 29: 799-802.

Smith, R. O. A., Lal, C., Mukerjee, S. and Halder, K. C., 1936. The transmission of kala-azar by the bite of the sand fly, *Phlebotomus argentipes*. Indian J. Med. Res., 24: 313-316.

Sridhar, Rakesh Kumar and Sharma, R. D., 1989. A note on an attempt to transmit *Trypanosoma evansi* infection through ticks (*Hyalomma anatolicum anatolicum*) in rabbits. Indian J. Anim. Hlth., 28: 173-174.

Sukhbir Kaur, Kaul, P. and Khullar, 2001. Programmed cell death in experimental visceral leishmaniasis. J. Parasit. Dis., 25: 30-33.

Swaminath, C. S., Shortt, N. E. and Anderson, L. A. P., 1942. Transmission of Indian kala-azar to man by the bites of *Phlebotomus argentipes*, Ann and Brun. Indian J. Med. Res., 30: 473-477.

Vijay Veer, Parashar, B. D. and Shri Prakash, 2002. Tabanid and muscoid haematophagous flies, vectors of trypanosomiasis or surra disease in wild animals and livestock in Nandankanan Biological Park, Bhubaneswar (Orissa, India). Curr. Sci., 82: 500-503.(Reviewed vectors of surra in India, collected and identified them, not indexed).

7.1 FAMILY: SARCOCYSTIDAE

Sarcocystis

Abraham, S. S., Jayakumar, C. S. and Alexander, J., 2009. Sarcocystosis in mithuns (*Bos frontalis*) - a report. J. Threat. Taxa, 1(4): 252.

Acharjyo, L. N. and Rao, A. T., 1988. Sarcocystosis in some Indian wild ruminants. Indian Vet. J., 65: 169-170.

Achutan, H. N., 1983. *Sarcocystis* and sarcocystosis in buffaloes (*Bubalis bubalis*) in calves. Indian Vet. J., 60: 344-346.

Achutan, H. N. and Ebenezer Raja, E., 1990. Occurrence of *Sarcocystis* sp. in horse (*Equus caballus*). Indian Vet. J., 67: 472.

Agnigotri, R. K., Juyal, P. D. and Bhatia, B. B., 1987. Microsarcocysts in pigs in Uttar Pradesh. J. Vet. Parasitol., 1: 69-70.

Agnigothri, P. K. Juyal, P. D. and Bhatia, B. B., 1988. Isolation of microsarcocysts in musculature of pigs. Second Natl. Parasitol., March 3-5, 1988, Dept. Parasitol., Vet. Coll., Bangalore, India, Souvenir, Stream 2, p.9.

Agarwal, M. C., Singh, K. P. and Shah, H. L., 1983. Comparative efficacy of three coccidiostats in dogs fed *Sarcocystis* of goats. Seventh Natl. Cong. Parasitol., Ravi Shankar Univ., Raipur, 26-28, Dec., 1986, Abstr. C-26, p.122.

Agarwal, M. C., Singh K. P and Shah H. L., 1991. Caprine sarcocystosis in Jabalpur area. J. Vet. Parasitol., 11: 108-112.

Agrawal, M. C., Shah, H. L., Sharma, R. K. and Singh, K. P., 1987. A tissue mince method for detection of intact cysts of *Sarcocystis* in musculature. J. Vet. Parasitol., 1: 67-68.

Ahamad Pandit, B., 1993. Epidemological studies in *Sarcocystis* infection in cattle. Ph. D. Thesis submitted to G. B. Pant Univ. Agri.& Tech., Pantnagar-263 145, U. P.

Aulakh, R. S. Joshi, D. V. and Juyal, P. D., 1996. Occurrence of *Sarcocystis* cysts in raw and commercially cooked goat meat. Global meet on parasitic Diseases, New Delhi, India 18-22 March, 1996. Abstr., J. Parasit. Dis., 20: 99.

Aulakh, R. S., Joshi, D. V. and Juyal, P. D., 1997. Prevalence of *Sarcocystis capracanis* in naturally infected goats: A preliminary report. J. Vet. Parasitol., 11: 99-100.

Avapal, R. S., 2001. Prevalence and immunodiagnosis of *Sarcocystis* species of public health significance in swine. M. V. Sc. Thesis submitted to Punjab Agri. Univ., Ludhiana.

Avapal, R. S., Sharma, J. K. and Juyal, P. D., 2003a. Comparison of techniques to detect *Sarcocystis* infection in slaughtered pigs. J. Vet. Parasitol., 17: 65-66.

Avapal, R. S., Sharma, S. K. and Juyal, P. D., 2003b. Prevalence of *Sarcocystis* species infection in slaughtered pigs. J. Vet. Parasitol., 17: 151-153.

Avapal, R. S., Sharma, J. K. and Juyal, P. D., 2003c. Comparative morphology of *Sarcocystis suihominis* and *S. miescheriana* in domestic pigs. Indian J. Anim. Sci., 73: 392 - 393

Avapal, R. S., Sharma, J. K. and Juyal, P. D., 2004. Pathological changes in *Sarcocystis* infection in domestic pigs (*Sus scrofa*). Vet. J. (Netherlands), 168: 358-361.

Avapal, S., Singh, B. B., Sharma, J. K. and Juyal, P. D., 2001. Changes in blood glucose and total serum protein levels naturally *Sarcocystis* spp. infected pigs. Twelfth Natl. Cong. Vet. Parasitol., August 26-27, 2001, Dept. Parasitol., Coll. Vet. Sci., Acharya Ranga Agri. Univ., Tirupati-517 502, India, Abstr. S-3: 2, p.90.

Avapal, R. S., Singh, B. B. Sharma, J. K. and Juyal, P. D., 2002. Sero-prevalence of *Sarcocystosis* in swine. J. Parasit. Dis., 26: 109-110.

Banerjee, P. S., 1998. Studies on pathogenesis of *Sarcocystis tenella* in sheep. Ph. D. Thesis submitted to G. B. Pant Univ. of Agri. & Tech., Pantnagar-263 145, U. P. (Vide Thesis Abstract vide Banerjee, P. S., 1998. J. Vet. Parasitol., 12: 65.).

Banothu Dasmabai, Udaya Kumar, M., Reddy, Y.N. and Anand Kumar, A., 2012a. Certain morphological studies on *Sarcocystis* spp. affecting bovines. XXII Natl. Cong. Vet. Parasitol., March 15-17, 2012, Coll. Vet. Sci. Anim. Husb., U. P. Pandit Deen Dayal Upadhyaya Pashu Chikitsa Vigyan Vishwavidyalaya Evam Go Anusandhan Sansthan (DUVASU), Mathura-281001(UP.), India, Abstr. No.I.02, pp.1-2.

Banothu Dasmabai, Udaya Kumar, M., Reddy, Y.N. and Anand Kumar, A., 2012b. Immunofluorescent antibody technique in the diagnosis of bovine sarcocystosis. XXII Natl. Cong. Vet. Parasitol., March 15-17, 2012, Coll. Vet. Sci. Anim. Husb., U. P. Pandit Deen Dayal Upadhyaya Pashu Chikitsa Vignan Vishwavidyalaya Evam Go Anusandhan Sansthan (DUVASU), Mathura-281001(UP.), India, Abstr. No.III.05, p.35.

Bineesh, Saket Sharma, Kansal, S. K., Banga and Brar, R. S., 2006. Sarcocystosis in a chicken. Indian Vet. J., 83: 809.

Biswas, S., Chakraborty, A. and Roy, S., 1990.Sarcocystis infection in buffalo meat (carabeef) in West Bengal. J.Vet. Anim.Sci., 21: 29-31.

Biswas, S., Chakraborty, A. and Roy, S., 1992. Sarcocystic infection in beef cattle in West Bengal. J.Vet. Anim.Sci., 23: 52-54.

Chhabra, M. B. and Mahajan, R. C., 1978. *Sarcocystis* sp. from the goat in India. Vet. Rec.,103: 562-563.

Chhabra, M. B. and Samantaray, S., 2013. *Sarcocystis* and sarcocystosis in India: status and emerging perspectives. J. Parasit. Dis., 37: 1-10. (Review article, not indexed).

Chaudhry, R. K., 1983. Biochemical studies on sarcocysts of *Sarcocystis fusiformis* of the buffaloes (*Bubalis bubalis*). Ph.D. Thesis submitted to Jawaharlal Nehru Krishi Vishwa Vidyalaya, Jabalpur.

Chaudhry, R. K. and Shah, H. L., 1987. Histochemical observation on sarcocysts of *Sarcocystis cruzi* of cattle. Indian J. Anim. Sci., 57: 688-691.

Chaudhry, R. K. and Shah, H, L., 1988. Role of man in the life-cycle of *Sarcocystis* sp. of the goat (*Capra hircus*). Indian Vet. J., 65: 742.

Chaudhry, R. K., Kushwah, H. S. and Shah, H, L., 1985. Biochemistry of the sarcocysts of *Sarcocystis fusiformis* of buffaloes, *Bubalis bubalis*. Vet. Parasitol., 17: 295-298.

Chaudhry, R. K., Kushwah, H. S. and Shah, H, L., 1986. Biochemical and histochemical studies of the sarcocysts of *Sarcocystis fusiformis* of buffaloes (*Bubalis bubalis*). Vet. Parasitol., 21: 271-273.

Chaudhry, R. K., Shah, H, Land Guru, V. K., 1986. Oxygen consumption by the sarcocysts of *Sarcocystis fusiformis* of buffalo. Indian J.Anim. Sci., 56: 53.

Chauhan, P. P. S. and Agrawal, R. D., 1977. Prevalence of *Sarcocystis* in goats (*Capra hircus*) from Western Uttar Pradesh. Indian Vet. J., 74: 1065-1066.

Chauhan, P. P. S., Agrawal, R. D. and Arora, G. S., 1978. Incidence of *Sarcocystis fusiformis* in Indian buffaloes. Indian J. Parasitol., 2: 123-124.

Chauhan, P. P. S., Bhatia, B. B., Agrawal, R. D., Kataria, R. P. and Ahluwalia, S. S., 1977. On the gametogonic development of bubaline *Sarcocystis fusiformis* in pups – an experimental study Indian J. Exp. Biol., 15: 492-494.

Chhabra, M. B. and Mahajan, R. C., 1978. *Sarcocystis* spp. from goats in India.Vet. Rec. 103: 562-563.

Chowdhury, S., 1999. Pathogenesis and treatment of lambs experimentally infected with *Sarcocystis tenella*. M.V.Sc., Thesis, submitted to G.B. Pant Univ. Agri. & Tech., Pantnagar, India.

Dafedar, A., 2005., Prevalence of Sarcocystosis in food animals in Bangalore with special emphasis on serodiagnosis. M.V.Sc., Thesis, submitted to the Karnataka Vet. Anim. Fisher. Sci. Univ., Bidar.

Dafedar, A. and D'Souza, P. E., 2008. Prevalence and morphological studies on *Sarcocystis* species infecting cattle in Bangalore. 20th Natl. Cong. Parasitol., November 3-5, 2008, Dept. Parasitol. and Indian Soc. Parasitol., North Eastern Hill Univ., Shillong-793 022, Abstr. P-33, p.109.

Dafedar, A. and D'Souza, P. E., 2009. Serodiagnosis of *Sarcocystis capracanis* infection in goats with ELISA and CIEP. XIX Natl. Cong. Vet. Parasitol. & Natl. Sym. Natl. impact on parasitic diseases of livestock health and production., February 3-5, 2009, IAAVP & Dept. Vet. Parasitol., Coll. Vet. Sci., Guru Angad Dev Vet. & Anim. Sci. Univ., Ludhiana-141 004, OI 33, p.114.

Dafedar, A., D'Souza, P. E. and Mamatha, G. S., 2011. Prevalence and morphological studies on *Sarcocystis* species infecting cattle in Bangalore. J. Vet. Parasitol., 25: 183-184.

Dafedar, A., D'Souza, P. E., Ananda, K. J. and Puttalakshmamma, G. C., 2008. Prevalence of sarcocystosis in goats slaughtered at an abattoir in Bangalore, Karnataka State. Vet. World, 1: 335-337.

Das, M. R and Borkakoty, M. R., 1978. Occurrence of *Sarcocytis* sp. in the diaphragm of cattle in Assam. Indian J. Parasitol., 2: 179.

Daya Shankar, 1991. Studies on epidemiological aspects of sarcocysts infection in domestic goat (*Capra hircus*) in Tarai (U. P) and its serodiagnosis with reference to eimerian infection. M. V. Sc. Thesis submitted to G. B. Pantnagar Univ. of Agri. & Tech., Pantnagar.

Daya Shankar and Bhatia, B. B., 1993a. Gametogonic and sporogonic development of *Sarcocystis capracanis* and *S. hircicanis* of goats in experimental pups with remarks on their pathogenicity. J. Parasit. Appl. Anim. Biol., 2: 75-79.

Daya Shankar and Bhatia, B. B., 1993b. *Sarcocystis* infection in goats of Uttar Pradesh. Indian J. Anim. Sci., 63: 284-287.

Deodar, N. S., Narasapur, V. S. and Ajinkya, M., 1968. A survey of *Trichinella* and *Sarcosporidium* in pigs in Bombay. Bombay Vet. Coll. Mag., 15: 38-40.

Deshpande, V. S. and Pethkar, D.K., 1995. *Sarcocystis* infection in buffaloes (*Bubalis bubalis*) in Palanpur, Gujarat. Seventh Natl. Cong. Vet. Parasitol., 19-21, Aug. 1995. Dept. Parasitol., Madras Vet. Coll., Madras -7, Abst. SII 11, p.99.

Deshpande, A. V. and Shastri, U. V., 1982. Sarcosporidiasis in buffaloes in Marathwada region (Maharashtra) India: A preliminary study. Proc. Fourth Natl. Cong, Parasitol., 28-30 March, Aligarh Muslim Univ., Aligarh, p.44.

Deshpande, A. V., Shastri, U. V. and Deshpande, M. S., 1982. Experimental infection of pups by feeding cysts of *Sarcocystis levinei* Dissanaike and Kan, 1978. Indian J. Parasitol., 6: 331-332.

Deshpande, A. V., Shastri, U. V. and Deshpande, M. S., 1983. Experimental infection of pups by feeding cysts of *Sarcocystis levinei* Dissanaike and Kan, 1978. Indian J. Parasitol., 6: 331-332.

Deshpande, A. V., Shastri, U. V. and Deshpande, M. S., 1987. Prevalence of sarcocysts in cattle and buffaloes in Marathwada (India). Trop. Vet. Anim. Sci. Res., 1: 92-93.

Devi, P. P., Sarmah, P. C. and Saleque, A., 1998. *Sarcocystis* sp. infection in carcasses of domestic pigs in Guwahati, Assam. J. Vet. Parasitol., 12: 56-57.

Dey, S., Gupta, S. L., Singh, R. P. and Verma, V. C. 1993. Clinico-pathological changes on goats experimentally infected with *Sarcocystis capracanis*. Indian J.Vet Med., 13: 5-8.

Dey, S., Khatkar, S. K., Ghosh, J. D. and Akbar, M. A., 1995. Clinicohaematologic and serum biochemical changes in piglets experimentally infected with *Sarcocystis meescheriana*. Indian Vet. J., 72: 126-130.

Gangadharan, B., Valsala, K.V., Nair, M. V. and Rajan, A., 1992. Sarcocystosis in a sambar deer (*Cervus unicolor*). Indian J. Anim. Sci., 62: 127-128.

George Verghese, C., Madhavan Pillai and Ouseph, C. K., 1986. Occurrence of *Sarcocystis* in buffaloes (*Bubalus bubalis*) in Trichur, Kerala. Kerala J. Vet. Sci., 17: 140-142.

Ghosal, S. B., 1978a. Studies on *Sarcocystis* of the buffalo (*Bubalus bubalis*). Ph. D. Thesis submitted to Jawaharlal Nehru Vishwa Vidyalaya, Jabalpur, India.

Ghosal, S. B., 1978b. Studies on *Sarcocystis* of buffaloes (*Bubalis bubalis*). Proc. Second Natl. Cong. Parasitol.,Varanasi, India, Sec II, pp.3-4.

Ghosal, S. B., 1989a. Morphological studies of oocysts and sporocysts of sarcocysts of bubaline *Sarcocystis*. Indian J. Parasitol., 13: 327-328.

Ghosal, S. B., 1989b. Comparative morphology of the sarcocysts of *Sarcocystis levinei* and *S. fusiformis* in experimentally infected water buffaloes (*Bubalus bubalis*). Indian J. Parasitol., 13: 329- 330.

Ghosal, S. B. and Shah, H. L., 1979. Studies on *Sarcocystis* of buffaloes (*Bubalus bubalis*) 2nd Natl. Cong. Parasitol. Banaras, Sect. II, pp.3-4.

Ghosal, S. B., Joshi, S. C. and Shah, H. L., 1986. A note on the natural occurrence of *Sarcocystis* of buffaloes (*Bubalus bubalis*) in Jabalpur region, M. P. Indian Vet. J., 63: 165-166.

Ghosal, S. B., Joshi, S. C. and Shah, H. L., 1987a. Development of sarcocysts of *Sarcocystis fusiformis* in water buffalo (*Bubalus bubalis*) experimentally infected with sporocysts from cats. Indian J. Anim. Sci., 57: 413-415.

Ghosal, S. B., Joshi, S. C. and Shah, H. L., 1987b. Development of sarcocysts of *Sarcocystis levinei* in the water buffalo (*Bos bubalis*) experimentally infected with sporocysts from dogs. Vet. Parasitol., 26: 165-167.

Ghosal, S. B., Joshi, S. C. and Shah, H. L., 1987c. Morphological studies of the sarcocysts of *Sarcocystis levinei* from naturally infected water buffaloes (*Bubalis bubalis*). Indian Vet. J., 64: 915-917.

Ghosal, S. B., Joshi, S. C. and Shah, H. L., 1987d. Sporocysts output in cats fed sarcocysts of *Sarcocystis fusiformis* of the buffalo. Indian J. Anim. Sci., 57: 1100-1101.

Ghosal, S. B., Joshi, S. C. and Shah, H. L., 1988a. Sporocysts output in dogs fed sarcocysts of *Sarcocystis levinei* of the buffalo (*Bos bubalis*). Vet. Parasitol., 28: 173-174.

Ghosal, S. B., Joshi, S. C. and Shah, H. L., 1988b. Studies on the morphology of the sarcocysts of *Sarcocystis fusiformis* from naturally infected water buffaloes (*Bubalis bubalis*). Indian Vet. J., 65: 196-199.

Gill, H. S., Singh, A., Vadehra, D. V. and Sethi, S. K., 1978. Shedding of unsporulated oocysts in faeces by dogs fed diaphragm muscles from water buffalo (*Bubalus bubalis*) naturally infected with *Sarcocystis*. J. Parasitol., 64: 549-551.

Goldy Srivastava and Jain, P. C., 2000. Incidence of *Sarcocystis* in sheep (*Ovis aries*) in Madhya Pradesh. Indian Vet. Med. J., 24: 41-45.,

Goldy Srivastava and Jain, P. C., 2001. Sporocyst output in dogs fed with sarcocysts of *Sarcocystis arieticanis* and *S. tenella* of the sheep. Indian Vet. J., 78: 533-535.

Gopalakrishna Rao, D. and Rama Rao, P., 1983. 5th Natl. Cong. Parasitol., June 25-27, 1983, Coll. Vet. Sci., Andhra Pradesh Agri. Univ., Tirupati, Andhra Pradesh and Indian Soc. Parasitol., Abstr. C-46, p.76.

Gopalakrishna Rao, D and Rama Rao, P., 1987. A note on sarcocystosis in domestic animals of Andhra Pradesh. Indian Vet. J., 64: 614-615.

Gupta, P. P. and Singh, S. P., 1988. Sarcocystosis in the brain of a rabid cow. Indian Vet. J., 65: 1130.

Gupta, R.S., 1990. Metabolic studies on various fractions of sarcocysts of *Sarcocysts fusiformis* of buffalo (*Bubalus babalis*). Ph.D. Thesis submitted to Jawaharlal Nehru Krushi Vishwa Vidyalaya, Jabalpur (Madhya Pradesh), India.

Gupta, R. S., Kushwah, H. S. and Ameeta Kushwah, 1992. Some glucose metabolic enzymes in various fractions of sarcocysts of *Sarcocystis fusiformis* of buffaloes (*Bubalus bubalis*). Vet. Parasitol., 44: 45-50.

Gupta, R. S., Kushwah, H. S. and Ameeta Kushwah, 1993. Enzymatic studies on various fractions of sarcocysts of *Sarcocystis fusiformis* of buffaloes. Indian J. Anim. Sci., 63: 20-21.

Gupta, R. S., Kushwah, H. S. and Ameeta Kushwah, 1993. *Sarcocystis fusiformis*: some protein metabolic enzymes in various fractions of sarcocysts of *Sarcocystis fusiformis* of buffalo (*Bubalus bubalis*). Vet. Parasitol., 45: 185-189.

Gupta, R. S., Kushwah, H. S. and Ameeta Kushwah, 1994. Some enzymes of glucogenesis of various fractions of sarcocysts of *Sarcocystis fusiformis* of buffaloes (*Bubalus bubalis*). Vet. Parasitol., 52: 145-149.

Gupta, R. S., Kushwah, H.S. and Ameeta Kushwah, 1995. *Sarcocystis fusiformis*: some Kerb cycle enzymes in various fractions of sarcocysts of buffalo (*Bubalus babalis*). Vet. Parasitol., 56: 1-5.

Gupta, S. C. and Iyer, P. K. R., 1984. Spontaneous sarcocystosis in the brain of pigs. Indian Vet. J., 61-738-739.

Gupta, S. L., 1983. Prevalence of *Sarcocystis* infection in animals of Northern India and its transmission to dogs. 5th Natl. Cong. Parasitol., June 25-27, 1983, Coll. Vet. Sci., Andhra Pradesh Agri. Univ., Tirupathi, Andhra Pradesh and Indian Soc. Parasitol., Abstr. E-15, pp.101-102.

Gupta, S. L. and Gautam, O. P., 1982. *Sarcocystis* infection in goats of Hisar and its transmission to dogs. Indian J. Parasitol., 6: 73-74.

Gupta, S. L. and Gautam, O. P., 1984. *Sarcocystis* infection in pigs of Hisar, Haryana, India and its transmission to dogs. Vet. Parasitol., 16: 1-3.

Gupta S. L., Gautam., O.P. and Bhardwaj, R. M., 1979. A note on the prevalence of *Sarcocystis* infection in sheep from Hisar area as studied by peptic digestion technique. Indian J. Anim. Sci., 49: 971-972.

Hemaprasanth., 1995a. *Sarcocystis* infection in domestic pigs in Uttar Pradesh. J. Vet. Parasitol., 9: 119-123.

Hemaprasanth., 1995b. Immunological and biochemical comparison of *Sarcocystis miescheriana* and *S. suihominis* in pigs. Ph. D. Thesis submitted to G. B. Pantnagar Univ. of Agri. & Tech., Pantnagar-263 145, U. P. Abstr., J. Vet. Parasitol., 9: 151-152.

Hemaprashanth and Bhatia, B. B., 1995a. Morphology of sarcocysts of three species of *Sarcocystis* from naturally infected pigs in India. Seventh Natl. Cong. Vet. Parasitol., 19-21 August, 1995, Dept. Parasitol., Madras Vet. Coll., Madras -7, Abstr. SII 17, p.103.

Hemaprashanth and Bhatia, B. B., 1996. On the species of *Sarcocystis* of the pig from Uttar Pradesh, India with a description of a sarcocyst of hitherto unrecognized species. J. Vet. Parasitol., 10: 57-61.

Hussain, M. M., Gupta, S. L. and Singh, R. P., 1986a. Lambs experimentally infected with *Sarcocystis ovicanis*. 1. Clinical haematological observations. Indian Vet. Med. J., 6: 110.

Hussain, M. M., Gupta, S. L., Singh, R. P. and Verma, P. C., 1986b. Pathological changes in lambs infected with *Sarcocystis ovicanis* during experiments. Indian J. Anim. Sci., 56: 169-173.

Hussain, M. M., Gupta, S. L. and Singh, R. L., 1986c. Prevalence of *Sarcocystis* infection in sheep and its transmission to dogs. Indian J. Anim Sci., 56: 341-342.

Hussain, M. M., Gupta, S. L., and Singh, R. P., 1987. Redischarge of sporocysts from pups fed with *Sarcocystis ovicanis* infected meat repeatedly. Indian J. Anim. Res., 21: 113-114.

Iyer, P. K. R., 1971. The occurrence of a cyst of *Sarcocysts* Lankester, 1882 in the cerebrum of a goat. Curr. Sci., 40: 135-136.

Jain, A. K., 1984. Studies on some aspects of *Sarcocystis levinei* in buffaloes. M. V. Sc. Thesis, Haryana Agri. Univ., Hissar, India.

Jain, A. K., Gupta, S. L. and Singh, R. P., 1985. Discharge of sporocysts from pups fed with *Sarcocystis levinei*. Indian J. Parasitol., 9: 77.

Jain, A. K., Gupta, S. L. and Singh, R. P., 1986a. Prevalence of *Sarcocystis* infection in buffaloes (*Bubalis bubalis*) and its transmission to dogs. Indian J. Parasitol., 10: 75-76.

Jain, A. K., Gupta, S. L. and Singh, R. P. and Mahajan, S. K., 1986b. Experimental *Sarcocystis levinei* infection in buffalo calves. Vet. Parasitol., 21: 51-53.

Jain, P. C., 1982. Studies on some aspect of *Sarcocystis* of cattle (*Bos indicus*) and buffaloes (*Bubalus bubalis*). Ph. D. Thesis submitted to Jawaharlal Nehru Krishi Viswa Vidyalaya, Jabalpur, India.

Jain, P. C. and Shah, H. L., 1985a. Cross-transmission of *Sarcocystis cruzi* of the cattle to buffalo calves. Indian J. Anim. Sci., 55: 27-28.

Jain, P. C. and Shah, H. L., 1985b. Prevalence and seasonal variations of *Sarcocystis* of cattle of Madhya Pradesh. Indian J. Anim Sci., 55: 29-31.

Jain, P. C. and Shah, H. L., 1986a. Gametogony and sporogony of *Sarcocystis fusiformis* of buffaloes in the small intestine of experimental infected cats. Vet. Parasitol., 21: 205-209.

Jain, P. C. and Shah, H. L., 1986b. Experimental study on gametogenic development of *Sarcocystis leveini* of buffaloes in the small intestine of dogs. Indian J. Anim. Sci., 56: 314-318.

Jain, P. C. and Shah, H. L., 1986c. Determination of sporocysts discharged during the patent period by dogs fed with sarcocysts of *Sarcocystis cruzi* of the cattle. Indian J. Anim. Sci., 56: 1005-1008.

Jain, P. C. and Shah, H. L., 1987a. *Sarcocystis hominis* in cattle in Madhya Pradesh and its public health importance. Indian Vet. J., 64: 650-654.

Jain, P. C. and Shah, H. L., 1987b. Comparative morphology of oocysts and sporocysts of bovine and bubaline *Sarcocystis* in Madhya Pradesh. Indian J. Anim. Sci., 57: 849-852.

Jain, P. C. and Shah, H. L., 1988. A comparative morphology of sarcocysts of three species of *Sarcocystis* infecting cattle in Madhya Pradesh. Indian J. Anim. Sci., 58: 8-13.

Jayashree Shit, Arpita Basu and Subhasish Biswas, 2010. Evaluation of sarcocyst infested bovine carcasses. Proc.22nd Natl. Cong. Parasitol., October 30 - November 1, 2010, Editor-in-Chief, Prof. P. K. Bandyopadhyay, Dept. Zool., Univ.of Kalyani, Kalyani-741235, West Bengal, India, pp.102-109.

Jog, M. M., and Watve, M. G., 2005. *Sarcocystis* of chital-dhole: conditions for evolutionary stability of a predator parasite mutualism. B. M. C. Ecol., 5: 3 doi: 10.1186/1472-6785-5-3.

Jog, M. M., Marathe, R. R., Goel, S. S., Ranade, S. P., Kunte, K. K. and Watve, M. G., 2003. *Sarcocystis* infection chital (*Axis axis*) and dhole (*Cuon alpinus*) in two Indian protected areas. Zoos' Print J., 18: 1220-1222.

Jog, M. M., Marathe, R.R., Goel, S. S., Ranade, S. P., Kunte, K.K. and Watve, M. G. 2005. Sarcocystosis of chital (*Axis axis*) and dhole (*Cuon alpines*): ecology of mammalian prey-predator-parasite system in Penninsular India. J. Trop. Ecol., 21: 479-482.

Jumde, P. D., Bhojne, G. R., Maske, D. K. and Kolte, S. W., 2000. Prevalence of sarcocytsosis in goats in Nagpur. Indian Vet. J., 77: 662-663.

Juyal, P. D., 1987. Studies on the incidence and clinico-pathology and immune response of *Sarcocystis* infection in goats. Ph. D. Thesis submitted to Haryana Agri. Univ., Hisar-125004. Abstr., J. Vet. Parasitol., 1: 81-82.

Juyal, P. D., 1991. *Sarcocystis* and sarcocystosis in India. Southeast Asian J. Trop. Med. Pub. Hlth. 22: 138-141.(Review article, not indexed).

Juyal, P. D., 1993. Animal Sarcocystis and sarcocystosis in India. Livestock Advisor, XIX (V): 5-8. (Review, not indexed)

Juyal, P. D. and Bali, H.S., 1991. Equine Sarcocystosis: Indian perspective. J. Rem. Vet. Cor., 4: 227-231.(Review, not indexed)

Juyal P. D. and Bhatia, B. B., 1987. A note on the occurrence of *Sarcocystis* infection in goats (*Capra hircus*) and buffaloes (*Bubalus bubalis*) of Tarai region of Uttar Pradesh. Indian Vet. Med. J., 11: 234-235.

Juyal P. D. and Bhatia, B. B., 1989a. Sarcocystosis: An emerging zoonosis. Indian Vet. Med. J., 13: 66-69. (Review, not indexed).

Juyal, P. D. and Bhatia, B. B., 1989b. Prevalence and intensity of *Sarcocystis fusiformis* infection in water buffaloes (*Bubalis bubalis*) in Tarai region of Uttar Pradesh. Indian Vet. Med. J., 13: 269-271.

Juyal, P. D. and Bhatia, B. B., 1992c. Buffalo oesophagi: A source of *Sarcocystis fusiformis* infection in cats. J. Vet. Anim. Sci., 23: 69-70.

Juyal, P. D., Bhatia, B.B. and Saleque, A., 1988a. Frequency of *Sarcocystis* infection in goats and buffaloes in Tarai region of Uttar Pradesh. Second Natl. Parasitol., March 3-5, 1988, Dept. Parasitol. Vet. Coll., Bangalore, India, Souvenir, Stream II, Abstr., p.15.

Juyal, P. D., Bhatia, B. B. and Saleque, A., 1989a. Experimentally induced *Sarcocystis capricanis* infection in pregnant Black Bengal goats. Indian Vet. Med. J., 13: 200-202.

Juyal, P. D., Chhabra, M. B., and Ruprah, N. S., 1987. Proc. Inaug. Symp., Indian Asso. Adv. Vet. Parasitol., (abst.) pp.15.

Juyal, P. D., Gupta, M. P, Gupta, P.P., Singh, H. and Kalra, I.S., 1994. Chronic sarcocystis infection in naturally infected mare. Indian Vet. Med. J., 18: 54-55. (Case report).

Juyal, P. D., Kalra, I. S. and Bali, H. S., 1990a. Occurrence of *Sarcocystis equicanis* in a horse (*Equus caballus*) in India. J. Vet. Parasitol., 5: 53-54.

Juyal, P. D., Kalra, I. S. and Gupta, P. P., 1993. A preliminary report on the development of *Sarcocystis equicanis* from eqine musculature in a dog. Indian Vet. Med. J., 17: 70-71.

Juyal, P. D., Ruprah, N. S. and Chhabra, M. B., 1988b. Assessment of efficacy of different techniques for the detection of microscopic *Sarcocystis* in oesophageal muscle. Indian J. Parasit., 12: 185-196.

Juyal, P. D., Ruprah, N. S. and Chhabra, M. B., 1989b. Incidence of micro-sarcocystis infection in slaughtered goats (*Capra hircus*) at Hisar, Haryana. Indian J. Parasitol., 13: 13-16.

Juyal, P. D., Ruprah, N. S. and Chhabra, M. B., 1989c. Evidence for a cellular immune response in caprine sarcocystosis. Indian J. Parasitol., 13: 197-200.

Juyal, P. D., Ruprah, N. S. and Chhabra, M. B., 1989d. Rapid isolation of intact micro-sarcocystis (Protozoa-Apicomplexa) cysts from muscular tissues. Indian J. Anim. Hlth., 28: 69-71.

Juyal, P. D., Ruprah, N. S. and Chhabra, M.B., 1989e. Epidemiology of *Sarcocystis* (microcyst) infection in slaughter goats (*Capra hircus*) and its transmission to dogs at Hisar (Haryana), India. Proc. 13th WAAVP Conf. 1989, Berlin, GDR., p.27.

Juyal, P. D., Ruprah, N.S. and Chhabra, M. B. 1989f. Experimentally induced *Sarcocystis capracanis* infection in pregnant goats. Indian Vet. Med. J., 13: 200-202.

Juyal, P. D., Ruprah, N. S. and Chhabra, M. B., 1990b. Evidence of cellular immune response in experimental caprine sarcocystosis. Indian J. Parasitol., 13: 197-200.

Juyal, P. D., Saleque, A and Bhatia, B. B., 1990. Double diffusion (DID) and single radial immunodiffusion (SRID) tests for detection of antibody responses against cystic antigen for *Sarcocystis fusiformis*. J. Vet. Parasitol., 4: 83-85.

Juyal, P. D., Sahai, B. N., Srivastava, P. S. and Sinha, S. R. P., 1982. Heavy sarcocystosis in ocular musculature of cattle and buffaloes. Vet. Res. Commun., 5: 337-342.

Juyal, P. D., Gupta, M. P., Gupta, P. P., Singh, H. and Kalra, I.S., 1994. Chronic sarcocystis infection in naturally infected mare. Indian Vet. Med. J., 18: 54-55.

Khatkar, K., Singh, R. P., Gupta, S. L. and Verma, P. C., 1993. Histopathology of experimental *Sarcocystis meischeriana* infection in pigs. Indian J. Anim. Sci., 63: 932-935.

Khulbe, D. C., Kushwah Ameeta and Kushwah, H. S., 1989a. Biochemistry of various fractions of sarcocysts of *Sarcocystis fusiformis* of buffaloes (*Bubalis bubalis*). Vet. Parasitol., 31: 1-5.

Khulbe, D. C., Kushwah Ameeta and Kushwah, H. S., 1990. Biochemical studies on the various fractions of *Sarcocystis fusiformis* in buffaloes. Indian J. Anim. Sci., 60: 799-800.

Khulbe, D. C., Kushwah, A., Kushwah, H. S. and Quadri, M. A., 1989b. Electrophorectic separation of the soluble proteins of different fractions of the sarcocysts of *Sarcocystis fusiformis* of buffaloes (*Bubalis bubalis*). Vet. Parasitol., 32: 127-131.

Kumar, A., 1988. Studies on the immunology and control of caprine sarcocystis. M. V. Sc., Thesis submitted to Bihar Vet. Coll., Patna, Bihar, India. (Not indexed).

Kumar, A., Srivastava, P. S. and Sinha, S. R. P., 1988. Chemotherapy of experimental caprine sarcocystosis in goats and young dogs with salinomycin (Coxistac Pfizer). J. Vet. Parasitol., 2: 129-132.

Lal Singh, 1991. Studies on *Sarcocystis capracanis*. M. V. Sc. Thesis submitted to Rajasthan Agri. Univ., Bikaner.

Lal Singh, Raisinghani, P. M., Kumar, D., Pathak, K. M. L. and Manohar, G. S., 1992a. Sporocyst output in dogs fed with sarcocyts of *Sarcocystis capracanis*. Indian J. Parasitol., 6: 35-36.

Lal Singh, Raisinghani, P. M., Pathak, K. M. L., Kumar, D and Manohar, G. S., 1992b. Epidemiology of *Sarcocystis capracanis* in goats at Bikaner, Rajasthan, India. Indian J. Anim. Sci., 62: 1044-1045.

Lal Singh, Raisinghani, P. M., Kumar, D., Pathak, K. M. L. and Sivasankar, C. P., 1993. Chemical and hematobiochemical changes in dogs experimentally infected with *Sarcocystis capracanis*. Indian J. Anim. Sci., 63: 1055-1057.

Mamatha, G. S., 2006. Immunodiagnosis and molecular characterization of *Sarcocystis* sp. in cattle. Ph. D. Thesis, Karnataka Vet. Anim. Fishr. Sci. Univ., Bidar, Karnataka. (Thesis Abstract, J. Vet. Parasitol., 20: 199).

Mamatha, G. S., 2008. Sarcocytsosis of domestic animals and man. XI. Natl. Trng. Prog.Vet. Parasitol.on Trends and perspectives in the detection, identification and control of parasitic.diseases.10th to 30th November, 2008. Centr Adv. Stud., Dept. Parasitol., Vet. College, Univ. Agri. Sci., Hebbal, Bangalore-560 024. pp.67-75.(Review article, not indexed).

Mamatha, G. S. and D'Souza, P. E., 2006a. Counter current immunoelectrophoresis for rapid detection of sarcocystosis in cattle. Proc. XVII Natl. Cong. Vet. Parasitol., November 15-17, 2006, Dept. Parasitol., Rajiv Gandhi Coll. Vet. & Anim. Sci., Kurumbapet, Puducherry- 605 009, Ab S.IV-12, p.127.

Mamatha, G. S. and D'Souza, P. E., 2006b. Serodiagnosis of sarcocystosis in cattle by ELISA. Proc. XVII Natl. Cong. Vet. Parasitol., November 15-17, 2006, Dept. Parasitol., Rajiv Gandhi Coll. Vet. & Anim. Sci., Kurumbapet, Puducherry- 605 009, Ab S.V-28, p.156.

Mamatha, G. S. and D'Souza, P. E., 2007. Identification of immunoreactive peptides in *Sarcocystis* species in cattle by enzyme immunotransfer blot (EITB). Natl. Conf. on Parasitic diseases and their control in the new millennium organized by Janardhana Foundation at Bangalore,13-15th April, 2007.

Mamatha, G. S., D'Souza, P. E. and Suryanarayana, V. V. S., 2008a. RAPD-PCR assay for differentiation of two *Sarcocystis* species in cattle. 20th Natl. Cong. Parasitol., November 3-5, 2008, Dept. Parasitol. and Indian Soc. Parasitol., North Eastern Hill Univ., Shillong 793 022, Abstr. OP-1, p.72.

Mamatha, G. S., D'Souza, P. E. and Suryanarayana, V. V. S., 2008b. Antigenic profile of *Sarcocystis* species in cattle. 20th Natl. Cong. Parasitol., November 3-5, 2008, Dept. Parasitol. and Indian Soc. Parasitol., North Eastern Hill Univ., Shillong 793 022, Abstr. PP-94, p.176.

Mamatha, G. S., D'Souza, P. E. and Suryanarayana, V. V. S., 2008c. Serodiagnosis of bovine sarcocystosis by enzyme immune transfer blot (EITB) in naturally infected cattle. J. Vet. Parasitol., 22: 69-72.

Mamatha, G. S., D'Souza, P. E. and Suryanarayana, V. V. S., 2009. RAPD-PCR assay for differentiation of *Sarcocystis bovicanis* and *S. bovifelis* in cattle. In: Currents Trends in Parasitolgy, Veena Tandon, Arun K. Yadav and Bishnupada Roy (Eds). Proc. 20ᵗʰ Natl. Cong. Parasitol, Shillong, India (November 3-5, 2008), Panima Publishing Corporation, New Delhi, pp.101-109.

Mithalesh Kumari, Arti Sachan, Parul, Dherendra Kumar and Daya Shankar, 2012. Prevalence of sarcocystosis in slaughtered buffaloes in and around Mathura (UP). XXII Natl. Cong. Vet. Parasitol., March 15-17, 2012, Coll. Vet. Sci. and Anim. Husb., U. P. Pandit Deen Dayal Upadhyaya Pashu Chikitsa Vignan Vishwavidyalaya Evam Go Anusandhan Sansthan (DUVASU), Mathura-281001(UP.), India, Abstr. No.VI.54, p.110.

Mohanty, B., Misra, S. C. and Rao, A. T., 1995a. Epidemiology of *Sarcocystis* infection in ruminants and dogs in Bhubaneswar. Sixth Nat. Cong. Vet. Parasitrol., 22-24 October, 1994, Dept. Parasitol., Coll. Vet. Sci., JNKVV, Jabalpur-482 001, M. P., India. Abstr. S-1: 6, p.3.

Mohanty, B., Misra, S. C. and Rao, A. T., 1995b. Pathology of *Sarcocystis* in naturally infected cattle, buffaloes, sheep and goats., Sixth Nat. Cong. Vet. Parasitol., 22-24 October, 1994, Dept. Parasitol., Coll. Vet. Sci., JNKVV, Jabalpur-482 001, M. P., India. Abstr. S-1: 7, pp.3-4.

Mohanty, B., Misra, S. C. and Rao, A. T., 1995c. Pathology of *Sarcocystis* in naturally infected cattle, buffaloes, sheep and goats. Indian Vet. J., 72: 569-571.

Mohanty, B., Misra, S. C., Panda, D. N. and Panda, M. R., 1995d. Prevalence of *Sarcocystis* infection in ruminants in Orissa. Indian Vet. J., 72: 1026-1030.

Mohanty, B. N., Panda, D. N. and Panda, M. R., 2005. Morphology and identification of *Sarcocystis* recorded from cattle, buffaloes, sheep and goats in and around Bhubaneswar. Proc. XVI Nat. Cong. Vet. Parasitol., Dec. 6-8, 2005 Dept. Parasitol., Coll. Vet. Sci. & Anim. Husb., Indira Gandhi Agri. Univ., Anjora, Durg-491 001, Chattisgarh, Ab-10, p.115.

Neelu Gupta, 2005. Sarcocystosis in a Sahiwal cow in Durg. Proc. XIV Natl. Cong. Vet. Parasitol., December 6-8, 2005, Dept. Parasitol., Coll. Vet. Sci.& Anim. Husb., Indira Gandhi Agri. Univ., Anjora, Durg-491 001., Chhattisgarh, Ab-29, p.124.

Nikita Sharma, Reena Mukherjee, Shivsharanappa and Asish Srivatava, 2012. *Sarcocystis muris*: A rare and incidental finding in laboratory mice. XXII National Congress of Veterinary Parasitology, March 15-17, 2012, Coll. Vet. Sci. Anim. Husb., U. P. Pandit Deen Dayal Upadhyaya Pashu Chikitsa Vigyan Vishwavidyalaya Evam Go Anusandhan Sansthan (DUVASU), Mathura-281001(UP.), India, Abstr. No.VI.51, p.108.

Palanivel, K. M., Muthuswamy, P., Sureshkumar, K. and Kumarasamy, P., 2011. Sarcocystosis in a organized farm. Indian Vet. J., 88(8): 66-67.

Pandit, B. A., 1993. Epidemological studies on *Sarcocystis* infection in cattle. Ph. D. Thesis. Submitted to G. B. Pant Univ. of Agri. &Tech., Pantnagar-263 145, U. P., India. Abstr., J. Vet. Parasitol. 7: 132-133.

Pandit, B. A. and Bhatia, B. B., 1994a. *Sarcocystis* infection in cattle in Uttar Pradesh. Sixth Nat. Cong. Vet. Parasitol., 22-24 October, 1994, Dept. Parasitol., Coll. Vet. Sci., JNKVV, Jabalpur-482 001, M. P., India. Abstr. S-1: 4, p.2.

Pandit, B. A. and Bhatia, B. B., 1994b. Experimental development of *Sarcocystis* species of cattle in pups and kitten. Sixth Natl. Cong. Vet. Parasitol., 22-24 October, 1994, Dept. Parasitol., Coll. Vet. Sci., JNKVV, Jabalpur-482 001, M. P., India, Abstr. S-1: 5, p.2-3.

Pandit, B. A. and Bhatia, B. B., 1996. Epidemology of *Sarcocystis* spp. in cattle of Uttar Pradesh. Indian J. Anim. Sci., 66: 435-442.

Pandit, B. A. and Bhatia, B. B., 2002. Gametogonic and sporogonic development of *Sarcocystis hirsuta* of cattle in experimental kitten with remarks of pathogenicity. J. Vet. Parasitol., 16: 175-177.

Pandit, B. A., Bhatia, B.B. and Banerjee, P. S., 1994a. Comparative morphology of sarcocysts of three species of *Sarcocystis* in cattle from Tarai region of Uttar Pradesh. J. Vet. Parasitol., 8: 59-63.

Pandit, B. A., Bhatia, B. B. and Banerjee, P. S., 1994b. Effect of temperature on the viability of sarcocysts of *Sarcocystis hirsuta* of cattle. J. Vet. Parasitol., 8: 85-87.

Pandit, B.A., Bhatia, B. B. and Garg, S.K., 1996. Detection of sarcocystis antibodies in the sera of naturally infected cattle. Indian J. Anim. Sci., 66: 431-434.

Pandit, B. A., Garg, S. K. and Bhatia.J., 1993. Preparation and assessment of precipitating antigens of *Sarcocystis cruzi* and *S. hirsuta*. J. Vet. Parasitol., 7: 5-15.

Patra, N. C., Bera, A. K., Bhattachary, D. and Karmarkar, P. K., 2006. Pathobiology and histochemistry of sarcocystosis in naturally infected bovines. Environ. Ecol., 24: 98-102.

Pethkar, D. K., 1979. Studies on *Sarcocystis* of goats (*Capra hircus*) Ph.D. Thesis. Jawaharlal Nehru Krishi Vishwa Vidyalaya, Jabalpur.

Pethkar, D. K., 1999. Sarcocystosisis. In: Review of Parasitic Zoonosis. S. C. Parija (Ed.), AITBS, New Delhi, India pp.44-52.

Pethkar, D. K. and Shaw, H. L., 1980. Sarcocyst output in dogs experimentally fed sarcocysts of *Sarcocystis capracanis* of the goat. Indian J. Anim. Sci. 58: 588-589.

Pethkar, D. K. and Shah, H. L., 1981. Indian J. Parasitol.,3: 85.

Pethkar, D. K. and Shah, H. L., 1982a. Prevalence of *Sarcocystis* in goats in Madhya Pradesh. Indian Vet. J., 59: 110-114. (Recorded two species, *S. capracanis* and one unnamed species, Type II as *Sarcocystis* sp.)

Pethkar, D. K. and Shah, H. L., 1982b. Study on pre-patent and patent periods of sporocysts of *Sarcocystis capracanis* of the goat and high infective potential of dogs. 5th Natl. Cong. Parasitol., June 25-27, 1983, Coll. Vet. Sci., A. P. Agri. Univ., Tirupati, A. P. and Indian Soc. Parasitol., Abstr. E-28, p.108.

Pethkar, D. K. and Shah, H. L., 1982c. Attempted cross transmission of *Sarcocystis capracanis* of the goat to the sheep. Indian Vet. J., 59: 766-768.

Pethkar, D. K. and Shah, H. L., 1982d. A simple method for detection of microscopic sarcocysts. Gujarat Coll.Vet. Sci. Mag., 14: 28-30.

Powar, K.V., Tripathi, S. A. and Chandhary, D. P., 2011. Incidence of *Sarcocystis* infection in a black buck. (*Antelope cervicapra*) in captivity. XII Natl. Cong. Vet. Parasitol., 5-7th Jan, 2011, IAAVP and Dept. Vet. Parasitol., & Bombay Vet. Coll., Mumbai, S3-O25, p.121.

Prasad, M. C., Sinha, B. K. and Verma, B. B., 1969. Malayasian Vet. J., 4: 325.

Pratibha Jumde, Bhilgaonkar, N. G., Maske, D. K. and Kolte, S. W., 2010. Seroprevalence of caprine sarcocystosis in Nagpur. XX NCVP Parasitology Today-Ecology to Molecular Biology, 18-20 February, 2010, Dept. Vet. Parasitol., CVS, CCS, Haryana Agri. Univ., Hisar-125004, p.20.

Purohit, S. K., 1970. M.V.Sc. Dissertation, Madras Univ., Madras.

Purohit, S.K. and D'Souza, B.A., 1973. An investigation into the of transmission of sarcosporidios. British Vet. J. 129: 230-235.

Rajat Garg, Saroj Kumar, Banerjee, P. S., Hira Ram, Krishendra Kundu, Mayura, P. S. and Rakesh, R. L., 2012. Molecular identification of *Sarcocystis* species in slaughtered water buffaloes of Bareilly, Uttar Pradesh. XXII Natl. Cong. Vet. Parasitol., March 15-17, 2012, Coll. Vet. Sci. Anim. Husb., U. P. Pandit Deen Dayal Upadhyaya Pashu Chikitsa Vignan Vishwavidyalaya Evam Go Anusandhan Sansthan (DUVASU), Mathura-281001(UP.), India, Abstr. No.III.22, p.45.

Ranga Rao, G. S. C., Sharma, R. L. and Shah, H. L., 1997. Occurrence of sarcocystis in the camel (*Camelus dromedaries*) in India. Indian Vet. J. 74: 426.

Rohini, K and Hafeez, Md., 2005. Serodiagnosis of sarcocystosis in naturally infected buffaloes. Indian Vet. J., 82: 330-331.

Saha, A. K., 1984. Experimental studies on *Sarcocystis levinei*. M. V. Sc., Thesis submitted to Bihar Veterinary College, Patna, Bihar, India.(not indexed).

Saha, A. K. and Ghosh, D., 1984. Prevalence of sarcocystis in Black Bengal goats in Tripura. Indian Vet. J., 69: 82-83.

Saha, A.K., Srivastava, P. S. and Sinha, S. R. P., 1985. Toxic effects of extract of *Sarcocystis fusiformis* to laboratory mice. Indian J. Anim. Sci., 55: 656-658.

Saha, A. K., Srivastava, P. S., Sinha, S. R. P. and Sahai, B. N., 1986. Morphological characteristics of the developmental stages of *Sarcocystis levinei* in canine and bubaline hosts. Rivista di Parasitologica., 3: 315.

Saha, A. K., Srivasatava, P. S., Sahai, B. N., Sinha, S. R. P. and Singh, S. P., 1986. Prevalence of sarcocystosis in slaughtered cattle and buffaloes in Bihar. Indian J. Parasitol., 10: 63-66.

Sahai, B. N., 1981. Paper presented in Summer Institute on "Epidemiology and host parasite relationship and control of parasitic diseases of livestock", Ranchi, June, 1981, pp.235-245.

Sahai, B. N., Singh, S. P., Sahay, M. N., Srivastava, P. S. and Juyal, P. D., 1982. A note on the incidence and epidemiology of *Sarcocystis* infection in cattle, buffaloes and pigs in Bihar. Indian J. Anim. Sci., 52: 1005-1006.

Sahai, B. N., Singh, S. P., Sahay, M. N., Srivastava, P. S. and Juyal, P. D., 1983. Role of dogs and cats in the epidemiology of bovine sarcosporidiasis. Indian J. Anim. Sci., 53: 84-85.

Saleque, A., 1990. Studies on some aspects of epidemiology of *Sarcocystis* in river buffaloes (*Bubalus bubalis*) and domestic pig with a reference to *Eimeria* infection in buffaloes. Ph.D Thesis. G. B. Pantnagar Univ. Agri. & Tech., Pantnagar-263 145, U. P., India. (Not indexed).

Saleque, A. and Bhatia, B. B., 1991. Prevalence of *Sarcocystis* in domestic pigs in India. Vet. Parasitol., 40: 151-153.

Saleque, A., Bhatia, B. B. and Daya Shanker, 1992. Prevalence of two species of *Sarcocystis* in sheep in Uttar Pradesh. Indian Vet. J., 69: 841-842.

Saleque, A., Bhatia, B. B. and Juyal, P. D., 1990a. Prevalence of sarcocystis in goats (*Capra hircus*) in Tarai region of Uttar Pradesh. Indian Vet. Med. J., 14: 276-277.

Saleque, A., Juyal, P. D. and Bhatia, B. B., 1990b. Effect of temperatures on the infectivity of *Sarcocystis miescheriana* cysts in pork. Vet. Parasitol., 36: 343-346.

Saleque, A., Bhatia, B. B., Juyal, P. D. and Rahman, H., 1991. Toxic effects of cyst extract of *Sarcocystis fusiformis* from buffalo in rabbits and mice. Vet. Parasitol., 38: 61-65.

Sandeep Dey, 1989. Studies on some aspects of experimental sarcocystis infection in goats. Punjab Agri. Univ., Ludhiana, Punjab. (Not indexed).

Sathia Singh, S. 1971. Investigation into the possible modes of transmission of sarcosporidiasis in animals. M. V. Sc. Dissertation, Univ. Madras, Madras.

Sen, M. R., 1951. Sarcosporidiasis in cattle in Bengal. Indian Vet. J.,27: 261-264.

Shah, H. L., 1981. *Sarcocystis* and sarcocystosis in buffaloes & goats in India. Paper presented at the Summer Institute on epidemiology, host parasite relationship and control of parasitic diseases of livestock., 22.6.81 to 12.7.81 at B. A. U., Ranchi.

Shah, H. L., 1983. Summer Institute on epidemiology, host-parasite relationship and control of parasitic diseases of livestock. Ranchi Vet. Coll., India, June 22 to July, 1983, pp.246-273. (Review article, not cited in Index).

Shah, H. L., 1983. Epidemiology of *Sarcocystis* in domestic animals. Haryana Vet. 22: 59-73. (Review article, not cited in Index).

Shah, H. L., 1984. *Sarcocystis* as zoonosis. In: Compendium of lectures. Summer Institute on animal parasitic infections of public health importance at APAU, Tirupati. 1984.(Not indexed).

Shah, H. L., 1990a. Human factors in dissemination and transmission of *Sarcocystis* infection. Proc. 1ˢᵗ Asian Cong.Vet. Parasitol., 26-28 November, 1990, Patna, Bihar, pp.1-8. (Review, not indexed)

Shah, H. L., 1990b. Indian J. Anim. Sci., 60: 1274.

Shah, H. L., 1995. Sarcocystosis as zoonosis with special reference to India. J. Vet Parasitol., 9: 57-61.

Shah, II. L. and Chowdhury, R. K., 1994. Recent advances in our knowledge of *Sarcocystis* and Sarcocystosis in India: A review. In: Current Advances in Veterinary Science and Animal Production in India. R. Somvanshi and R. R. Lokeshwar, (Eds.), International Book Distributing Co., Lucknow, India. pp.1-48. (Review article, not cited in Index).

Shah, H. L., Sharma, R.K., Singh, K. P. and Agrawal, M. C. 1986. Attempted separation of *Sarcocystis capracanis* from *S. hircicanis* from in dogs. Seventh Natl. Cong. Parasitol., Ravi Shankar Univ., Raipur, 26-28, Dec., 1986, Abstr. A-43, p. 43.

Shahardar, R. A., and Pandit, B. A., 2010. Studies on *Sarcocystis* infection in food animals of Kashmir Valley. XX NCVP Parasitology Today-Ecology to Molecular Biology, 18-20 February, 2010, Dept. Vet. Parasitol., CVS, CCS Haryana Agri. Univ., Hissar-125004, p.29.

Sharma, B. K. and Chaudhury, R. K., 1994. Effect of various stains on the sarcocysts of *Sarcocystis capracanis* the naturally infected goats. Sixth Natl. Cong. Vet. Parasitol, 22-24, October, 1994, Dept. Parasitol., Coll. Vet. Sci., JNKVV, Jabalpur-482 001, M. P., India, Abstr.S-1: 8, p.4.

Sharma, R. K. and Shah, H. L. 1992. Occurrence of *Sarcocystis hircicanis* in goat. Indian J. Anim. Sci., 62: 561-563.

Sharma, R.K., Chaudhry, R. K., Shah, H. L., and Kohli, S., 1991. Histochemical observations on intact sarcocysts of *Sarcocystis capracanis* of the goat. Indian J. Anim. Sci., 61: 1181-1182.

Shastri, U. V., 1988a. *Sarcocystis* infection in goats in Maharastra. J. Vet. Parasitol., 2: 189-192.

Shastri, U. V., 1988b. Prevalence of *Sarcocystis* and other coccidial infections in stray dogs in and around Prabhani town (Maharastra). Second Natl. Parasitol., March 3-5, 1988, Dept. Parasitol., Vet. Coll., Bangalore, India, Stream 1, p.8.

Shastri, U. V., 1990. *Sarcocystis* infection in goats in Maharashtra. Indian Vet. J., 67: 70-71.

Shrivastava, A. B., Sharma, R. K., Chaudhary, R. K.and Malik, P., 1999. Sarcocystosis in a Barasingha deer (*Cervis duvauceli branderi*) J. Zoo Wild., 30: 454-455.

Shukla, D. C. and Victor, D.A., 1974. Incidence of sarcosporidial infection in bovine carcasses at Madras slaughter house. Gujarat Vet., 7: 84-87.

Shukla, D. C. and Victor, D. A., 1976. The complement fixation test in the diagnosis of sarcosporidiosis in bovines. Indian Vet. J., 53: 852-854

Singh, B. B., Sharma, A., Aradhana and Gumber, S., 2004. *Sarcocystis capracanis* and *Sarcocystis hircicanis* from goats in Punjab. Indian Vet. Med. J. 28: 155-156.

Singh, B. B., Sharma, J. K., Juyal, P. D. and Aulakh, R. S., 2003. Prevalence and comparative morphology of *Sarcocystis* species of cattle in Punjab. J. Res. Punjab agri. Univ., 40: 456-458.

Singh, B. B., Sharma, J. K., Juyal, P. D. and Gill, J. P. S., 2004. Seroprevalence of *Sarcocystis* species of cattle in Punjab. J. Vet. Parasitol., 18: 75-76.

Singh, B. B., Sharma, R., Sharma, J. K. and Juyal, P. D., 2010. Parasitic zoonosis in India: An overview. Rev. sci. tech. Off. Int. Epiz., 29: 629-637. (Sarococystis; Review, not indexed).

Singh, B. B., Juyal, P. D., Sharma, J. K., Aulakh, R. S. and Avapal, R.S., 2002. Public health significance of sarcocystosis. Punjab Vet. J., 2 (1): 18-20. (Review, not indexed).

Singh, K. P. and Singh, C. M., 1970. Parasitic infestations of bovine heart in Uttar Pradesh. Indian Vet. J. 47: 1023.

Singh, K. P. and Shah, H. L., 1990a. Viability and infectivity of sarcocysts of *Sarcocystis capracanis* of the goats. Indian J. Anim. Sci., 60: 429-430.

Singh, K. P. and Shah, H. L., 1990b. Offals from slaughtered goats as source of infection *Sarcocystis capracanis* for dogs. Indian J. Anim. Sci., 60: 1315.

Singh, K.P., Agrawal, M.C. and Shah, H. L., 1987. Prevalence of sarcocystis sporocysts in stray dogs. Indian J. Anim. Sci., 57: 1102-1107

Singh, K. P., Agrawal, M. C. and Shah, H. L., 1990a. Prevalence of sarcocysts of *Sarcocystis capracanis* in oesophagus and tail muscles of naturally infected goats. Vet Parasitol., 36: 153-155.

Singh, K. P., Shah H. L. and Agrawal, M. C., 1990b. Trials with three coccidiostats against *Sarcocystis capracanis* of the goat in experimentally infected dogs. J. Vet. Parasitol., 4: 13-16.

Singh, L., (vide Lal Singh supra).

Solanki, P. K., 1989. Studies on *Sarcocystis* of domestic pigs (*Sus scrofa domestica*). M. V. Sc. Thesis. Jawarhalal Nehru Krishi Vishwa Vidyalaya, Jabalpur (Not indexed).

Solanki, P. K., Srivastava, H. O. P.and Shah H. L., 1988. Life cycle of *Sarcocystis miescheriana* of the pig (*Sus scrofa domestics*). Proc. 3rd Natl. Congr. Vet. Parasitol., 1988, NN21-23, A-16.

Solanki, P. K., Shrivastava, H. O. P. and Shah, H. L., 1990. Endogenous life-cycle of *Sarcocystis meischeriana* in experimentally infected pigs. Indian J. Anim. Sci., 60: 1274-1278.

Solanki, P. K., Shrivastava, H. O. P. and Shah H. L., 1991a. Gametogenous and sporogonous life-cycle of *Sarcocystis meischeriana* of the pig in experimentally infected dogs. Indian J. Anim. Sci., 61: 374-378.

Solanki, P. K., Srivatava, H. O. P. and Shah H. L., 1991b. Prevalence of *Sarcocystis* in naturally infected pigs in Madhya Pradesh with an experimental explanation for the higher prevalence of *Sarcocystis suihominis*. Indian J. Anim. Sci., 61: 820-821.

Solanki, P. K., Srivatava, H. O. P. and Shah H. L., 1991c. Morphology of sarcocyst of *Sarcocystis meischeriana* (Khun, 1865) Labbe, 1889 *S. suihominis* (Tadros and Laarman, 1976; Heydorn, 1977) from naturally infected pigs (*Sus scrofa*) Indian J. Anim. Sci., 61: 1030-1033.

Solanki, P. K., Shrivastava, H. O. P., Shah, H. L. and Awadhiya, R. P., 1992. Pathology of *Sarcocystis miescheriana* in experimentally infected pigs. Indian J. Anim. Sci., 62: 797-801.

Somvanshi, R., Koul, G. L. and Biswas, J. C., 1987. *Sarcocystis* in a leopard (*Panthera pardus*). Indian Vet. Med. J., 11: 174-175.

Sreemannarayana, O. and Christopher, K. J., 1977. A case of combined infection of cysticercosis and sarcosporidiosis in bullock. Indian J. Anim. Hlth., 17: 188.

Sreenivas Rao, K, and Hafeez, Md., 2001. Prevalence of sarcocystosis in domestic animals of Andhra Pradesh. Twelfth Natl. Cong. Vet. Parasitol., August 26-27, 2001, Dept. Parasitol., Coll. Vet Sci., Acharya Ranga Agri. Univ., Tirupati-517 502., India, Abstr. S-2: 64, p.81.

Sreenivas Rao, K, and Hafeez, Md., 2002a. Incidence of porcine sarcocystosis in Andhra Pradesh. J. Parasit. Dis., 26: 69-71.

Sreenivas Rao, K, and Hafeez, Md., 2002b. Efficacy of amprolium and maduramicin against sarcocystosis in experimentally infected pups. J. Parasit. Dis., 26: 111-113.

Srivastava, C. P., Sinha, B. K. and Sahai, B. N., 1977. Observations on sarcosporidiasis in cattle and pigs. Indian J. Anim. Hlth., 16: 105-107.

Srivastava, P. S., Juyal, P. D. and Sinha, S.R.P. 1982. Dual infection of *Gongylonema pulchrum* and *Sarcocystis cruzi* in the tongue of cattle in Bihar-Case report. Indian J. Parasitol. 6 (1): 79.

Srivastava, P. S., Saha, A. K. and Sinha, S. R. P., 1985a. Spontaneous sarcocystosis in indigenous goats in Bihar, India. Acta Protozool., 24: 339-345.

Srivastava, P. S., Sahai, B. N., Sinha, S. R. P. and Saha, A. K., 1985b. Some differential features of the developmental cycle of bubaline *Sarcocystis* spp. in canine and feline definitive hosts. Protisitologica., 21: 385-390.

Srivastava, P. S., Kumar, A., Sinha, S. R. P. and Sinha, A. K., 1990. Morphological differentiation of caprine Sarcocystis species: Evidence of occurrence of *S. hircicanis* in India. First Asian Cong. Vet. Parasitol., Patna 26-28, Nov, Abstr. No 51: 38, p.55.

Srivastava, P. S., Kumar, A., Sinha, S. R. P. and Sinha, A. K., 1991. Morphological differentiation of caprine sarcocystis species: evidence of occurrence of *S. hircicanis* in India. Acta Protozool., 30: 61-62.

Srivastava, P. K., Saha, A. K and Sinha, S. R. P., 1986. Effects of heating and freezing on the viability of sarcocysts of *Sarcocystis levinei* from cardiac tissues of buffaloes. Vet. Parasitol., 19: 329-332.

Srivastava, P. S., Sinha, S. R. P., Juyal, P. D. and Saha, A. K., 1987. Host resistance and faecal sarcocyst excretion in dogs exposed to repeated infection with *Sarcocystis levinei*. Vet. Res. Comm., 11: 185-190.

Srivastava, P. S., Juyal, P. D., Sinha, S. R. P., Singh, S. P. and Sahai B. N., 1988. Attempt to infect monkeys with *Sarcocystis cruzi* sporocysts. J. Vet. Parasitol., 2: 63-65.

Suresh Kumar Khatkar, 1990. Studies on prevalence and clinical pathological aspects of sarcocystis infection in swine. M. V. Sc. Thesis, HAU, Hissar, India. (Not Indexed).

Swapna Susan Abraham (Vide supra, Abraham, S.S.)

Swarnkar, C. P., Khan, F. A., Juyal, P. D. and Bhagwan, P. S. K., 1999. *Sarcocystis tenella* in sheep in and around farms of Rajasthan. J. Small Rum., 5: 28-30.

Thirunavukkarasu, P. R., 1964. M. V. Sc. Dissertation, Madras Vet. Coll., Submitted to Univ. of Madras, Madras.

Venu, R. and Hafeez Md. 1999. Prepatent periods in dogs experimentally infected with *Sarcocystis* spp. Indian Vet. J., 76: 574-576.

Venu, R. and Hafeez Md., 2000. Prevalence of sarcocystic infections in the slaughtered domestic ruminants in Tirupati (A.P.). Indian Vet. J., 77: 165-166.

Venu, R., Hafeez Md. and Reddy, P. R., 2000. Morphometry of sporocysts of *Sarcocystis* spp. developing in the dog. Indian Vet. J., 77: 354-356.

Wadajkar, S. V., Shastri, U. V. and Narladkar, B. W., 1993. A note on the development of *Sarcocystis capracanis* in pigs. J. Vet. Parasitol., 7: 121-123.

Wadajkar, S. V., Shastri, U. V. and Narladkar, B. W., 1994. Prevalence of caprine *Sarcocystis* spp. In Marathwada region. J. Vet. Parasitol., 8: 43-46.

Wadajkar, S. V., Shastri, U. V. and Narladkar, B. W., 1995a. Caprine sarcocystosis: clinical signs gross and microscopic pathology. Indian Vet.J., 72: 224-228.

Wadajkar, S. V., Shastri, U. V. and and Narladkar, B. W., 1995b. Caprine sarcocystosis: haematological and biochemical studies. Indian Vet. J., 72: 392-393.

Toxoplasma

Ali, M.S., Singh, B. and Chhabra, M. B., 1983. Serological survey of toxoplasmosis in sheep and goats in Marathwada region of Maharashtra. Livestock Adviser, 8: 49-52.

Bhandari, S. K., Ahuja, S., Sharma, V. K. and Saxena, S. N., 1980. *Toxoplasma* antibodies in mules and horses in India. Indian J. Parasitol., 4: 155-156.

Bharadwaj, R. M., 1974. Studies on the prevalence, some aspects of pathogenesis, immunity and diagnosis of toxoplasmosis. Ph. D. Dissertation, Haryana Agri. Univ., Hissar.

Bhoop Singh and Msolla, 1986. Seroprevalence and pathogenesis of *Toxoplasma gondii* in sheep and goats in tropical region. Bull. Anim. Hlth. Prod. Afr., 34: 236-240.

Chhabra, M. B., 1978. *Toxoplasma gondii*–Natural infection sought in earthworms. Haryana agri. Univ. J. Res., 8: 759-76.

Chhabra, M. B. and Gautam, O. P., 1980. Antibodies to *Toxoplasma gondii* in equids in north India. Equine Vet. J., 12: 146-148.

Chhabra, M. B. and Mahajan, R. C., 1978a. A serological study of toxoplasmosis prevalence in buffaloes in North India. Rivista Parassitol., 39: 39-43.

Chhabra, M. B. and Mahajan, R. C., 1978b. Isolation of *Toxoplasma gondii* in slaughter pigs. Indian J. Parasitol., 2: 125-126.

Chhabra, M. B. and Mahajan, R. C., 1979a. Occurrence of *Toxoplasma gondii* in slaughter pigs in India. Trop. Geogr. Med., 31: 123-126.

Chhabra, M. B. and Mahajan, R. C., 1979b. Prevalence of *Toxoplasma* antibodies in dogs in Chandigarh territory, North India. Trop. Geogr. Med., 31: 499-502.

Chhabra, M. B. and Mahajan, R. C., 1979c. Prevalence of *Toxoplasma* antibodies in zebu cattle in India. Trop. Anim. Hlth. Prod., 11: 27-28.

Chhabra, M. B. and Mahajan, R. C., 1979d. Virulence of some isolates of *Toxoplasma gondii*. Indian J. Med. Res., 69: 752-755.

Chhabra, M. B. and Mahajan, R. C., 1982. Toxoplasmosis in India. Prevalence of serum antibodies in sheep and goats. Indian J. Anim. Hlth., 21: 5-8.

Chhabra, M. B., Ganguly, N. K., and Mahajan, R. C., 1980. Antigenic characterization of *Toxoplasma gondi* strains by gel diffusion method. Indian J. Med. Res., 72: 206-209.

Chhabra, M. B., Ganguly, N. K., and Mahajan, R. C., 1981. Polyacrylamide gel electrophoresis of Indian isolates of *Toxoplasma gondii*. Indian J. Microbiol., 21: 340-342.

Chhabra, M. B., Gupta, S. L. and Goutam, O. P., 1985. *Toxoplasma* seroprevalence in animals in northern India. Int. J. Zoon., 12: 136-142.

Chhabra, M. B., Mahajan, R. C. and Ganguly, N. K., 1978. Indirect fluorescent and indirect haemagglutination tests in the serodiagnosis of toxoplasmosis. Indian J. Parasitol., 2: 87-90.

Chhabra, M. B., Mahajan, R. C. and Ganguly, N. K., 1979a. Value of double vaccination by irradiated tachyzoites in experimentally induced murine toxoplasmosis. Haryana agri. Univ. J. Res., 9: 59-62.

Chhabra, M. B., Mahajan, R. C. and Ganguly, N. K., 1979b. Effects of ⁶⁰Co irradiation on virulent *Toxoplasma gondii* and its use in experimental immunization. Int. J. Radiat. Biol., 35(5): 433-440.

Chhabra, M. B., Mahajan, R. C. and Mahajan, M. K., 1979c. Isolation of *Toxoplasma gondii* from suspected human cases. Indian J. Med. Res., 69: 746-751.

Chhabra, M. B., Mahajan, R. C. and Ganguly, N. K., 1980. Antibody response and serum protein alterations in experimental toxoplasmosis. Indian Vet. J., 57: 627-631.

Chhabra, M. B., Mahajan, R. C. and Malik, P. D., 1982a. Comparative evaluation of indirect latex agglutination (Toxotest- MT) indirect haemagglutination tests in the detection of *Toxoplasma gondii* antibodies in animals. Indian J. Parasitol., 6: 95-96.

Chhabra, M. B., Bharwaj, R.M., Goutam, O. P. and Gupta, R. P., 1981. *Toxoplasma* infection and abortion in dairy goats. Trop. Anim. Hlth. Prod., 13: 222-226.

Chhabra, M. B., Mahajan, R. C., Ganguly, N. K. and Chtikara, N. L., 1976. Prevalence of *Toxoplasma* antibodies in rhesus monkeys in India. Trop. Geogr. Med., 28: 101-103.

Chhabra, M. B., Mahajan, S. K., Gupta, S. L. and Goutam, O. P., 1982b. Experimental toxoplasmosis in pregnant goats. Indian J. Anim. Sci., 52: 661-664.

Devada, K., 1996. Studies on the possible source of infection of *Toxoplasma gondii* to cats and its seroprevalence in chickens. Ph. D. Thesis, Tamil Nadu Univ. of Vet. Anim. Sci., Chennai- 600 007, India. vide Abstract, J. Vet. Parasitol., 1998, 12: 151.

Devada, K. and Anandan, R., 1998. Prevalence of *Toxoplasma gondii* in some common rodents and birds. Indian Vet. J., 75: 581-582.

Devada, K. and Anandan, R., 2000a. Detection of antibodies to *Toxoplasma gondii* in chicken by Enzyme linked immunosorbent assay. Indian Vet. J., 77: 60-61.

Devada, K. and Anandan, R., 2000b. Suitability of modified direct agglutination test (MDAT), Enzyme linked immunosorbent assay (ELISA) and avidin-biotin ELISA in the detection of antibodies to *Toxoplasma gondii* in chicken. Indian Vet. J., 77: 196-198.

Devada, K., Anandan, R. and Dubey, J. P., 1998. Serological prevalence of *Toxoplasma gondii* in chickens in Madras, India. Research note. J. Parasitol., 84: 621-622.

Dhananjay Kumar, Surajit Baidya, Soumitra Pandit and Bandyopadhyay, M. C., 2010. Comparative histopathological changes in experimentally induced *Toxoplasma gondii* (RH strain) infection in mice. XX Natl. Cong. Vet. Parasitol., February 18-20, 2010, Dept. Vet. Parasitol., Coll. Vet. Sci., Chaudhary Charan Singh Haryana Agri. Univ., Hisar-125 004, Abstr., p.75.

Dubey, J. P., 1986. Toxoplasmosis in cattle. Vet. Parasitol., 22: 20-22. (Review).

Dubey, J. P., 1986. Toxoplasmosis in pigs. Vet. Parasitol., 19: 181-223. (Review).

Dubey, J. P., 1987a. Toxoplasmosis in domestic animals in India. J. Vet. Parasitol., 1: 13-18.(Review, not indexed)

Dubey, J. P., 1987b. Toxoplasmosis in goats. Agri. Pract., 7: 43-52. (Review).

Dubey, J. P., 1988. Toxoplasmosis in India. In: Perspectives in Parasitology, Vol. 2, A. B. Sen, J. C. Katiyar, and P. Y. Guru (Eds.). CBS Publishers and Distributors, Delhi, India, pp.131–152. (Not indexed).

Dubey, J. P., 2009a. *Toxoplasma gondii* infection in chicken (*Gallus domesticus*): Prevalence, clinical disease, diagnosis and public health significance. Zoon. and Pub. Hlth., 57: 60-73.(Review).

Dubey, J. P., 2009b. Toxoplasmosis in sheep-The last 20 years. Vet. Parasitol.,163: 1-14. (Review).

Dubey, J. P. and Towle, A., 1986. Toxoplasmosis in Sheep: A Review and Annotated Bibliography. Commonwealth Institute of Parasitology, St. Alberts, Herts, UK, pp.1–152.(Review).

Dubey, J. P., Somvanshi, R., Jithendran, K. P. and Rao, J. R., 1993. High seroprevalence of *Toxoplasma gondii* in goats from Kumaon region of India. J. Vet. Parasitol. 7, 17–21.

Gaikwad, A. V., 1993. Pathological alterations in experimental toxoplasmosis in poultry. M. V. Sc. Thesis, submitted to Marathwada Agri. Univ., Parbhani.

Gautam, O. P. and Chhabra, M. B., 1983. Public health aspects of toxoplasmosis. Indian J. Trop. Agric., 1: 159-165. (Not indexed).

Gautam, O. P., Chhabra, M. B. and Bharadwaj, R. M., 1979. Current status of toxoplasmosis in India. Haryana Vet., 18: 89-102. (Not indexed)

Gautam, O. P., Chhabra, M. B., Gupta, S. L. and Mahajan, S. K., 1982a. Experimental toxoplasmosis in buffalo calves. Vet. Parasitol., 11: 293-299.

Gautam, O. P., Chhabra, M. B., Gupta, S. L. and Mahajan, S. K., 1982b. Incidence, diagnosis and control of toxoplasmosis in animals. Final report of ICAR research scheme (1978-1982), Dept. Vet. Med., Coll. Vet. Sci., Haryana Agri. Univ., Hisar.

Gill, H. S., 1972. Toxoplasmosis in India: A note on the prevalence of dye test in water buffaloes. Indian J. Anim. Sci., 42: 1027-1028.

Gill, H. S. and Prakash, O., 1969. Toxoplasmosis in India. Prevalence of antibodies in camels. Ann. Trop. Med. Parasitol., 63: 265-267.

Gill, H. S. and Prakash, O., 1970a. Occurrence of toxoplasmosis antibodies in goats in India. Trop. Geogr. Med., 22: 364-366.

Gill, H. S. and Prakash, O., 1970b. Toxoplasmosis in India. Survey of antibodies in sheep. J. Trop. Med. Hyg., 73: 77-78.

Gill, H. S. and Prakash, O., 1971a. Prevalence of antibodies against *Toxoplasma gondii* in slaughtered pigs in India. Vet. Rec., 89: 130.

Gill, H. S. and Prakash, O., 1971b. Prevalence of *Toxoplasma* antibodies in cattle in India. Trop. Geogr. Med., 23: 204-207.

Goyal, M., Ganguly, N. K. and Mahajan, R. C., 1988. Cytotoxic activity of monocytes against *Toxoplasma gondii* in acute, chronic and reactivated murine toxoplasmosis. Med. Microbiol. Immunol., 177: 339-348.

Goyal, M., Ganguly, N. K. and Mahajan, R. C., 1989. Immunological response in experimentally reactivated toxoplasmosis in mice. Med. Microbiol. Immunol., 178: 269-278.

Gupta, S. L., 1971. Studies on some aspects of pathogenesis and therapy of toxoplasmosis in domestic animals.M. V. Sc. Thesis, Haryana Agri. Univ., Hissar. (Not indexed).

Gupta, S.L., Chhabra, M. B. and Gautam, O. P., 1985. *Toxoplasma* prevalence and human occupational groups. Int J. Zoon., 12: 143-146.

Gupta, S. L., Gautam, O. P. and Bhardwaj, R. M., 1979. Isolation of *Toxoplasma gondii* from sheep. Indian J. Parasitol., 2: 103-105.

Gupta, S. L., Gautam, O. P. and Bhardwaj, R. M., 1980. Cultivation of *Toxoplasma gondii* in lamb testicular cell culture. Indian J. Med. Res., 71: 217-220.

Gupta, S. L., Gautam, O. P. and Bhardwaj, R.M., 1981a. Note on toxoplasmosis: Serological survey of antibodies in sheep of Hisar area by indirect haemoagglutination test (microtitre system). Indian J. Anim. Sci., 51: 381-382.

Gupta, S. L., Gautam, O. P. and Chhabra, M. B., 1981b. Chemotherapeutic efficacy of some agents against experimental toxoplasmosis in mice. Indian J. Med. Res., 71: 767-771.

Gupta, S. L., Gautam O. P., Bharwaj, R. M. and Bhardwaj, S., 1980a. A rapid card agglutination for toxoplasmosis. Trop. Anim. Hlth. Prod., 12: 95-96.

Gupta, S. L., Gautam O. P., Bharwaj, R. M. and Banerjee, D.P., 1980b. Cell mediated immunity in experimental toxoplasmosis. Indian J. Med. Res., 71: 526-530.

Hira Ram, Rao, J. R., Tewari, A. K. and Banerjee, P. S., 2009. A real-time PCR assay for sensitive detection of *Toxoplasma gondii*. J.Vet. Parasitol., 23: 111-114.

Jani, R. G., Bhuva, C. N., Katara, R. D., Bhanderi, G. and Vadaliya, D., 2006. Study of seroprevalence of toxoplasmosis in workers of zoological gardens of prevention and control: Gujarat. Intas Polivet, 7(2): 452-454.

Jithendran, K. P. and Rao J. R., 1996. Parasitic zoonoses: Role of migratory sheep and goats in Himachal Pradesh. Himaparyavaran,8: 6-8. (Review, not indexed).

Jithendran, K. P. and Vaid, J., 1996. Infectious abortion in sheep and goats in Himachal Pradesh- Facts and trends. Vet. Rev., 11: 24-26.

Korwar, D. A. and Bhoop Singh, 1982. Note on intradermal test for the diagnosis of caprine toxoplasmosis. Indian J. Anim. Sci., 52: 1167-1169.

Korwar, D. A., Bhoop Singh and Chhabra, M. B., 1982. Note on the prevalence of *Toxoplasma* antibodies in sheep and goats in Maharashtra. Indian J. Anim. Sci., 52: 1125-1127.

Krishnan, K. V. and Lal, J. C., 1933. A note on the finding of *Toxoplasma cuniculi* in two experimental rabbits. Indian J. Med. Res. 20: 1049-1059.

Kumar, D., Baidya, S., Pandit, S. and Bandyopadhyay, M. C., 2009. Purification and cryopreservation of *Toxoplasma gondii* (RH strain) tachyzoite from mice peritoneal exudates. J. Interacad.

Kumar, D., Baidya, S., Bandyopadhyay, M. C., Pandit, S., Dasgupta, C. K. and Ghosh, J. D. 2010. Study of acute *Toxoplasma gondii* (RH strain) infection in albino mice. Environ. Ecol., 28(2A): 1002-1004.

Kumar, D., Baidya, S., Pandit, S. Pravin, P. K., Bordolor, G., Bandyopadhyay, M. C., Tewari, A. K. and Banerjee, P. S., 2011. Infectivity and pathogenicity of experimentally induced *Toxoplasma gondii* (RH Strain) infection in mice. XII Natl. Cong. Vet. Parasitol., 5-7th January, 2011, IAAAP and Dept. Vet. Parasitol., & Bombay Vet. Coll., Mumbai, S4-O3, pp.139-140.

Kumar, M. U. (Vide Udayakumar, supra.).

Mahajan, R. C. and Ganguly, N. K., 1980. Antibody response and serum protein alterations in experimental toxoplasmosis. Indian Vet. J., 57: 627-631.

Mahajan, S. K., Chhabra, M. B. and Goutam, O. P., 1982. Laboratory maintenance of virulent *Toxoplasma gondii* a convenient alternative. Indian J. Microbiol., 22: 139-140.

Mahajan, S. K., Chhabra, M. B., Ganguly, N. K. and Singh, R. P., 1977. Fluorescent antibody tests in immunodiagnosis of toxoplasmosis. Indian J. Med. Res., 66: 29-32.

Malik, S. V. S., Rao, J. R., Samanta, S., Barbuddhe, S. B., Shakuntala, I., Rawool, D. B., 2005. Seroprevalence of toxoplasmosis in humans, animals and poultry by modified agglutination test. Indian J. Comp. Microbiol. Immunol. infect. Dis., 29 (2): 86-88.

Mandhalikar, M. V., 1992. Studies on characteristics of Prabhani isolate of from cat. M. V. Sc. Thesis submitted to Marathwada Agri. Univ., Parbhani.

Mandhalikar, M. V., Shastri, U. V., Narladkar, B. W. and Moregaonkar, S. D., 1994a. Further studies on Parbhani isolate of *Toxoplasma*: Infection in mice by parenteral routes. Indian Vet. J., 71: 329-332.

Mandhalikar, M. V., Shastri, U. V., Narladkar, B. W. and Moregaonkar, S. D., 1994b. Effects of oral infection of Parbhani isolate of *Toxoplasma* in rabbits. Indian Vet. J., 71: 930-931.

Mandhalikar, M. V., Shastri, U. V., Narladkar, B. W. and Moregaonkar, S. D., 1994c. Studies on oral infection of Parbhani isolate of *Toxoplasma* in mice and rats. Indian Vet. J., 71: 1152-1154.

Mir, N. A., Chhabra, M. B., Bhardwaj, R. M. and Goutam. O. P., 1979. *Toxoplasma* infection and some other protozoan parasites of the wild rat in India. Indian Vet. J., 59: 60-63.

Mirdha, B. R. and Samantray, J. C., 1996. Latent toxoplasmosis in animals. Global Meet on parasitic Diseases, New Delhi, 18-22 March, 1996. Abstr., J. Parasit. Dis., 20: 99.

Mirdha, B. R., Samantray, J. C. and Pandey, A., 1999. Seropostivity of *Toxoplasma gondii* in domestic animals. Indian J. Pub. Hlth., 43: 91-92.

Narladkar, B.W., Kulkarni, R. R., Deshpande, A. R. and Deshpande, P. D., 2006. Toxoplasmosis- Public health significance.Intas Polivet, 7(2): 444-451.

Nazir Ahmed, 1979. Studies on some aspects of pathogenesis and prophylaxis of experimental toxoplasmosis in rodents. M. V. Sc. Thesis, Haryana Agri. Univ., Hissar, India. (Not indexed).

Nene, S. S., Joshi, B. N. and Patki, J., 1986. *Toxoplasma* antibodies in local domestic animals. Int. J. Zoon., 13: 187-189.

Pande, P. G. Shukla, R. R. and Sekariah, P. C., 1961. *Toxoplasma* from the eggs of the domestic fowl (*Gallus gallus*). Science, 133: 648.

Pillai, M. T. and Khader, T. G. A., 1983. Survey of ovine toxoplasmosis in Tamil Nadu by "Tox HA-test". Cheiron, 12: 57-59.

Pravin, P. K., Baidya, S., Pandit, S, Kumar, D., Sar, T. K. and Bandyopadhay, M. C., 2011. Comparative efficacy of sulphadoxine-pyrimethamine, arthromycin and clindamycin in experimentally induced acute *Toxoplasma gondii* (RH strain) infection in mice. XII Natl. Cong. Vet. Parasitol., 5-7th January, 2011, IAAAP and Dept. Vet. Parasitol., & Bombay Vet. Coll., Mumbai, S4-O4, pp.140-141.

Pravin, P. K., Pandit, S., Kumar, D., Baidya, S., Sar, T. K., Jas, R., Bordoloi, G., Bandyopadhyay, M. C., 2012. Comparative histopathological changes in experimentally induced *Toxoplasma gondii* (RH strain) infected and treated mice. XXII Natl. Cong. Vet. Parasitol., March 15-17, 2012, Coll. Vet. Sci. Anim. Husb., U. P. Pandit Deen Dayal Upadhyaya Pashu Chikitsa Vigyan Vishwavidyalaya Evam Go Anusandhan Sansthan (DUVASU), Mathura-281001(UP.), India, Abstr. No.VI.57, pp.111-112.

Rajkhowa, S., Rajkhowa, C. and Chamuah, J., 2008. Seroprevalence of *Toxoplasma gondii* antibodies in free-ranging mithuns (*Bos frontalis*) from India. Zoon. Pub. Hlth., 55: 320-322.

Raote, Y. V., Gaikwad, A. V. Bharkad, G. P. and Jayraw, A. K. 2007. Immune response analysis in *Toxoplasma gondii* infected chicks. J. Parasit. Dis., 31: 74-75.

Ray, H.N. and Raghavachari, K., 1941.Toxoplasmosis. Indian J. Vet.Sci.,11: 28-32.

Reddy, P. M., 2006. Toxoplasmosis in domestic animals and man with reference to epidemiology and control. J. Parasitol., 83-86.(Review, not indexed).

Selvaraj, J., Murali Manohar, Sarman Singh and Balachandran, C., 2007. Seroprevalence of *Toxoplasma gondii* in buffaloes. J. Vet. Parasitol., 21: 41-42.

Shahrooz, M., 1988.Toxoplasmosis in animals together with some observations on its epidemiology. P. G. Thesis submitted to Bombay Vet. Coll., Bombay.(Not indexed).

Sharma, S., Sindhu, K. S., Bal, M. S., Kumar, H., Verma, S. and Jubey, J. P., 2008. Serological survey of antibodies to *Toxoplasma gondii* in sheep, cattle and buffaloes in Punjab, India. J. Parasitol., 94: 1174-1175.

Sharma, S. P., 1971 Studies on the incidence, some aspects of pathogenesis and therapy of toxoplasmosis septicemia in cattle and buffaloes.M. V. Sc Thesis, HAU, Hissar, India.

Sharma, S. P. and Gautam, O. P., 1972. Prevalence of *Toxoplasma* antibodies in sheep and goats in the area of Hissar, Haryana, India. Trop. Anim. Hlth. Prod., 4: 245-248.

Sharma, S. P. and Gautam, O. P., 1974a. Toxoplasmosis: Serological incidence in dogs. Indian J. Anim. Hlth., 13: 77-80.

Sharma, S. P. and Gautam, O. P., 1974b. A note on the prevalence of *Toxoplasma* antibodies among camels and pigs in Hissar. Indian J. Anim. Sci., 44: 214-215

Shastri, U. V., 1990. Isolation of *Toxoplasma* oocysts from a cat, *Felis catus* at Parbhani (Maharashtra). J. Vet. Parasitol., 4: 45-47.

Shastri, U. V. and Mandakhalikar, M. V., 1992. Some observations on Parbhani isolate of *Toxoplasma* from cat. Indian Vet.J., 69: 1088-1089.

Shastri, U. V. and Ratnaparkhi, 1992. *Toxoplasma* and other intestinal coccidia in cat in Maharashtra (Parbhani). Indian Vet.J., 69: 14-16.

Shastri, U. V., Narladkar, B. W. Deshpande, P. D., Kulkarni, G. B., Bhikane, A. U. and Mir Salbat Ali, 1993. Infectivity of *Toxoplasma* (Parbhani isolate) in sheep and goats. Agresco Report, 1993-94, Marathwada Agri. Univ., Parbhani.

Singh, B. B., Sharma, R., Sharma, J. K. and Juyal, P. D., 2010. Parasitic zoonosis in India: An overview. Rev. sci. tech. Off. Int. Epiz., 29: 629-637.(Toxoplasmosis; Review, not indexed).

Singh, H., Tewari, A. K., Mishra, A. K., Maharana, B. R.and Rao, J. R., 2011. A purified tachyzoite protein based ELISA for the sero-surveillance of *Toxoplasma gondii* specific antibodes in domestic ruminants. Indian J. Anim. Sci., 83: 20-23.

Singh, H., Tewari, A. K., Mishra, A. K., Maharana, B. R., Rao, J. R. and Raina, O. K., 2011. Molecular cloning comparative sequence analysis and prokaryote expression of GR A5 protein of *Toxoplasma gondii*. Indian J. Anim. Sci., 81: 209-215.

Singh, S., 1998. Toxoplasmosis an emerging disease in man and animals. In: Zoonotic Parasites – Their Diagnosis and Control. III Trng Prog., 9th -23rd March, 1998. Centr. Adv. Stud., Dept. Parasitol., Vet. Coll., Univer. Agri. Sci., Hebbal, Bangalore-560 024.(review, not indexed).

Somvanshi, R., and Singh, S., 1999. Spontaneous toxoplasmosis in laboratory rats. In: Toxoplasmosis in India. Singh, S., (Ed.), Second edition, Pragati Publishing Co. Gazhiabad & New Delhi.

Sonar, S. S. and Brahmbhatt, M. N., 2010.Toxoplasmosis: An Important Protozoan Zoonosis. Vet. World, 3 (9): 436-439.(Reveiw, not indexed).

Srivatsava, A. K., Singh, B. and Gupta, S. L., 1983. Prevalence of *Toxoplasma* antibodies in sheepand goats in India. Trop. Anim. Hlth. Prod., 15(4): 207-208.

Sreekumar, C., 2001. PCR based diagnosis and genotyping of *Toxoplasma gondii*. Ph. D. Thesis (Vet. Parasitol.), Deemed Univ., Indian Vet. Res. Inst., Izatnagar-243 122, India.(Not indexed).

Sreekumar, C., Rao, J. R., Mishra, A. K., Joshi, P. and Singh, R. K., 2004. Detecting of toxoplasmosis in experimentally infected goats by PCR.Vet. Rec., 154: 632-635.

Sreekumar, C., Rao, J. R., Mishra, A. K., Ray, D and Joshi, P., 2001a. Isolation and *in vitro* cultivation of *Toxoplasma gondii*. Twelfth Nat. Cong. Vet. Parasitol., August 26-27, 2001, Dept. Parasitol., Coll. Vet. Sci., Acharya Ranga Agri. Univ., Tirupati-517 502, India, Abst. S-2: 62,p.80.

Sreekumar, C., Rao, J. R., Mishra, A. K., Ray, D and Joshi, P., 2001b. First isolation of *Toxoplasma gondii* from chicken in India. J. Vet. Parasitol., 15: 103-106.

Sreekumar, C., Graham, D. H., Dahl, E., Lehman, T., Raman, M., Bhalerao, D. P., Vianna, M. C. and Dubey, J. P., 2003. Genotyping of *Toxoplasma gondii* isolates from chicken from India. Vet. Parasitol., 118: 187-194.

Srivastava, A. K., Bhoop Singh and Gupta, S. L., 1983a. Prevalence *Toxoplasma* antibodies in sheep.and goats in India. Trop. Anim. Hlth. Prod., 15: 207-208.

Srivastava, A. K., Bhoop Singh and Mir Salabat-Ali., 1983b. Therapeutic activity of Mitomycin-C and Fluracil against experimental *Toxoplasma gondii* in mice. 5th Natl. Cong. Parasitol., June 25-27, 1983, Coll. Vet. Sci., A. P. Agri. Univ., Tirupati, A. P. and Indian Soc. Parasitol., Abstr. C-46, p.76.

Srivastava, A. K., Bhoop Singh and Gupta, S. L., 1984. Toxoplasmosis in sheep and goats and its correlation to postnatal mortality and abortion. Livestock Adviser, 9: 55-57.

Sucilathangam, G., Palaniappan, N., Sreekumar, C. and Anna, T., 2012. Seroprevalence of *Toxoplasma gondii* in southern districts of Tamil Nadu using IgG-ELISA. J. Parasit. Dis.,36: 159-164.

Sudan, V., Sharma, R. L., Borah, M. K. and Mishra, R., 2012. Molecular characterization of surface antigen 3 (SAG3) of gene of *Toxoplasma gondii* RH- IVRI strain. J. Parasit. Dis.,36: 207-209.

Sudan, V., Tewari, A. K., Patyal, A., Maharana, B. R. and Rao, J. R., 2011a. Cellular immune response in mice againt *Toxoplasma gondii*. XII Natl. Cong. Vet. Parasitol., 5-7th January, 2011, IAAAP and Dept. Vet. Parasitol., & Bombay Vet. Coll., Mumbai, S4-O13, pp.145.

Sudan, V., Tewari, A. K., Maharana, B. R., Kundu, K., Raina, O. K., Garg, R., Patyal, A. and Rao, J. R., 2011b. Histopathological observations on experimental *Toxoplasma gondii* infection in mice. XII Natl. Cong. Vet. Parasitol., 5-7th January, 2011, IAAAP and Dept. Vet. Parasitol., & Bombay Vet. Coll., Mumbai, S4, p.2.

Syamala, K. and Devada, K., 1999. Prevalence of caprine to toxoplasmosis in Kerala. Proc. 11th Kerala Sci. Conf., Kasargod, February, 27 - March, 1, 1999.

Syamala, K., Devada, K. and Nair, G. K., 2005. Detection of *Toxoplasma gondii* by modified agglutination test. J. Vet. Anim. Sci., 36: 104-106.

Syamala, K., Devada, K. and Pillai, K. M., 2007. Carbon immuno assay- a sensitive and rapid serodiagnosis for caprine toxoplasmosis. J. Vet. Anim. Sci.,38: 1195-1196.

Syamala, K., Devada, K. and Pillai, K. M., 2008. Diagnosis of caprine toxoplasmosis by latex agglutination test. J. Vet. Anim. Sci., 39: 53-54.

Tewari, A. K., Harkirat Singh, Sudan V. and Rao, J. R., 2010a. Recombinant source antigen 2 based serodetection of toxoplasmosis in cattle. XX Natl. Cong. Vet. Parasitol., February,18-20, 2010, Dept. Vet. Parasitol., Coll. Vet. Sci., Chaudhary Charan Singh Haryana Agri. Univ., Hisar - 125 004, Abst., p 42.

Tewari, A. K., Sudan,V., Sankar, M., Saravanan, B. C. and Bansal, G. C., 2012. Critical determination of serum concentration of Th1 major cytokines in mice experimentally infected with *Toxoplasma gondii*. XXII Natl. Cong. Vet. Parasitol., March 15-17, 2012, held at College of Veterinary Science and Animal Husbandry, U. P. Pandit Deen Dayal Upadhyaya Pashu Chikitsa Vigyan Vishwavidyalaya Evam Go Anusandhan Sansthan (DUVASU), Mathura-281001(UP.), India, Abst. No.III.17, p.42.

Tewari, A. K., Velmurugan, G. V., Sudan,V., Rao, J. R., Dalal, S., Raina, O. K. and Banerjee, P. S., 2010b. Recombinant surface antigen 2 (SAGI) based seroproduction of toxoplasmosis in cattle. XX NCVP, Parasitology Today-Ecology to Molecular Biology, 18-20 February, 2010, Dept. Vet. Parasitol., CVS, CCS, Haryana Agri. Univ., Hissar-125004, p.41.

Thimma Reddy, P. M., 2003. Studies on toxoplasmosis with special reference to serodiagnosis. Ph. D. Thesis submitted to Acharya N. G. Ranga University, Hyderabad, India.

Udayakumar, M., 2004. Immunological and nucleic acid based diagnosis of toxoplasmosis in goats. Ph. D. Thesis (Vet. Parasitol.), Deemed Univ., Indian Vet. Res. Inst., Izatnagar-243 122, India.

Udayakumar, M., Mishra, A. K., Rao, J. R. and Tewari, A. K., 2009. Antibody response to oocyst and trachyzoite induced *Toxoplasma gondii* infection in goats. XIX Natl. Cong. Vet. Parasitol. & Nat. Sym. National Impact on Parasit. Dis, Livestock Hlth. Prod., February, 3-5, 2009, IAAVP& Dept. Vet. Parasitol., Coll. Vet. Sci., Guru Angad Dev Vet. & Anim. Sci. Univ., Ludhiana-141 004, OI 37, p.114.

Udayakumar, M., Mishra, A. K., Rao, J. R. and Tewari, A. K., 2010a. Polymerase chain reaction based copro-detection of *Toxoplasma gondii* infection in cats. XX NCVP, Parasitology Today-Ecology to Molecular Biology, 18-20 February, 2010, Dept. Vet. Parasitol., CVS, CCS, Haryana Agri. Univ., Hissar-125004, p.43.

Udaya Kumar, M., Mishra, A. K., Rao, J. R. and Tewari, A. K., 2010b. Analytical sensitivity of 35 copy B1 PCR Assay in detecting *Toxoplasma gondii* infection in mouse. J. Appl. Anim. Res. (India), 38: 65-67.

Udayakumar, M., Mishra, A. K., Rao, J. R. and Tewari, A. K., 2011. Slide enzyme linked immunosorbant assay (SELISA) in the detection of *Toxoplasma gondii* in goats. XII Natl. Cong. Vet. Parasitol., 5-7th Jan, 2011, IAAAP and Dept. Vet. Parasitol., & Bombay Vet. Coll., Mumbai, S4 -011, pp.144-145.

Velmurugan, G.V., Tewari, A. K., Rao, J. R., Baidya, S., Udaya Kumar, M. and Mishra, A. K., 2008. High level expression of SAG1 and GRA7 gene of *Toxoplasma gondii*

(Izatnagar isolate) and their application in serodiagnosis of goat toxoplasmosis. Vet. Parasitol., 154: 185-192.

Verma, S. P., Bharadwaj, R. M. and Gautam, O. P., 1988a. Efficacy of enzyme linked immunosorbent assay (ELISA) in the serodiagnosis of experimental *Toxoplasma gondii* in sheep. Second Nat. Parasitol, March 3-5, 1988, Dept. Parasitol., Vet. Coll., Bangalore, India, Souvenir, Stream III, Abstr., p.28.

Verma, S. P., Bharadwaj, R. M. and Gautam, O. P. 1988b. Application of counter-current immunoelectrophorosis (CIEP) in the serodiagnosis of experimental *Toxoplasma gondii* in sheep. Second Natl. Parasitol March 3-5, 1988, Dept. Parasitol., Vet. Coll., Bangalore, India, Souvenir, Stream III, Abstr., p.28.

Verma, S. P., Bharadwaj, R. M. and Gautam, O. P., 1988c. Seroprevalence of *Toxoplasma gondii* antibodies in aborted ewes. Indian J. Vet. Med., 8: 40-41.

Verma, S. P., Bharadwaj, R. M. and Gautam, O. P., 1988d. Comparative histopathological changes in experimentally induced *Toxoplasma gondii* (RH strain) infection in mice. XX Natl. Cong. Vet. Parasitol., February,18-20, 2010, Dept. Vet. Parasitol., Coll. Vet. Sci., Chaudhary Charan Singh Haryana Agri. Univ., Hisar - 125 004, Abstr., p.75.

Verma, S. P., Bharadwaj, R. M. and Gautam, O. P., 1989a. *In vitro* cultivation of *Toxoplasma gondii*. J. Vet. Parasitol., 3: 57-59.

Verma, S. P., Bharadwaj, R.M. and Gautam, O. P. 1989b. Application of countercurrent immunoelectrophorosis (CIEP) in the sero-diagnosis of *Toxoplasma* antibodies. J. Vet. Parasitol., 3: 61-62.

Verma, S. P., Bharadwaj, R. M. and Gautam, O. P., 1989c. Isolation of *Toxoplasma gondii* from foetal brain of aborted ewes. Indian J. Vet. Med., 9: 40-41.

Vijaya Bharathi, M., Mohamed Basheer, A., Nedunchelliyan, S., Sarman Singh and Meenachi Selvan, M. S., 2003. Comparison of modified direct agglutination and indirect haema-agglutination tests for detection of *Toxoplasma gondii* antibodies in goats in Chennai. J. Vet. Parasitol., 17: 49-51.

Vijayakumar, K., 1999. Diagnostic and immunization studies in *Toxoplasma* in sheep. Ph. D. Thesis, Tamil Nadu Vet. and Anim. Sci. Univ., Chennai, India.

Wagle, N. M., Deshpande, C. K. and Bhave, G. G., 1984. Immunodiagnosis of toxoplasmosis by indirect fluorescent antibody test. Indian J. Parasitol., 211-215.

7.2 FAMILY: HAEMOGREGARINIDAE

Christophers, S. R., 1906. *Leucocytozoon canis*. Sci. Mem. Off. Med. Sanit. Dept., Govt. of India. 26: 1-18.

Christophers, S. R., 1907. The sexual life cycle of *Leucocytozoon canis* in the tick. Sci. Mem. Off. Med. Sanit. Dep. Gov. India, 28: 1-14.

Christophers, S. R., 1912. The development of *Leucocytozoon canis* in the tick with reference to the development of *Piroplasma*. Parasitology, 5: 37-48.

Harikrishnan, T. J. and Ponnudurai, G., 2006. *In vitro* propogation of *Hepatozoon canis* in a canine macrophage cell line. Proc. XVII Natl. Cong. Vet. Parasitol., November

15-17, 2006. Dept. Parasitol., Rajiv Gandhi Coll. Vet. Sci.& Anim. Sci., Kurumbapet, Puducherry- 605 009. Abstr., S.II: 40, p.73.

Harikrishnan, T. J., Edith, S. and Chellappa, D. J., 2001. Development of *Hepatozoon canis* in the tick, *Rhipicephalus sanguineus*. Twelfth Natl. Cong. Vet. Parasitol., August 26-27, 2001, Dept. Parasitol., Coll. Vet. Sci., Acharya Ranga Agri. Univ. Tirupati-517 502., India, Abstr., S-2: 16, p.53.

Harikrishnan, T. J., Edith, R. and Ponnudurai, G., 2008a. Observations on *Hepatozoon canis* in *Rhipicephalus sanguineus*. J. Vet. Parasitol., 22(2): 1-4.

Harikrishnan, T. J., Pazhanivel, N. and Chellappa, D. J., 2008b. Observation on development of *Hepatozoon canis* in a dog. J. Vet. Parasitol., 22: 35-40.

Juyal, P. D. and Singla, L. D., 1994. Canine hepatozoonosis (Protozoa: Apicomplexa): A review. Indian Vet. Med. J., 18 (3): 125-130. (Review, not indexed)

Rau, M. A. N., 1925. *Haemogregarina canis*. Vet. J. 81: 293-307.(Not indexed).

Megat Adb Rani, P. A., Irwin, P. J., Gatne, M. L., Colman, G. T., Mcinnes, L. M. and Tromb, R. J., 2010. Canine vector-borne diseases in India: a review of the literature and identification of existing knowledge gaps. Parasites and Vectors 3: 28. (Review, not cited in Index)

7.3 FAMILY: PLASMODIIDAE

Helen Adie, 1915. The sporogony of *Haemoproteus columbae*. Indian Vet. Med. J., 2: 671-680.

Ponnudurai, G., Rajendran, K., Ravi, N., and Harikrishnan, T. J., 2010 A note on the occurrence of *Haemoproteus* spp. and its new vector in Tamil Nadu. XX NCVP, Parasitology Today-Ecology to Molecular Biology, 18-20 February, 2010, Dept. Vet. Parasitol., CVS, CCS, Haryana Agri. Univ., Hisar-125004, p.20.

7.4 FAMILY: BABESIIDAE

Achutan, H. N., Mahadevan, S. and Lalitha, C. M., 1980. Studies on the developmental forms of *Babesia bigemina* and *Babesia canis* in ixodid ticks. Indian Vet. J. 57: 181-184.

Christophers, S. R., 1907. *Piroplasma canis* and its life cycle in the tick. Sci. Mem. Med. Ind., Calcutta, 29: p.1-83.

Chaudhuri, R. P., Gill, B. S. and Khan, M. H., 1975. Studies on transmission of *Babesia bigemina*. Ann. Soc. Bel. Med.Trop., 55: 327-332.

Gautam, O. P., 1980. Babesiosis in India. In: Haemoprotozoan Diseases of Domestic Animals. Gautam, O. P., Sharma, R. D. and Dhar, S. (Eds.), Proc. Semr. Haemoprotozoan Dis., Dept. Vet. Med., Haryana Agri. Univ., Hissar 125004 (Haryana) India, October 27-November 1, 1980, B-4 pp. 115-131. (Review, not indexed, suggestive vectors name are given for all species of *Babesia*).

Megat Adb Rani, P. A., Irwin, P. J., Gatne, M. L., Colman, G. T., Linda, M. M. and Rebeca, J. T., 2010. Canine vector-borne diseases in India: a review of the literature and identification of existing knowledge gaps. Parasites and Vectors, 3: 28. (Review, not cited in Index).

Rajamohanan, K., 1982. Identification of the vector for babesiosis of cattle in Kerala. In: Vectors and Vector-borne Diseases. Proc. All India Sym., Trivandrum, Kerala, India. February 26-28, 1982, pp.125-128.

Rau, M. A. N., 1926. Experimental infection of the jackal (*Canis aureus*) with *Piroplasma canis*. Indian J. Med. Res. 14: 243-244.

Rau, M. A. N., 1926. *Piroplasma gibsoni*, Patton, 1910. Indian J. Med. Res., 14: 785-800.

Ravindran, R., Mishra, A. K. and Rao, J. R., 2006. Clinico-parasitological observations in experimentally induced bovine babesiosis. J. Parasit. Dis., 30: 178-180.

Ravindran, R., Mishra, A. K. and Rao, J. R., 2008. Randomly amplified polymorphic DNA-polymerised chain reaction fingerprinting of *Babesia bigemina* isolates of India - Short communication. Vet Arhiv., 78: 545-551.

Sanjeev Kumar, Malhotra, D. V., Sangwan, A. K., Goel, P., Kumar, A. and Kumar, S., 2007. Infectivity rate and transmission potential of *Hyalomma anatolicum anatolicum* ticks for *Babesia equi* infection. Vet. Parasitol.,144: 338-343.

Sanjeev Phogat, 1999. Studies on transmission of *Babesia equi* in *Hyalomma anatolicum anatolicum* from experimental infected donkey. M. V. Sc. Thesis, Punjab Agri. Univ., Ludhiana.

Sen, S. K., 1933.The vectors of canine piroplasmosis due to *Piroplasma gibsoni*. Indian J. Vet. Sci., 3: 356-363.

Swaminathan, C. S. and Shortt, H. E., 1937. The arthropod vector of *Babesia gibsoni*. Indian J. Med. Res., 25: 499-503.

7.5 FAMILY: THEILERIIDAE

Anandan, R., Koshy, T. J., Abdul Basith, S., Ganeshmoorthy, M., Lalitha John, Lalitha, C. M., 1988. Experimental transmission of two strains of *Theileria annulata* through infective blood and tick stages. Indian Vet. J., 65: 567-573.

Anon., 1970-73. Final Report of research scheme on investigation of theileriasis in cattle (ICAR scheme). Dept. Vet. Med., Coll. Vet. Sci. Haryana Agri. Univ., Hissar.

Anon., (1988-89). Annual report of Animal Disease Research Laboratory, National Dairy Development Board, Anand, India.(Not Indexed).

AnishaTiwari, Singh, N. K., Harkirat Singh, Jyoti, Bhat, S. A. and Rath, S. S., 2013. Prevalence of *Theileria annulata* infection in *Hyalomma anatolicum anatolicum* collected from crossbred cattle of Ludhiana, Punjab. J. Parasit. Dis., from Internet.

Ashok Kumar, Swarup, S., Sharma, R. D., Nichani, A. K. and Parveen Goel., 1991. Chemoprophylactic efficacy of buparvaquone against bovine tropical theileriosis. Indian Vet. J., 68: 54-56.

Bansal, G. C. and Gaur, G. N. S., 1977. Note on biochemical changes in experimental bovine theileriosis. Pantnagar J. Res., 2: 221-222.

Bansal, G. C. and Ray, D., 1997. A simple method for cryopreservation of *Theileria annulata*. Indian Vet. J., 74: 813-814.

Bansal, G. C. and Ray, D., 1998. Antibody response to stage specific antigens of *Theileria annulata*. J. Vet. Parasitol., 12: 109-111.

Bansal, G. C. and Sharma, N. N., 1986. Efficacy of parvaquone and long acting oxytetracycline in *Theileria annulata* infection. Vet. Parasitol., 21: 145-149.

Bansal, G. C. and Sharma, N. N., 1991. Prophylactic efficacy of buparvaquone in experimentally induced *Theileria annulata* infection in calves. Vet. Parasitol., 33: 219-224.

Bhattacharyulu, Y., Chaudhuri, R. P. and Gill, B. S., 1975a. Studies on the development of *Theileria annulata* Dschunskowsky and Luhs, 1904 in the tick–*Hyalomma anatolicum anatolicum* Koch, 1844. Ann. Parasitol. (Paris), 50: 397-408.

Bhattacharyulu, Y., Chaudhuri, R. P. and Gill, B. S., 1975b. Trans stadial transmission of *Theileria annulata* through common ticks infesting Indian cattle. Parasitology, 71: 1-7.

Bhattacharyulu, Y., Dhar, S. and Gautam, O. P., 1972. Experimental theileriosis in crossbred calves. 14th conf. res. workers in anim. dis., 25-27th November, 1972, Mathura, India.

Bhattacharyulu, Y., Singh, A. and Grewal., A. S., 1990. Some experiments on the transmission of *Theileria annulata* through ticks. Indian J. Anim. Sci., 60: 1313-1314.

Chaudhari, S. S. and Subramanian, G., 1991. Immunoprophylaxis against *Theileria annulata* with protein from plasma membrane of infected lymphocytes. Vet. Parasitol, 39: 53-60.

Chaudhari, S. S. and Subramanian, G., 1992a. *In vitro* infectivity of sporocyst of *Theileria annulata* to viable lymphocytes of cattle and buffalo. Indian J. Anim. Sci., 62: 561-563.

Chaudhari, S. S. and Subramanian, G., 1992b. Cell-mediated immune response to sporozoites and macroschizonts of *Theileria annulata*. Vet. Parasitol, 41: 23-24.

Das, G., 2002. Transmission of *Theileria annulata* in cattle with reference to acquired resistance to vector ticks. Ph. D. Thesis (Vet. Parasitol.), Deemed Univ., Indian Vet. Res. Inst., Izatnagar-243 122, India.

Das, G. and Ray, D., 2003a. Differentiation of *Theileria annulata* in salivary gland of *Hyalomma anatolicum anatolicum*. Blue Cross Book, 21: 21-22.

Das, G and Ray, D., 2003b. PCR-based detection of *Theileria annulata* infection in ticks collected from cattle of West Bengal, India. J. Vet. Parasitol., 17: 11-14.

Das, G., Ray, D. D. and Ghosh, S., 2005. Different stages of *Theileria annulata* in the salivary glands of ticks *Hyalomma anatolicum anatolicum*. Proc. XVI Nat. Cong. Vet. Parasitol, December 6-8, 2005, Dept. Vet. Parasitol., Coll. Vet. Sci. & Anim. Husb., Indira Gandhi Agri. Univ. Anjora, Durg-491 001, Chhattisgarh, Ab-12, p.23.

Das, G., Ray, D. D. and Ghosh, S., 2007. Reduced infectivity of *Theileria annulata* in ticks fed on calves immunised by purified tick antigen. J. Vet. Parasitol., 21: 59-61.

Das, S. S., 1994. Correlation of *Theileria* parasites masses in salivary gland acini of *Hyalomma anatolicum anatolicum* with levels of hosts parasitaemia. J. Vet. Parasitol., 8: 39-42.

Das, S. S. and Sharma, N. N., 1975. Study of *in vitro* development of sporozoites in cattle and buffalo lymphocytes. Indian Vet. J., 72: 460-462.

Das, S. S. and Sharma, N. N., 1991a. Prevalence of *Theileria* infection in *Hyalomma anatolicum anatolicum* ticks in north district of Tripura (India). J. Vet. Parasitol., 5: 25-27.

Das, S. S. and Sharma, N. N., 1991b. Effect of temperature on transtadial transmission of *Theileria annulata* infection in *Hyalomma anatolicum anatolicum* ticks. Vet. Parasitol., 40: 155-158.

Das, S. S. and Sharma, N. N., 1992. Effect of radiation on sporozoites of *Theileria annulata* in the salivary glands of ticks *Hyalomma anatolicum anatolicum*. J. Vet. Parasitol., 6: 37-38.

Das, S. S. and Sharma, N. N., 1993a. Study on the prevalence and *in vitro* development of *Theileria* sporozoites in the salivary gland acini of *Hyalomma anatolicum anatolicum* in the west district of Tripura (India). J. Parasitol. Appl. Anim. Biol., 2: 71-74.

Das, S. S., and Sharma, N. N., 1993b. *In-vitro* anti-schizontal effects of parvaquone and buparvaquone on *Theileria annulata* in lymphoblastoid cells. Indian J. Anim. Sci., 63: 1021-1024.

Das, S. S. and Sharma, N. N., 1994a. Correlation of *Theileria* parasite masses in salivary gland acini of *Hyalomma anatolicum anatolicum* with levels of host's parasitaemia. J. Vet. Parasitol., 8: 39-42.

Das, S. S. and Sharma, N. N., 1994b. Study of intracellular development of *Theileria annulata* sporozoites. Indian Vet. J., 71: 547-549.

Datta, C. S., 1985. Studies on host-parasite relationship and chemoprophylaxis of theileriasis in exotic and cross-bred cattle. M. V. Sc. Thesis submitted to Bihar Vet. Coll., Patna, Bihar, India. (Not indexed).

Datta, C. S., Srivastava, P. S. and Sinha, S. R. P., 1988. Prevalence and epidemiology of a virulent strain of *Theileria annulata* in cattle in and around Patna., Bihar, India. Indian J. Anim. Hlth., 27: 151-158.

Davra, M. L., 1973. Studies on some aspects of pathogenesis, diagnosis and treatment of experimental theileriasis in sheep and goats. M.V. Sc. Thesis, Punjab Agri. Univ., Ludhiana, India. (Not indexed).

Dhar. S. and Gautam, O. P., 1977a. Some biochemical aspects of *Theileria annulata* infection in cattle. Indian J. Anim. Sci., 47: 169-172.

Dhar, S. and Gautam, O. P., 1977b. *Theileria annulata* infection in cattle. 2. Capillary–tube agglutination test for serodiagnosis. Indian J. Anim. Sci., 47: 458-462.

Dhar, S and Gautam, O. P., 1978. A note on the use of hyperimmune seruim in bovine tropical theileriosis. Indian Vet. J., 55: 738-740.

Dhar, S and Gautam, O. P., 1979a. Observations on anaemia in experimentally induced *Theileria annulata* infection in calves. Indian J. Anim. Sci., 49: 122-126

Dhar, S and Gautam, O. P., 1979b. Serum proteins in experimental *Theileria annulata* infection of cattle. Indian J. Anim. Sci., 49: 511-516.

Dhar, S., Malhotra, D. V., Chandra Bhushan and Gautam, O. P., 1986. Chemotherapy of *Theileria annulata* infection with buparvaquone. Vet. Rec., 119: 635-636.

Dhar, S., Malhotra, D. V., Chandra Bhushan and Gautam, O. P., 1987. Buparvaquone (BW720 C) for the treatment of bovine tropical theileriosis. Proc. Inaugaral Sym., Indian Asso. Vet. Parasitol., February 12-13th, 1987, pp.31-32.

Dhar, S., Malhotra, D. V., Chandra Bhushan and Gautam, O. P., 1988. Treatment of experimentally induced *Theileria annulata* infection in cross bred calves buparvaquone. Vet. Parasitol., 27: 267-275.

Dhar, S., Malhotra, D. V., Chandra Bhushan and Gautam, O.P., 1990a. Chemotherapy of different stages of experimentally induced bovine tropical theileriosis with buparvaquone. Indian Vet. J., 67: 598-602.

Dhar, S., Malhotra, D. V., Bhushan, C. and Gautam, O. P., 1990b. Chemoprophylaxis against bovine tropical theileriosis in young calves: a comparison between with buparvaquone and long acting oxytetracycline. Res. Vet. Sci., 49: 110-112.

Dhar, S., Chandra Bhushan, Malhotra, D. V., Mallick, K. P. and Gautam, O. P., 1982. Quantitative assessment of *Theileria annulata* (Piroplama: Theileridae) infection in the salivary glands of the tick, *Hyalomma anatolicum anatolicum*. Prog. Acarol., 7: 81-84.

Dhar, S., Mallick, K. P., Chandra Bhushan, Malhotra, D. V. and Gautam, O. P., 1987. Observations on *Theileria annulata* infection in *Hyalomma anatolicum anatolicum* adult ticks. Indian Vet. J., 64: 370-373.

Datta, C. S., Srivastava, P. S. and Sinha, S. R. P., 1988. Prevalence and epidemiology of a virulent strain of *Theileria annulata* in cattle in and around Patna (Bihar, India). Indian J. Anim. Hlth., 27: 151-157.

Dinesh Patel, Misraulia, K. S., Gopal Reddy, A., Garg, U. K., Sharma, R. K., Bagherwal, R. K. and Brajesh Kumar Gupta., 2001. Effectiveness of Artemether in induced bovine tropical theileriosis in cross bred calves. Indian Vet. J., 78: 386-389.

Ebenezer Raja, Joseph, S. A. and Lalitha, C. M., 1983. Vector potential in relation to the incidence of theileriasis in Tamilnadu. Ticks and Tick-borne Diseases. 21st-22nd March, 1983, (Tamil Nadu Agri. Univ.), Dept. Parasitol., Madras Vet. Coll., Madras-600 007, p.20.

Gautam, O. P., 1974. Theileriosis in India. 3rd Int. Cong. Parasit. Munich, West Germany.(Not Indexed).

Gautam, O. P., 1976. *Theileria annulata* infection in India. International Conference on Tick-Borne Diseases and their Vectors. J. H. Wilde (Ed.), Centr.Trop. Vet. Med., Univ. Edinburgh, U.K. (Not indexed).

Gautam, O. P. and Dhar, S., 1980. Bovine tropical theileriosis in India. In: Haemoprotozoan Diseases of Domestic Animals. Gautam, O. P., Sharma, R. D. and Dhar, S. (Eds.), Proc. Seminar Haemoprotozoan Dis., Dept. Vet. Med., Haryana Agri. Univ., Hissar 125004 (Haryana) India, October, 27-November 1, 1980, A-1, pp.11-29. (Review, not indexed, vectors details are given).

Gautam, O. P., Sastry, K. N. V., Dhar, S. and Singh, R. P., 1982. Effect of irradiation on *Theileria annulata* particles derived from ticks. Indian Vet. J., 59: 581-584.

Gautam, O.P.and Dhar, S., 1983. Bovine tropical theileriosis—a review. 1. Prevalence, transmission and symptoms. Trop. Vet. Anim. Sci. Res. 1: 1–18.(Not indexed).

Gill, B. S., Bansal, G. C., Bhattacharylu, Y., Kaur, D. and Singh, A., 1980. Immunological relationships between strains of *Theileria annulata* (Dschunkowsky and Luhs, 1904). Res. Vet. Sci., 29: 93-97.

Gill, B. S. and Bhattacharyulu, Y., 1976. Theileriosis – Rept. workshop,Nairobi, Kenya, 7-9 December 1976. Henson, J. B. and Campbell, M. (Eds.), Internatl. Devp. Res. Centr., Ottawa, Canada, pp.8-11.

Gill, B. S., Bhattacharyulu, Y. and Kaur, D., 1976a. Immunization against bovine tropical theileriasis (*Theileria annulata* infection). Res. Vet. Sci., 21: 146-149.

Gill, B. S., Bhattacharyulu, Y. and Kaur, D., 1976b. Annual Report 1975 of the ad-hoc research scheme, Investigation into bovine tropical theileriasis. ICAR, New Delhi. (Not indexed).

Gill, B. S., Bhattacharyulu, Y. and Kaur, D., 1977a. Studies on the relationship between the quantum of infection and ensuing reaction of cattle infected with *Theileria annulata*. Ann. Soc. Bel. Med. Trop., 57: 557-567.

Gill, B. S., Bhattacharyulu, Y. and Kaur, D., 1977b. Symptoms and pathology of experimental bovine tropical theileriosis (*Theileria annulata* infection). Ann. Parasitol., 52: 597-608.

Gill, H. S., Bhattacharyulu, Y. and Gill, B. S., 1980. Attempt to transmit *Theileria ovis* through tick, *Haemaphysalis bispinosa* and *Rhipicephalus haemaphysaloides*. Trop. Anim. Hlth. Prod., 12: 61.

Gill, B. S., Bhattacharyulu, Y., Kaur, D. and Singh, A., 1976c. Vaccination against bovine tropical theileriasis (*Theileria annulata* infection). Nature (London), 264: 355-356.

Gill, B. S., Bhattacharyulu, Y., Kaur, D. and Singh, A., 1977c. Immunization of cattle against tropical theileriasis (*Theileria annulata* infection) by 'infection and treatment' method. Ann. Rech. Vet., 8: 285-292.

Gill, B. S., Bhattacharyulu, Y., Kaur, D. and Singh, A., 1977d. Annual Report 1976, ad-hoc research scheme, Investigation into bovine tropical theileriasis. ICAR, New Delhi. (Not indexed).

Gill, B. S., Bhattacharyulu,Y., Singh, A. Kaur, D. and H. S., 1981. Chemotherapy against *Theileria annulata*. Curr. Topics Vet. Med. Anim. Sci., 14: 218-222.

Gill, B. S., Kaur, D. and Bhattacharyulu, Y., 1974. Transmission of *Theileria annulata* through the tick *Hyalomma detritum* (Schulze, 1919). Bull. Off. Int. Epiz., 81: 805-811.

Gill, H. S., Bhattacharyulu, Y., Kaur, D. and Singh, A., 1978. Chemoprophylaxis with tetracycline drugs in the immunization of cattle against *Theileria annulata* infection. Int. J. Parasitol., 8: 467-469.

Ghosh, S., Azhahianambi, P., Vanlalhmuaka, Rajendran, C., Suryanarayana, V. V. S., Ray, D. D., Chowdhary, Pallabh, Bansal, G. C. and Gupta, S. C., 2006. Prospects for the development of multivalent chimeric antigen vaccine for integrated control of *Hyalomma anatolicum anatolicum* and *Theileria annulata*. Proc. XVII Natl. Cong. Vet. Parasitol., November 15-17, 2006, Dept. Parasitol., Rajiv Gandhi Coll.Vet. & Anim. Sci., Kurumbapet, Puducherry- 605009, Ab. S.V-10, p.145.

Harikrishnan, T. J., Joseph, S. A. and Anandan, R., 2000. Experimental transmission of *Theileria annulata* through ixodid ticks. Indian Vet. J., 77: 382-384.

Harikrishnan, T. J., Chellappa, D. J. and Balasundaram, S., 2001. A note on erythrophagocytosis in experimental bovine theileriosis. Indian Vet. J., 78: 1162-1163.

Haque, M., Jyoti, Singh N. K. and Rath, S. S., 2010. Prevalence of *Theileria annulata* infection in *Hyalomma anatolicum anatolicum* in Punjab State, India. J. Parasitic Dis., 34: 48-51.

Haque, M., Jyoti, Singh, N. K. and Rath, S. S., 2011. Concurrent detection of *Theileria* organism from definitive host and vector tick. Indian Vet. J., 88: 144-145.

Jagadish, S., 1977. Studies on certain aspects of *Theileria annulata* in cattle with particular reference to chemotherapy and chemoprophylaxis and some serum enzyme changes. Ph. D. Thesis, Haryana Agri. Univ., Hissar. (Not indexed).

Jagadish, S., 1979. Chemoprophylactic immunization against bovine tropical theileriosis.Vet. Rec., 104: 140-142.

Jagadish, S., Singh, D. K. and Gautam, O. P., 1980a. Chemoprophylactic immunization against bovine tropical theileriasis. Indian Vet. J., 57: 177-178.

Jagadish, S., Singh, D. K. and Gautam, O. P., 1980b. A simple method for the demonstration of *Theileria annulata* in the salivary glands of *Hyalomma anatolicum anatolicum* by modified Feulgen technique. In: Haemoprotozoan Diseases of Domestic Animals. Gautam, O. P., Sharma, R. D. and Dhar, S (Eds.), Proc. Semr. Haemoprotozoan Dis., Dept. Vet. Med., Haryana Agri. Univ., Hissar 125004 (Haryana) India, October 27-November 1, 1980, A-3, pp.32-42.

Jeyabal, L., Kumar, B., Ray, D., Azahahianambi, P., Ghosh, S., 2012.Vaccine potential of recombinant antigens of *Theileria annulata* and *Hyalomma anatolicum anatolicum* against vector and parasite.Vet. Parasitol., 188: 231-138.

Khan, M. H., 1997. Factors affecting transmission of *Theileria annulata* by *Hyalomma* ticks. Indian J. Anim. Hlth., 36: 105-109.

Khanna, B. M., Shruti Dhar and Gautam, O. P., 1980a. Chemotherapy of experimental *Theileria annulata* infection in bovine calves. In: Haemoprotozoan Diseases of Domestic Animals. Gautam, O. P., Sharma, R. D. and Dhar, S.(Eds), Proc. Semr. Haemoprotozoan Dis., Dept. Vet. Med., Haryana Agri. Univ., Hissar 125004 (Haryana) India, October 27-November 1, 1980, A-11, pp.74-80.

Khanna, B. M., Shruti Dhar and Gautam, O. P., 1980b. Immunization against bovine tropical theileriosis by using infection and treatment method. In: Haemoprotozoan Diseases of Domestic Animals. Gautam, O. P., Sharma, R. D. and Dhar, S.(Eds), Proc. Semr. Haemoprotozoan Dis., Dept. Vet. Med., Haryana Agri. Univ., Hissar 125004 (Haryana) India, October 27-November 1, 1980, A-20, pp.91-99.

Khanna, B. M., Dhar, S and Gautam, O. P., 1983a. Immunization against bovine tropical theileriosis by using infection and treatment method. Indian Vet. J., 60: 257-261.

Khanna, B. M., Shruti Dhar and Gautam, O. P., 1983b. Chemotherapy of experimental *Theileria annulata* infection in bovine calves. Indian Vet. J., 60: 603-606.

Khurana, K. L., Chhabra, M. B. and Samantaray, J., 1988. Comparative *Theileria* transmission potential of *Hyalomma* spp. ticks. Proc. 2nd Sym. Vectors and Vector-borne Dis., pp.39-46.

Kumar, R. and Malik, J. K., 1999a. Influence of experimentally induced theileriosis (*Theileria annulata*) on the pharmacokinetics of a long-acting formulation of oxytetracycline (OTC-LA) in calves. J. Vet. Pharmacol. Ther., 22(5): 320-326.

Kumar, R. and Malik, J. K., 1999b. Effects of experimentally induced *Theileria annulata* infection on the pharmacokinetics of oxytetracycline in cross-bred calves. Vet. Res., 30(1): 75-86.

Lal, H. and Soni, J. L., 1983. Auto-immune reaction associated with experimental *Theileria annulata* infection. Indian J. Anim. Sci., 53: 654-658.

Mallick, K. P., Dhar, S., Malhotra, D. V., Chandra Bhushan and Gautam, O. P., 1987. Immunization of neonatal bovine against *Theileria annulata* by an infection and treatment method. Vet. Parasitol., 24: 169-173.

Manickam, R., Dhar, S. and Singh, R. P. 1983. Protection of cattle against *Theileria annulata* infection using *Corynebacterium parvum*. Trop. Anim. Hlth. Prod.,15: 209-213.

Mehta, H. K., Sisodia, R.S. and Misraulia, K. S., 1988. Clinical and haematological observations in experimentally induced cases of bovine tropical theileriasis. Indian J. Anim. Sci., 58: 584-587.

Momin, R. R., Banerjee, D. P. and Samantaray, S., 1991. Attempted immunization of crossbred calves (*Bos taurus* x *Bos indicus*) by repeated natural attachment of ticks *Hyalomma anatolicum anatolicum* Koch (1884). Trop. Anim. Hlth Prod., 23: 227-231.

Naithani, R. C. and Subramanian, G., 1980. Role of *Hyalomma anatolicum anatolicum* in transmission of bovine Theileriosis. In: Haemoprotozoan Diseases of Domestic Animals. Gautam, O. P., Sharma, R. D. and Dhar, S. (Eds.), Proc.Semr. Haemoprotozoan Dis., Dept. Vet. Med., Haryana Agri. Univ., Hissar 125004 (Haryana) India, October 27-November 1, 1980, A-7, p.56.

Nirmal Sangwan and Sangwan, A.K., 2007a. Zinc distribution or loss in relation to progression of *Theileria annulata* infection in crossbred calves. J. Vet. Parasitol., 21: 63- 65.

Nirmal Sangwan and Sangwan, A.K., 2007b. Changes in blood and tissue copper levels due to *Theileria annulata* infection in cross-bred calves. Indian J. Anim. Res., 41: 278- 281.

Patil, A. I. and Avasathi, B. L., 1983. Resistance of cross bred bull calves having 75 per cent Kankrej blood and to *Theileria* annulata. Semr, Ticks and Tick-borne Diseases. 21st-22nd March, 1983, (Tamilnadu Agri. Univ.) Dept. Parasitol., Madras Vet. Coll., Madras-600 007, p.16.

Prem, S., Gupta, S. L., Malhotra, D. V. and Sagar, P., 2002. Erythrocyte associated haemato-biochemical changes in cross-bred calves experimentally infected with *Theileria annulata*. Indian J. Vet. Res., 11: 18-24.

Prem Sagar, 1992. Some haematological and scanning electron microscopic students of erythrocytes in experimental *Theileria annulata* infection in crossbred calves. M.V.Sc.Thesis, Punjab Agri. Univ., Ludhiana. (Not indexed).

Prem Sagar, Gupta, S. L., Malhotra, D. V. and Sagar, P., 2002. Erythrocyte associated haemato-biochemical changes in cross-bred calves experimentally infected with *Theileria annulata*. Indian J. Vet. Res., 11: 18-24.

Raghorte, S. D. Maske, D. K. Jayraw, A. K. and Bhaviskar, B. S., 2006a. Haematological observations in cattle infested *Rhipicephalus* ticks harbouring *Theileria annulata* infection. Proc. XVII Natl. Cong. Vet. Parasitol., November 15-17, 2006. Dept. Parasitol., Rajiv Gandhi Coll. Vet. Sci. & Anim. Sci, Kurumbapet, Puducherry-605 009. Abstr., S.II.36, p.71.

Raghorte, S. D. Maske, D. K. Jayraw, A. K. and Gawande, P. J., 2006b. Biochemical observations in cattle infested *Rhipicephalus* ticks harbouring *Theileria annulata* infection. Proc. XVII Natl. Cong. Vet. Parasitol., November 15-17, 2006. Dept. Parasitol., Rajiv Gandhi Coll. Vet. Sci. & Anim. Sci, Kurumbapet, Puducherry-605 009, Abstr., S.II.34, p.70.

Raghav, P.R.S. *et al.*, 1991. In Orientation and Coordination of Research on Tropical Theileriosis. Singh, D.K. and Varshney, B.C.(Eds.), National Dairy Development Board, Anand, India. p.39. (Not indexed).

Rajiv Kumar and Malik, J. K., 1999a. Influence of experimentally induced theileriosis (*Theileria annulata*) on the pharmacokinetics of a long-acting formulation of oxytetracycline (OTC-LA) in calves. J. Vet. Pharmacol. Therap., 22: 320-326.

Rajiv Kumar and Malik, J. K., 1999b. Effect of experimentally induced *Theileria annulata* infection in pharmacokinetics of oxytetracycline in cross-bred calves. Vet. Res., 30: 75-86.

Ram, R., Gupta, S. K. and Sangwan, A. K., 2004. Comparative infection rates of *Theileria annulata* in *Hyalomma anatolicum anatolicum* ticks in and arid and semi-arid regions of North-West India. J. Vet. Parasitol., 18: 109-114.

Rao, J. R., Misra, R. K., Ramprakash, V. and Sharma, N. N., 1989. Biogeneic amine levels in *Theileria annulata* infected bovine calves. 1989. J. Vet. Parasitol., 3: 77-79.

Ray, D. and Bansal, G. C., 1993. Use of schizont antigen of *Theileria annulata* in enzyme immunoassay. J. Vet. Parasitol., 7: 93-97.

Ray, D., Bansal, G.C., Srivastava, R.V. N. and Subramaniyan, G., 1987. Characterization of an isolate of *Theileria annulata* from Panagarh in the West Bengal region of India. J. Vet. Parasitol., 1: 31-34.

Ray, D., Patra, G. and Bansal, G.C., 1994. Studies on "lymphocytic stimulation test" with fractionated schizont antigens of *Theileria annulata*. Sixth Natl. Cong. Vet. Parasitol., 22-24 Oct., 1994, Dept. Parasitol., Coll. Vet. Sci., JNKVV, Jabalpur-482 001, M. P., India, Abstr. S-3: 16, pp.36-37.

Ray, D. D., Bhar, R., Chaturvady,V. B., Pattnaik, A. K., Lalu, K., Pathak, A. K. and Bansal, G. C., 2006. Growth performance of cross bred bovine infected with *Theileria annulata*. Proc. XVII Natl. Cong. Vet. Parasitol., November 15-17, 2006. Dept. Parasitol., Rajiv Gandhi Coll. Vet. Sci.& Anim. Sci., Kurumbapet, Puducherry-605 009. Abstr., S.II.40, p.73.

Ray, H. N., 1950. Hereditary transmission of *Theileria annulata* infection in the tick, *Hyalomma aegyptium* Neum. Trans. R. Soc. Trop. Med. Hyg., 44: 93-104.

Rup Ram, Gupta, S. K., Sangwan, A. K. and Nichani, A. K., 2001. Comparative sero-positive prevalence of bovine tropical theilerosis in arid and semiarid regions of north-west India. Proc. Natl. Conf. Vet. Parasitol., August 25-27, 2001,Tirupati-517 502, India, Abstr. No. S-2: 43, p 69.

Rup Ram, Gupta, S. K. and Sangwan, A. K., 2004. Comparative infection rates of *Theileria annulata* in *Hyalomma anatolicum anatolicum* ticks in arid and semi-arid regions of north-west India. J. Vet. Parasitol., 18: 109-114.

Sahoo, P. K. and Mishra, S. C., 1994. Natural infection rate of *Theileria annulata* in *Hyalomma anatolicum anatolicum* at Bhubaneswar. Sixth Natl. Cong. Vet. Parasitol., 22-24 October, 1994, Dept. Parasitol., Coll. Vet. Sci., JNKVV, Jabalpur-482 001, M. P., India. Abstr. S-1: 15, p.7.

Sahoo, P. K. and Mishra, S. C., 1995. Natural infection rate of *Theileria annulata* in *Hyalomma anatolicum anatolicum* at Bhubaneswar. Indian Vet. J., 72: 687-689.

Saluja, P. S., Gupta, S. L., Malhotra, D. V. and Ambawsi, H. K., 1999. Status of plasma malondialdehyde in experimental *Theileria annulata* infection in crossbred bovine calves. Indian Vet. J., 76: 379-381.

Samantary, S. N., Bhattacharylu, Y. and Gill, B. S., 1980. Immunization of calves against bovine tropical theileriosis (*Theileria annulata*) with graded doses of sporozoites and irradiated sporozoites. International J. Parasitol., 10: 355-358.

Sandhu, G. S., 1996. Histopathlogical, biochemical and haematological studies in crossbred calves suffering from experimental tropical theileriosis infection. Thesis, Punjab Agri. Univ. (Not indexed).

Sandhu, G. S., Grewal, A. S., Singh, A., Kendal, J. K., Singh, J and Brar, R. S., 1998. Haematological and biochemical studies on experimental *Theileria annulata* infection in crossbred calves.Vet. Res. Comm., 22: 347-354.

Sangwan, A. K., 2007. Impact of cattle migration from Rajasthan in Haryana state in India in transmission of tropical theileriosis. J. Vet. Parasitol., 21: 87-88.

Sangwan, A. K. and Goel, M. C., 1999. Distribution of tick vectors of bovine tropical theileriosis in India. Proc. XXII Intl. Conf. on Sustainable Animal Production, Health and Environment: Future Challenges, CCSHAU, Hisar, India November 24-27, 1999, pp.144-145.

Sangwan, A. K. and Malhotra, D.V., 2001. Trends in prevalence of *Theileria annulata* in *Hyalomma* ticks in Haryana, Proc. Natl. Conf. Vet. Parasitol., August 25-27, 2001, Tirupati-517 502, India, Abstr. S-2: 48, pp.72-73.

Sangwan, N and Sangwan, A. K., 1999. Thiamin, riboflavin and ascorbic acid in relation to tropical theileriosis in cattle. Indian J. Anim. Sci., 69: 316-317.

Sangwan, N and Sangwan, A. K., 2007. Changes in blood and tissue copper levels due to *Theileria annulata* infection in cross-bred calves. Indian J. Anim. Res., 41 (4): 278-281.

Sangwan, A. K., Chhabra, M. B. and Samantaray, S., 1986. *Theileria* infectivity of *Hyalomma* ticks in Haryana, India. Trop. Anim. Hlth. Prod., 18: 149-154.

Sangwan, A. K., Chhabra, M. B. and Samantaray, S., 1989. Relative role of male and female *Hyalomma* ticks in *Theileria* transmission. Vet. Parasitol., 31: 83-87.

Sangwan, A. K., Chaudhri, S. S., Sangwan, N. and Gupta, R. P., 1994. Seasonal distribution of *Theileria* in *Hyalomma anatolicum anatolicum* ticks of an endemic area in Haryana, India. Trop. Anim. Hlth. Prod., 26: 241-246.

Sanjay Kumar, Rajender Kumar and Chihiro Sugimoto., 2009. A perspective on *Theileria equi* infections in donkeys. Japanese J. Vet. Res., 56(4): 171-180.(Review, not indexed, mention has been made on IH recorded from India, vide under *Babesia equi*, supra).

Saravanan, B. C., Bansal, G. C., and Ray, D., 2003. A comparative study on the clinico-parasitogical responses of Izatnagar and Parbhani isolates of *Theileria annulata* in cattle. J. Vet. Parasitol., 17: 161-162.

Saravanan, B. C., Sankar, M., Bansal, G. C., Sreekumar, C., Tewari, A. K., Rao, J. R. and and Ray, D., 2010. Random amplified polymorphic DNA profiles in two Indian strains of *Theileria annulata*. J. Vet. Parasitol., 24: 39-49.

Sastry, K. N. V., 1980. Studies on certain aspects of immunization against tropical bovine Theileriosis. Ph. D. Thesis, Haryana Agri. Univ., Hissar. (Not indexed).

Sastry, K. N. V., Dhar, S. and Singh, R. P., 1980. Infectivity of *Theileria annulata* infected *Hyalomma anatolicum anatolicum* ticks incubated at 37⁰C and 95 per cent relative humidity. In: Haemoprotozoan Diseases of Domestic Animals. Gautam, O. P., Sharma, R. D. and Dhar, S., (Eds).Proc. Seminar Haemoprotozoan Dis., Dept. Vet. Med., Haryana Agri. Univ., Hissar 125004 (Haryana) India, October, 27-November,1, 1980, A-6 pp.50-55.

Sharma, N. N., and Mishra, A. K., 1990. Treatment of bovine tropical theileriosis with buparvaquone (BW 720C). Trop. Anim. Hlth. Prod., 22: 63-65.

Shastri, U. V., Jadhav, K. V., Pathak, S.V. and Deshpande, P. D., 1985. Studies on *Theileria* sp. from buffalo, *Bubalus bubalis,* from Maharashtra. Indian J. Parasitol., 9: 275-279.

Shastri, U. V., Pathak, S.V., Jadhav, K. V. and Deshpande, P. D., 1988. Occurrence of *Theileria orientalis* in bovines from Maharashtra state and its transmission by *Haemaphysalis bispinosa* ticks. Indian J. Parasitol., 12: 173-177.

Sharma, L. D., Sharma, N. N., Sabir, M., and Bhattacharya, N. K., 1987. Clinco-haematological and biochemical changes in experimental *Theileria* infection in crossbred calves. Indian J. Vet. Med., 7: 154-156.

Shukla, P. C., 1991. Ph. D. Thesis submitted to Haryana Agri. Univ., Hissar 125004 (Haryana) India. (Not indexed).

Shukla, P. C. and Sharma, R. D., 1988. Immunization trials in bovine calves inoculated with supernatant of *in-vitro* cell culture infected with *Theileria annulata*. Indian J. Vet. Med., 8: 58-60.

Shukla, P. C. and Sharma, R. D., 1991. Immunization of bovine calves with cell culture vaccine against *Theileria annulata*. Acta Vet. Brno., 60: 79-86.

Singh, A. 1998. Clinicopathological studies on experimental *Theileria annulata* infection in crossbred calves. M.V.Sc. Thesis. Punjab Agri. Univ., Ludhiana.(Not indexed).

Singh D. K., Thakur, M., Raghav, P.R. and Varshney, B.C., 1993. Chemotherapeutic trials with four drugs in crossbred calves experimentally infected with *Theileria annulata*. Res. Vet. Sci., 54(1): 68-71.

Singh, A., Singh, J., Grewal, A. S. and Brar, R.S., 2001. Studies on some blood parameters of crossbred calves with experimental *Theileria annulata* infection. Vet. Res. Commn., 25: 289-300.

Singh, D. K., 1977. Studies on *Theileria annulata* with special reference to its attenuation for immunization of calves. Ph. D. Thesis, Haryana Agri. Univ., Hisar.(not indexed).

Singh, D. K., 1986. Tropical theileriosis in India: In: Orientation & Coordination of Research in Tropical theileriosis (Report of EEC Workshop Centr.Trop. Vet. Med., Edinburgh, pp.40-42.

Singh, D. K., 1990. Recent development in research and control of *Theileria annulata* in India. In: Recent developments in the research and control of *Theileria annulata*. Proc. Workshop, ILRAD, Nairobi, Kenya, 17-19 september, 1990, T.T.Dolan (Ed.), Internatl. Lab. Res. Anim. Dis., Box 30709, Nairobi,Kenya, 1992, pp.11-14.

Singh, D. K., 1991. Theileriosis in India. In: Orientation and Coordination of Research on Tropical Theileriosis. Singh, D.K. and Varshney, B.C. (Eds.), National Dairy Development Board, Anand, India. p.23. (Not indexed).

Singh, D. K. and Varshney, B. C. (Eds). 1991. Proc. Second EEC Workshop on Orientation and Coordination Res.Tropical Theileriosis: March 18-22, 1991. EEC and NDDB Anim. Dis. Res. Lab., Anand, India. p.175. (Not indexed).

Singh, D. K., Jagadish, S. and Gautam. O. P., 1979a. Immunization against bovine tropical theileroisis using co-irradiated infective particles of *Theileria annulata* (Dschunkowsky and Luhs, 1904). Am. J. Vet. Res., 40: 767- 769.

Singh, D. K., Jagadish, S. and Gautam. O. P., 1979b. Cell mediated immunity in tropical theileriosis (*Theileria annulata* infection). Res. Vet. Sci., 23: 391-392.

Singh, D. K., Jagadish, S., Gautam. O. P. and Dhar, S., 1979c. Infectivity of ground-up tick supernates prepared from *Theileria annulata* infected *Hyalomma anatolicum anatolicum*. Trop. Anim. Hlth. Prod., 11: 87-90.

Singh, D. K., Jagadish, S. and Gautam, O. P., 1980. Observations on the development of *Theileria annulata* in the salivary glands of *Hyalomma anatolicum anatolicum*. In: Haemoprotozoan Diseases of Domestic Animals. Gautam, O. P., Sharma, R. D. and Dhar, S. (Eds.), Proc. Semr. Haemoprotozoan Dis., Dept. Vet. Med., Haryana Agri.Univ., Hisar-125004 (Haryana) India, October 27-November 1, 1980, A-5, pp.45-49.

Singh, S., Grewal, A. S. and Mangat, A. P. S., 1991. Acquired immunity to the tick vector (*Hyalomma anatolicum anatolicum*) of bovine tropical theileriosis. Proc. 2nd ECC Workshop on Orientation Coordination of Res. Tropical Theileriosis, 18-22 March, 1991, Anand, India.

Sisodia, R. S. and Gautam. O. P., 1980. Clinical and experimental studies on sheep theileriosis. In Haemoprotozoan Diseases of Domestic Animals. Gautam, O. P., Sharma, R. D. and Dhar, S. (Eds.), Proc. Semr. Haemoprotozoan Dis., Dept. Vet. Med., Haryana Agri. Univ., Hisar 125004 (Haryana) India, October 27-November 1, 1980, A-13, p.82.

Srivastava, P. S. and Sharma, N. N., 1976a. Effects of ^{60}Co irradiation of the early developmental stages of an ixodid tick, *Hyalomma anatolicum*. International J. Radiation Biol., 29: 165-168.

Srivastava, P. S. and Sharma, N. N., 1976b. On the rearing of *Hyalomma* (*Hyalomma*) *anatolicum anatolicum* Koch their infection with *Theileria annulata* (Daschunkowsky and Luhs, 1904) in the laboratory. Indian J. Anim. Res., 10: 21-26.

Srivastava, P. S. and Sharma, N. N., 1976c. Infectivity and immunogenicity of washed bovine erythrocytes in crossbred calves infected with *Theileria annulata* (Daschunkowsky and Luhs, 1904). Pantnagar J. Res.,1: 70-72.

Srivastava, P. S. and Sharma, N. N., 1976d. Characteristics of a tick transmitted virulent strain of *Theileria annulata* (Daschunowsky and Luhs, 1904) in crossbred calves. Pantnagar J. Res., 1: 83-88.

Srivastava, P. S. and Sharma, N. N., 1978. Studies on the infectivity of *Theileria annulata* infected nymphs, adults and ground tissues of the *Hyalomma anatolicum*. Vet. Parasitol., 4: 83-89.

Srivastava, S. C. and Khan, M. H., 1997. Infection of *Theileria annulata* in *Hyalomma anatolicum anatolicum* Koch. J. Vet. Parasitol.,11: 207-210.

Sudan, N. A., Sinha, B. P. and Verma, S. P., 1988. Haematological and biochemical changes in crossbred calves infected experimentally with *Theileria annulata*. J. Vet. Parasitol., 2: 25-30.

Subramaniyan, G., Naithani, R. C. and Ray, D., 1987. Cross-reaction among four isolates of *Theileria annulata* from India. Vet. Parasitol., 25: 75-77.

Sundar, N., Balasundaram, S., Ramadass, P and Anandan, 1993. Usefulness of enzyme-linked immunosorbant assay (ELISA) for the diagnosis of *Theileria annulata* antigen. Indian J. Anim. Sci., 63: 1219-1221.

7.6 ORDER: RICKETTSIALES

Anon, 1974-1979. Final Report of ICAR Scheme for Investigation on the biology and control of *Anaplasma*, Dept. Parasitol., Vet. Coll., Univ. Agri. Sci., Hebbal, Bangalore 560 024, p.1-28.

Ashuma, Kaur Paramjit and Juyal P. D., 2005. Canine ehrlichiosis: An emerging silent tick borne disease of dogs. Intas Polivet, 6 (11): 346-349. (Review, not indexed).

Ebenezer Raja, E., Joseph, S. A. and Lalitha, C. M., 1983. Experimental transmission of *Anaplasma marginale*, Theiler, 1910 in calves by the argasid tick *Ornithodoros savignyi* Audouin, 1827. Semnr, Ticks and Tick-borne Diseases. 21st-22nd March, 1983, (Tamilnadu Agri. Univ., Coimbatore), Dept. Parasitol., Madras Vet. Coll., Madras-600 007, p.25.

Jagannath, M. S., 1988. Transmission of *Anaplama marginale* by *Boophilus annulatus* in calves. Indian J. Anim. Sci., 58: 6-7.

Manohar, B. M. and Ramakrishnan, R., 1984. Experimental ehrlichiosis in dogs. Cheiron, 13: 144-150.

Megat Adb Rani, P. A., Irwin, P. J., Gatne, M. L., Coleman, G. T., Linda, M. M. and Rebeca, J. T., 2010. Canine vector-borne diseases in India: a review of the literature and identification of existing knowledge gaps. Parasites and Vectors 3: 28. (Review, not cited in Index).

Misraulia, K. S., Sisodia, R. S. and Mehta, H. K., 1987. Experimental transmission of anaplasmosis in crossbred male calves. Indian J. Anim. Sci., 57: 802-803.

7.7 GENERAL REFERENCE

Gautam, O. P. Sharma, R. D and Dhar, S. (Eds.), 1980. Haemoprotozoan Diseases of Domestic Animals. (Ed. Gautam, O. P., Sharma, R. D. and Dhar, S.), Proc. Semnr. Haemoprotozoan Dis. (CwVA Asian/Australian regions). October, 27 to November, 1980, Dept. Vet. Med., Haryana Agri. Univ., Hissar-125004 (Haryana) India.

SECTION III–INTERMEDIARIES OF ZOONOTIC PARASITES
(Reviews)

Agarwal, M. C., 2000. Final Report on National Fellow Project on "Studies on the strain identification, epidemiology, diagnosis, chemotherapy and zoonotic potentials of Indian schistosomes". ICAR, New Delhi. (Not indexed).

Anandan, R., 1995. Summer institute lecture notes on recent advances in the epidemiology of zoonotic diseases. Dept. Vet. Prev. Med., Madras Vet. Coll., Chennai.

Anantaraman, M., 1984. Animal parasitic infection of public health importance. Summer Institute, Tirupati, India, p.107.

Anil Ahuja, Galhot, A. K. and Purohit, S. K., 2006. Zoonotic significance of cutaneous leishmaniasis- An important zoonosis of Western Rajasthan. Intas Polivet, 74: 437-443.

Anon., 1998. Zoonotic Parasites – Their Diagnosis and Control. III National Training Programme, 9th to 23rd March, 1998, Centr. Adv. Stud., Dept. Parasitol., Vet. Coll., Univ. Agri. Sci., Bangalore-560 024, pp.1-164.

Banerjee, D. P., 1988. Serological aspects and its role in the diagnosis of parasitic zoonotic diseases. Proc. VIII Natl. Cong. Parasitol., Calcutta, 10-12 February, 1988, Abstr, pp.129-130.

Banerjee, P. S., Bhatia, B. B. and Pandit, B. A., 1994. *Sarcocystis suihominis* in human beings in India. J. Vet. Parasitol., 8: 57-58.

Bhandyopadhyay, A. K., 1988. Parasitic zoonosis: Helminths. Proc. VIII Natl. Cong. Parasitol., Calcutta, 10-12 February, 1988, Abstr, pp.129.

Bhatia, B. B., 1991. Current status of food-borne parasitic zoonoses in India. In: Cross, J. H. ed. Emerging problem in food-borne parasitic zoonoses impact on agriculture and public health. Proc. 33rd SEAMED – TROPMED Reg. Semr. – Southeast Asian Trop. Med. Publ. Hlth., 22 (Suppl): pp.36-41.

Bhatia, B. B., 2006. Food- borne parasitic zoonosis. In: Bhatia, B. B., Pathak, K. M. L. and Banerjee, D. P. (Eds.), Text Book of Veterinary Parsitology. Kalyani Publishers, Ludhiana, New Delhi. pp.425-446.

Bhatia, B. B. and Pathak, K. M. L., 1990. Echinococcosis. In: Parija, S. C. (Ed), Review of Parasitic Zoonosis. AITBS Publishers, New Delhi, India. p.268.

Chatterjee, A., 1988. Parasitic zoonosis: An Overview. Proc. VIII Natl. Cong. Parasitol., Calcutta,10-12 February, 1988, Abstr, p.129.

Chhabra, M. B. and Pathak, K. M. L., 2008a. Helminth zoonoses in India: a resurgent problem. J. Vet. Parasitol., 32: 77-86.

Chhabra, M. B. and Pathak, K. M. L., 2008b. Protozoan zoonoses in India. J. Vet. Parasitol., 32: 87-96.

Chhabra, M. B. and Pathak, K. M. L., 2008c. Arthropod zoonoses in India. J. Vet. Parasitol., 32: 97-103.

Chhabra, M. B. and Singla, L. D., 2009. Food-borne parasitic zoonosis in India. Review of recent reports of human infection. J. Vet. Parasitol., 23: 103-110.

Das S. S., Kumar, D. and Sreekrishnan, R., 2003. Hydatidosis in animals and man. In: Helminthology in India. Sood, M. L. (Ed.). International Book Distributor, Dehradun, pp.425- 451.

Dubey, J. P., 1987. Toxoplasmosis in domestic animals in India: Present and future. J. Vet. Parasitol., 3: 13-18.

Dubey, J. P., 1988. Toxoplasmosis in India. In: Perspectives in parasitology, Vol. 2, Sen, A. B. Katiyar, J. C. and Guru, P. Y. (Eds.). CBS Publishers and Distributors, Delhi, India.

Dubey, J. P., 1990.Toxoplasmosis in sheep -The last 20 years.Vet. Parasitol., 163: 1-14. (Review).

Dubey, J. P., 2009. *Toxoplasma gondii* infection in chicken (*Gallus domesticus*): Prevalence, clinical disease, diagnosis and public health significance. Zoonosis and Pub. Hlth., 57: 60-73.(Review).

Gautam, O. P. and Chhabra, M. B., 1983. Public health aspects of toxoplasmosis. Indian J. Trop. Agric., 1: 159-165.

Ghosh, T. N., 1988. Parasitic zoonosis: Protozoa. Proc. VIII Natl. Cong. Parasitol., Calcutta, 10-12 February, 1988, Abstr, p.129.

Gill, B. S., 1972. A study on parasitic zoonosis in India. Proc. 59th Indian Sci. Cong. (Calcutta), Part III, Abstr.,37, p.549.

Hafeez, Md., 2003. Helminth parasites of public health importance - Trematodes. J. Parasit. Dis., 27: 69-75.

Jagannath, M. S., 1998 (Ed.). Zoonotic parasites – their diagnosis and control. In: III Natl. Trng. Prog., 9th to 23rd March, 1998, Centr. Adv. Stud., Dept. Parasitol., Vet. Coll., Univ. Agri. Sci., Bangalore-560 024. (Review articles contributed by many authors).

Jani, R., Bhuva, C. N. and Katara, R. D., 2006. Study of prevalence of toxoplasmosis in workers of zoological gardens of Gujarat. Intas Polivet,7: 452-454.

Joseph, S. A. and Karunamoorthy, G., 1990. Studies on haematophagus arthropods of zoonotic importance in Tamilnadu. Proc. Symp. Entomol. Def. Serv., 12-14 September 1990, DRDE, Gawlior. pp.186-192.

Juyal, P. D.and Cross, J. H., 1991., Sarcocystis and sarcocystosis in India. Southeast Asian J. Trop. Med. Pub. Hlth., 22: Suppl., 138-141.

Juyal, P. D., Singh, N. K. and Kaur, P., 2005. Hydatidosis in India: A review on veterinary perspective. J. Parasit. Dis., 29: 97-102.

Madhavi, R., 2008. Fish-borne zoonotic haplorchine trematodes: the question of differentiation of species. In: Current Trends in Parasitology. Veena Tandon, Arun K.Yadav and Bishnupada Roy (Eds.), Proceeding of the 20th National Congress of Parasitology, Shillong, India. (November, 3-5, 2008). Panima Publishing Corporation, Bangalore and New Delhi.p.277.

Mahajan, R. C., 2005. Paragonimiasis: an emerging public health problem in India. Indian J. Med. Res., 121: 716-718.

Mandal, S. and Manisha Deb Mandal, 2012. Human cystic echinococcosis: epidemiologic, zoonotic, clinical, diagnostic and therapeutic aspects. Asian Pacific J. Trop. Med., 5: 253-260. (Review).

Nancy Malla and Mahajan, R. C., 2010 (Oct10-Nov,10). Cysticercosis and epilepsy: need for better sanitation, personal hygiene and swine husbandry practices. Editorial. Everyman's Sci., 45: 205-206.

Narain, K., Mahanta, J., Dutta, R. and Dutta, P., 1994. Paddy field dermatitis in Assam: A cerarial dermatitis. J. Comm., Dis., 26: 26-30.

Narain, K., Rajguru, S. K. and Mahanta, J., 1998. Incrimination of *Schistosoma spindale* as a causative agent of farmer's dermatitis in Assam with a note n liver pathology in mice. J. Comm., Dis., 30: 1-6.

Narasapur, V. S. and Gatne, M. L., 2004. In: Elements of Veterinary Public Health. ICAR, pp. 420-427.

Neelima Gupta, Gupta, D. K. and Said Shalaby, 2008. Parasitic zoonotic infections in Egypt and India: an overview. J. Parasit. Dis., 32: 1-9.

Panda, A. K., Thakur, S. D. and Katoch, R., 2007. Meat and meat–borne parasitic zoonoses: an Indian perspective. XVIII Natl. Cong. Vet. Parasitol., Natl. Symp. 'Emerging Parastic Zoonoses and Advances in Herbal Medicines against Parasites of Veterinary importance', IAAVP., Dept. Parasitol., Vet. Faculty, Vet. Anim. Sci., SKUVAST, R. S. Pura, Jammu and Kasmir, p.53.

Parija, S. C. (Ed.), 1990. Review of Parasitic Zoonosis. AITBS Publishers, New Delhi, India.

Parija, S. C., 2002. Emerging Parastic Zoonoses in India. In: Natl. Sym. 'Veterinary public health in containing bioterrorism, Proc. 5th All India Ann. Conf. Vet. Pub. Hlth., 21-23 December, 2002, Tirupati, India, Andhra Pradesh, pp.103-121.

Parija, S. C., 2004. Review of emerging parasitic zoonoses in India. Current trends and strategies in diagnosing and control of parasitic diseases of domestic animals and poultry. 25 August to 14 September, 2004, Vet. Coll. Res. Inst., Namakkal, Tamil Nadu, India, pp.140-159.

Parija, S. C.and Swarna, S. R., 2005. Echinococcosis in India. In: Asian Monograph. The Federation of Asian Parasitologists, Japan, Vol 2, pp. 283-324.

Pathak, K. M. L., 1987. Parasitic Zoonoses. Agro Botanical Publishers, Bikaner.

Pathak, K. M. L., 1991. Fundamentals of Parasitic Zoonoses. Kalyani Publishers, Ludhiana, New Delhi.

Pathak, K. M. L. and Gaur, S.N., 1989. Prevalence and economic implications of *Taenia solium* taeniasis and cysticercosis in Uttar Pradesh State of India. Acta Leidensia, 57: 197-200.

Petkar, D. K., 1990. Sarcocystosis. In: Parija, S. C. (Ed.), Review of Parasitic Zoonosis. AITBS Publishers, New Delhi, India. pp.44-52.

Prasad, K. N., Chawla, S., Jain, D., Pande, C. M., Pal., L., Pradhan, S. and Gupta, R. K., 2002. Human and porcine *Taenia solium* infection in rural Northern India. Trans. Roy. Soc. Trop. Med. Hyg., 96: 515-516.

Prasad, K. N., Prasad, A. Verma, A. Gupta, R. K. and Pandey, C. M., 2008. *Taenia solium* taeniasis and neurocysticercosis (NCC): experience from North Indian pig farming community. In: Current Trends in Parasitology. Veena Tandon, Arun K.Yadav and Bishnupada Roy (Eds.), Proceeding of the 20th National Congress of Parasitology, Shillong, India. (November, 3-5, 2008). Panima Publishing Corporation, Bangalore and New Delhi.p.277.

Rathore, V. S., 2004. Parasitic Zoonosis. Pointer Publishers, Vyas Building, S.M.S. Highway, Jaipur, 392 003.(Rajasthan), India.

Rao, B. V., 1968. Certain biological and epidemiological considerations of hydatidosis in India with special reference to buffalo material. Bull. Indian Soc. Comm. Dis., 5: 257-261.

Roy, B. and Tandon, V., 1992. Seasonal prevalence of some zoonotic trematode infections in cattle and pigs in the north-east montane zone in India. Vet. Parasitol., 41: 69-76.

Sabu, L., Devada, K. and Subramanian, H., 2005. Dirofilariosis in dogs and humans in Kerala. Indian J. Med. Res., 121: 691-693.

Seghal, S. and Bhatia, R., 1981. Manual of zoonoses developed at the workshop on zoonoses of public health importance. Natl. Inst. Comm. Dis., New Delhi.

Shah, H. L., 1984. *Sarcocystis* as zoonosis. In: Compendium of lectures. Summer institute on animal parasitic infections of public health importance at APAU, Tirupati.

Shah, H. L., 1987. An integrated approach to study of zoonoses. J. Vet. Parasitol., 1: 7-12.

Shah, H. L., 1995. Sarcocystosis as a zoonosis with special reference to India. J. Vet. Parasitol., 9: 57-61.

Shah, H. L. and Agrawal, M. C., 1990. Schistosomiasis: In: A Review on Parasitic Zoonosis. Parija, S. C. (Ed.), ATIBS Publishers, Delhi-143-172.

Shah, H. L. and Chaudhury, R. K., 1994. Recent advances in our knowledge of *Sarcocysts* and sarcocystosis in India In: Current Advances in Veterinary Science and Animal Production in India. Somvanshi, R. and Lokeshwar, R. R. (Eds.)., Interenational Book Publishing Co., Lucknow.

Singh, B. B., Sharma, R., Sharma, J. K. and Juyal, P. D., 2010. Parasitic zoonosis in India: An overview. Rev. sci. tech. Off. Int. Epiz., 29: 629-637.

Singh, S., 1997. Toxoplasmosis in animal health. In: Prospects of Livestock and Poutry Development in 21st century. Verma, R. and Johri, D. C. (Eds.), Indian Vet. Res. Inst., Bareilly. pp. 204-212.

Singh, S., 1998. Toxoplasmosis-An emerging disease in man and animals. In: III National Training Programme, 9th to 23rd March, 1998. Centr. Adv. Stud., Dept. Parasitol., Vet. Coll., Univ. Agri. Sci., Bangalore-560 024.(Not indexed).

Singh, S., 2001. Immunology and Immunological Diagnosis of Toxoplasmosis. *In*: Advances in Immunology and Immunopathology. Chauhan, R. S., Singh G. K., and Agarwal, D. K. (Eds.). Soc. Immunology and Immunopathology, Pantnagar. pp.190-201.

Singh, S., 2010. Laboratory Diagnosis of Toxoplasmosis. In: Toxoplasmosis in India. Singh, S. (Ed.). Third Edition, Pragati Publishing Company, Ghaziabad & New Delhi.

Singh, S. and Banerjee, D. P., 1997. Role of wildlife in parasitic disease of man and animals. Zoos' Print,12(4): 14-16.

Singh, S. and Niti Singh, 1994. Sero-prevalence of toxoplasmosis in sheep and goats of Rajasthan State and their butchers. In: Current Advances in Veterinary Science and Animal Production in India. Somvanshi, R. and Lokeshwar, R. R., (Eds.)., Interenational Book Publishing Co., Lucknow. pp.204-212.

Singh, T. S. and Sugiyama, H., 2008. Paragonimiasis in India: A newly emerging food borne parasitic disease. Clinc. Parasitol., 19: 95- 98.

Somvanshi, R. and Singh, S., 1999. Spontaneous toxoplasmosis in laboratory rats. In: Toxoplasmosis in India. Singh, S.,(Ed.), Second edition, Pragati Publishing Co. Gazhiabad & New Delhi.

Srivastava, H. O. P. and Shah, H. L., 1970. On *Gastrodiscoides hominis* (Lewis and Mc Connell,1876) Leiper, 1911 from pigs (*Sus scrofa domestica*) in Madhya Pradesh, its pathology and public health significance. Indian J. Path. Bacteriol., 13: 68-72.

Tandon, V. and Dhawan, B. N., 2005. Infectious diseases of domestic animals and zoonoses in India. Proc. Natl. Acad. Sci., India 75 (B), Special issue, Natl. Acad. Sci., India, Allahabad.

Thapliyal, D. C., 1999. Diseases caused by parasites. In: Diseases of Animals Transmissible to Man, 1st Edn., International Book Distributing Co., pp. 275-327.

Traub, R. J., Robertson, Irwin, P., Mencke, N., Thompson, R. C. A., 2005. Canine gastrointestinal parasitic zoonosis in India. Trends Parasitol., 21: 42-48.

Veena Tandon, Bishnupada Roy and Prasad, P. K., 2012. Chapter 32. Fasciolopsis. In: Molecular Detection of Human Parasitic Pathogens. Dongyou Liub (Ed), Boca Raton, CRC Press, USA.

Vinay Kumar, V., 1980a. Zoonotic trematodiasis in south-east and far-east asian countries. Curr. Top. Vet. Med. Anim. Sci., 43: 106-118.

Vinay Kumar, V., 1980b. The digenetic trematodes, *Fasciolopsis buski*, *Gastrodiscoides hominis* and *Artyfechinostomum malayanum* as zoonotic infection in South Asian countries. Ann. Soc. Bel. Med. Trop., 60: 331-339.

Yadav, A. K. and Tandon,V., 1989. Helminths of public health significance in domestic animals of Meghalaya. Indian J. Anim. Hlth., 28(2): 169-171.

SECTION IV–BOOKS OF INDIAN AUTHORS

Abdul Rahman, S., 1985. Helminth Parasites of Veterinary Importance in India. Intervet Publishers 124/ 21, 7th B Main Road, 4th Block, Jayanagar, Bangalore. p.119.

Agrawal, M. C., 2012. Schistosomes and Schistosomiasis in South Asia. Springers, New Delhi, India

Anurudha Ray Ramakrishna, 2007. Handbook on Indian Freshwater Molluscs. Zoological Survey of India, Calcutta.p.400

Bhalerao, G. D., 1935. Helminth Parasites of Domesticated Animals in India. Scientific Monograph Series No. 6, Imperial Council of Agricultural Research Publication, New Delhi, p.365.

Bhatia, B, B. and Chauhan, P. P. S., 1984. Parasitic Dieases of Buffaloes in India.1st Ed. Cosmos, New Delhi. India, p.118.

Bhatia, B, B. and Shah, H. L., 2001. Protozoa and Protozoan Diseases of Domestic Animals. Indian Council of Agricultural Research, Division of Information and Publicity, New Delhi. p.202.

Bhatia, B, B., 2000. Text Book of Veterinary Protozoology. Division of Information and Publicity, Indian Council of Agricultural Research, New Delhi.p.368.

Bhatia, B. B., Pathak, K. M. L. and Banerjee, D. P., 2006. Text Book of Veterinary Parasitology. 2nd Edition, Kalyani Publishers, Ludhiana, New Delhi.

Bhatia, B. B., Pathak, K. M. L. and Juyal Banerjee, P. D., 2010. Text Book of Veterinary Parasitology. 3rd Edition, Kalyani Publishers, Ludhiana, New Delhi.

Chaudhuri, R. P., 1982. Insect Parasites of Livestock and Their Control, Research. Series 29, Indian Council of Agricultuaral Research, New Delhi.

Chauhan, R. S., Singh G. K. and Agarwal D. K. (Eds.), 2001. Advances in Immunology and Immunopathology. Society for Immunology and Immunopathology, Pant Nagar.

Chowdhary, N and Tada N. J., 1994. Helminthology. Narosa Publishing House, New Delhi.

Chowdhury, N. and Alonso Aguirre, A., (Eds.)., 2001. Helminths of Wild life. Science Publishers, Inc., Enfield (NH), USA, Plymouth, UK.

Chowdhury, N. and Tada. I., 2001. (Eds.). Perspective on Helminthology. Oxford & IBH Publishing Co. Pvt. Ltd., New Delhi. p.531.

Deo, P.G., 1964. Roundworms of Poultry. Animal Husbandry Series No. 3. ICAR, New Delhi.

Dubey, J. P. and Towle, A., 1986. Toxoplasmosis in Sheep: A Review and Annotated Bibliography. Commonwealth Institute of Parasitology, St. Alberts, Herts, UK, pp.1–152.

Dubey, J. P., 2010a. Toxoplasma and Toxoplasmosis in Animals and Man. CRC Press, Inc. Boca Raton, Florida, USA. pp.1-220.

Dubey, J. P., 2010b. Toxoplasmosis of Animals and Humans. Second edition. CRC Press, 1-313 pages.

Dutt, S. C., 1980. Paramphistomes and Paramphistomiasis of Domestic Ruminants of India. Punjab Agricultural University, Ludhiana, pp.1-168.

Gautam, O. P., Sharma, R. D. and Dhar, S., 1980. Haemoprotozoan Diseases of Domestic Animals Proceedings of the Seminar on Haemoprotozoan Diseases, October 27-November 1, 1980, Department of Veterinary Medcine, Haryana Agricultural University, Hissar-125004 (Haryana) India.

Geeverghese, G. and Misra, A.C., *Haemaphysalis* Ticks of India. Elsevier, London and Waltham, USA.1-268.

Gill, B. S., 1977. Trypanosomes and Trypanosomiasis of Indian Livestock.1ˢᵗ Edn., Indian Council of Agricultural Research, Division of Information and Publicity, New Delhi, pp.1-191

Gill, H. S. and Gill, B. S., 1977. Ixodid Ticks of Domestic Animals of Punjab State. Punjab Agricultural University, Ludhiana, India, p.69.

Greets, S., Kumar,V. and Brandt, K. J. (Eds.), 1987. Helminth Zoonoses (Current Topics in Veterinary Medicine). Martinus Nijhoff Publishers, Dordrecht, The Netherlands.

Gupta, N. K., 1993. Amphistomes: Systemics and Biology. Oxford and IBH Publishing company Private Limited, New Delhi.

Jain, P. C., 2006. Text Book of Veterinary Entomology and Acarology. Jaypee Publishers pp.1-360.

Mukherjee, R. P., 1992. Fauna of India and Adjacent Countries: Larval Trematodes, Part I – Amphistome Cercariae (pp. 1-89), Zoological Survey of India, Kolkatta. pp.1-158.

Parija, S. C. (Ed.), 1990. Review of Parasitic Zoonosis. AITBS Publishers, New Delhi, India.

Pathak, K. M. L., 1987. Parasitic Zoonoses. Agro Botanical Publishers, Bikaner.

Pathak, K. M. L., 1991. Fundamentals of Parasitic Zoonoses. Kalyani Publishers, Ludhiana, New Delhi.

Rathore, V. S., 2004. Parasitic Zoonosis. Pointer Publishers, Vyas Building, S.M.S. Highway, Jaipur, 392 003. (Rajasthan), India.

Roy, R. N. and Brown, A. W. A., 1987. Entomology (Medical and Veterinary) including Insecticides and Insect and Rat Control., Bangalore Printing and Publishing Co. pp.1-855.

Ruprah, N. S., 1985. A Text Book of Clinical Parasitology, Oxonian Press Pvt. Ltd, New Delhi.

Seghal, S. and Bhatia, R., 1981. Manual of zoonoses developed at the workshop on Zoonoses of Public Health Importance. National Institute of Communicable Diseases, New Delhi.

Sen, A. B., Katiyar, J. C. and Guru, P. Y. (Eds.)., 1985. Perspectives in Parasitology, Vol. 1. Print House (India), Lucknow, India.

Sen, A. B., Katiyar, J. C. and Guru, P. Y. (Eds.)., 1988. Perspectives in Parasitology, Vol. 2. CBS Publishers and Distributors, Delhi, India.

Sen, S.K. and Fletcher, T. B., 1962. Veterinary Entomology and Acarology for India Indian Council of Agricultural Research, New Delhi.

Sherikar, A.T., Bachhil, V. N. and Thapliyal, D. C., 2004. Elements of Veterinary Public Health. Directorate of Information and Publications of Agriculture, Indian Council of Agricultural Research, pp.1-572.

Singh, G. and Prabhakar, S., 2002. *Taenia solium* Cysticercosis: From Basic to Clinical Sciences.CAB International, U. K.

Singh, K. S., 2003. Veterinary Helminthology. Division of Information and Publicity, Indian Council of Agricultural Research, New Delhi.

Singh, M. L. (Ed.), 2001. Text Book of Helminthology in India. International Book Distributors, Dehradun,

Singh, S. (Ed.), 2010. Toxoplasmosis in India. Third Edition, Pragati Publishing Company, Ghaziabad & New Delhi.

Singh, S., 2001. Immunology and Immunological Diagnosis of Toxoplasmosis. *In*: Advances in Immunology and Immunopathology. Chauhan, R. S, Singh, G. K. and Agarwal, D. K. (Eds). Society for Immunology and Immunopathology, Pant Nagar. pp. 190-201.

Singh, S., Mahajan, R.C. and Dubey, J. P.(Eds.), 1997. Toxoplasmosis in India. Pragati Publishing Co., New Delhi.

Somvanshi, R. and Lokeshwar, R. R., (Ed.)., 1994. Current Advances in Veterinary Science and Animal Production in India. Interenational Book Publishing Co., Lucknow.

Sood, M. L., 2003. Helminthology in India, International Book Distributers, Dehradun, India.

Srivastava, H. D. and Dutt, S. C., 1962. Studies on *Schistosoma indicum*. Indian Council of Agricultural Research, Division of Information and Publicity, Series 24., New Delhi.pp.1-91.

Subba Rao, N. V., 1989. Handbook on Freshwater Molluscs of India. Zoological Survey of India, Calcutta. pp.1-281.

Thapliyal, D. C., 1999. Diseases of Animals Transmissible to Man, 1st Edn., International Book Distributing Co., Lucknow.

Veena Tandon, Arun K.Yadav and Bishnupada Roy (Ed.), 2008. Current Trends in Parasitology. Proceedings of the 20th National Congress of Parasitology, Shillong, India. (November, 3-5, 2008). Panima Publishing Corporation, Bangalore and New Delhi.pp.1-277.

Veer Singh Rathore and Yogesh Singh Senger, 2005. Diagnostic Parasitology. Pointer Publishers, Jaipur-302 003, India.

Verma, R. and Johari, D. C. (Eds.), 1997. Prospects of Livestock and Poultry Development in 21st Century. IVRI, Bareilly, 1997.

Vinay Kumar, V., 1999. Trematode Infections and Diseases of Man and Animals. Kluwer Academic Publishers Dordreht, P. O. Box 17, 0300 AA, Netherlands. pp.1-328.(See also Greets *et al.*, 1987, supra)

PART III
PARASITES INDEX

PART IV
AUTHORS INDEX

2. CESTODA

3. NEMATODA

4. ARTHROPODA

5. PROTOZOA (PROTISTA)

6. ORDER: RICKETTSIALES

www.ingramcontent.com/pod-product-compliance
Lightning Source LLC
Chambersburg PA
CBHW020139240326

41458CB00120BA/322

9 7 8 9 3 5 1 2 4 2 9 6 3